C000093009

Archaeological Oceanography

Archaeological Oceanography

Robert D. Ballard, EDITOR

PRINCETON UNIVERSITY PRESS *Princeton & Oxford*

Copyright © 2008 by Princeton University Press

Published by Princeton University Press, 41 William Street,
Princeton, New Jersey 08540
In the United Kingdom: Princeton University Press, 3 Market Place,
Woodstock, Oxfordshire OX20 1SY

All Rights Reserved

ISBN-13: 978-0-691-12940-2
ISBN-10: 0-691-12940-1

LIBRARY OF CONGRESS CATALOGING-IN-PUBLICATION DATA
Archaeological oceanography / Robert D. Ballard, editor.
p. cm.
Includes bibliographical references and index.
ISBN 978-0-691-12940-2 (cloth : alk. paper)
1. Underwater archaeology. 2. Coastal archaeology. 3. Excavations (Archaeology). 4. Shipwrecks. I. Ballard, Robert D.
CC77.U5A67 2008
930.1'028'04—dc22 2007017246

British Library Cataloging-in-Publication Data is available

This book has been composed in Adobe Garamond and Myriad
Printed on acid-free paper. ∞
press.princeton.edu
Printed in China

10 9 8 7 6 5 4 3 2 1

The frontispiece shows ROV Hercules investigating the bow of RMS Titanic,
copyright Institute for Exploration

CONTENTS

PART THREE DEEP-WATER SHIPWRECKS IN THE MEDITERRANEAN AND BLACK SEAS

PART FOUR SUBMERGED LANDSCAPE ARCHAEOLOGY

PART FIVE TELEPRESENCE AND SUBMERGED CULTURAL SITES

ACKNOWLEDGMENTS

This book has taken more than four years to complete so we will more than likely fail to thank many people who made it happen. Since its contents are based on a series of expeditions spanning in excess of 20 years, we want most importantly to thank the various teams that supported these expeditions, including those at the Deep Submergence Laboratory at the Woods Hole Oceanographic Institution, the Institute for Exploration in Mystic, Connecticut, and the Institute for Archaeological Oceanography at the University of Rhode Island's Graduate School of Oceanography.

We also want to thank the captains and crews of numerous research vessels associated with these expeditions, including the R/V *Starella,* R/V *Star Hercules,* R/V *Knorr,* SSV *Carolyn Chouest,* M/V *Northern Horizon,* R/V *Endeavor,* NOAA Ship *Ronald H. Brown,* and the U.S. Navy nuclear research submarine *NR-1*.

Once an expedition ends, the long months and years of analysis begin and a book such as this one begins to emerge. A new team takes over—people such as Cathy Offinger, Janice Meagher, Paul Oberlander, Sandra Witten, and Laurie Bradt, who helped type, edit, draft, organize, and assemble all of the chapters and illustrations contained within this volume. The authors of each chapter in this book thank them. Special thanks to Kathleen Cantner, the newest member of our research team, who not only compiled the glossary and index, but read through and corrected an earlier draft of the book. And thanks to George Bass, who reviewed the book for Princeton University Press and recommended its publication.

And let's not forget to thank the people and organizations that sponsored these expeditions, including the Office of Naval Research, NOAA's Office of Ocean Exploration, and the National Geographic Society's Expedition Council.

We also want to thank the team at Princeton University Press who had the patience to wait as deadlines passed and new ones rose on the horizon: Heath Renfroe, Ingrid Gnerlich, Linny Schenck, Virginia Dunn, Mark Bellis, Dimitri Karetnikov, and Elizabeth Byrd.

INTRODUCTION

Robert D. Ballard

Why the title *Archaeological Oceanography*? Why not *Marine Archaeology* or *Nautical Archaeology* or *Oceanographic Archaeology*? Good question.

The ocean covers 72% of the earth's surface, with an average depth of 4000 m. Since the early 1950s and before, archaeologists have been discovering, exploring, mapping, and excavating ancient shipwrecks beneath the sea. The depths they have been working at, however, are shallow by oceanographic standards, restricted until recently to the inner portions of the continental margins, to depths of less than 100 m. The choice has been ambient diving techniques, particularly the use of self-contained underwater breathing apparatus, or scuba.

Until recently, it was widely believed that the ancient mariner hugged the coastline, seldom venturing into the open waters of the world's oceans. But recent discoveries in deep water far from shore have shown that many ancient mariners were either driven out to sea by storms before sinking or chose for a variety of reasons to take the shortest routes from one destination to the next.

Not only are there many ancient shipwrecks to be found in the deep sea, but various conditions result in those shipwrecks being well preserved. These factors include total darkness, cold temperatures, low rates of sedimentation, and limited human activity, since fishing and diving rarely occur in such remote regions of the world. Ships sink in the deep sea generally due to storm action, which leads to them being swamped instead of striking the bottom, which frequently can tear their hulls open. Deep-water shipwrecks tend to sink in an upright position, coming to rest in a low-energy environment instead of being further damaged in shallow water during subsequent storm periods.

Another critical factor leading to their preservation is the thick layers of soft mud into which they settle when reaching the bottom. Deep-sea sediments commonly consist of fine-grained clay that is saturated with salt water. As a result, shipwrecks sinking into such a bottom commonly come to rest with their main deck near the bottom/water interface. Since deep-sea mud quickly becomes anoxic, the majority of a shipwreck and its contents, including its human occupants, are thrust into highly preserving anoxic conditions minutes after sinking.

Although this is all interesting, let us return to the question of the title: *Archaeological Oceanography*. Oceanography, like marine archaeology, is a relatively new field of research, a child of the 20th century. Unlike marine archaeology, oceanography is expensive and reliant upon costly resources, such as large research ships, submersibles, and advanced undersea vehicles, including remotely operated vehicles. It is not uncommon for an oceanographic expedition to cost \$30,000 to \$40,000 a day. As a result, one month at sea can cost \$1 million.

To the world of marine archaeology this is prohibitively expensive and even more so when compared to archaeological programs carried out on land. If one were to use these comparisons in making decisions regarding the allocation of scarce resources, the shipwrecks of the deep sea would never be studied—that is, if you expect the traditional archaeological community and its traditional sources of funding to finance work in the deep sea. But that need not be the case since there is no reason why archaeological oceanography could not be supported by the same sources of funding that support other fields of oceanographic research.

It is important to point out that oceanography is not a separate discipline such as physics, chemistry, or geology. It is an arena in which these disciplines work, bonded together by common needs such as the need for unique facilities that are required to carry out these separate lines of research. It is common for these various disciplines to work together on oceanographic field programs, no different than multidisciplinary programs carried out on land or in outer space. Oceanographers come from all fields of research in the physical and engineering sciences, fields of research that could easily be expanded to include the social sciences of maritime history, archaeology, and anthropology.

More importantly, it is a young enough science to be inclusive, commonly accepting new disciplines into its fold. The history of marine geology is an excellent example. It started in the 1930s and was dominated by sedimentologists concentrating on the continental margins of the world. But the evolving theory of plate tectonics in the 1960s took the earth sciences into the deeper ocean basins, bringing in petrologists, volcanologists, and structural geologists. The discovery of hydrothermal vents on the Mid-Ocean Ridge in 1977 saw an influx of chemists, geochemists, and a broad range of biologists entering the field, placing increasing demands on access to the expensive tools of oceanography.

When we first began to discover ancient shipwrecks in the central Mediterranean Sea in 1988, it was thought to be a rare occurrence. But in subsequent years, as this book documents, more and more ancient shipwrecks were discovered in other deep-water locations. More recently, professional salvage companies have obtained the necessary technology to carry out commercial recovery programs.

It has become increasingly clear that the deep sea can be of great importance to the social fields of archaeology, anthropology, history, and art, to name a few. But how can this interest turn into a meaningful and viable research program?

The term "archaeological oceanography" sounds like the former is subordinate to the latter but that is not the case. A geological oceanographer is a geologist working in the ocean. An archaeological oceanographer is an archaeologist working in the ocean as well.

It is encouraging to see the willingness on the part of the leadership of the National Oceanic and Atmospheric Administration's Ocean Exploration Program to support this budding field of archaeological oceanography and our only hope is that other federal funding agencies that support oceanography will follow suit.

PART ONE

The Technology and Techniques of Archaeological Oceanography

Oceanographic Methods for Underwater Archaeological Surveys

1

Dwight F. Coleman and Robert D. Ballard

Geophysical prospecting techniques for land-based archaeological studies are fairly well established. For the most part this is true for marine archaeological studies as well (Oxley and O'Regan 2001). Oceanographic survey techniques that focus on mapping and exploring the marine environment are also well established, but traditional oceanographic methodologies are not typically applied to marine archaeology. A major limiting factor that influences this is the high cost. For example, the current operational cost for using an ocean-class research vessel can be more than $20,000 per day. Deep submergence vehicle systems and advanced geophysical survey equipment that are used with these research vessels can cost more than $10,000 per day. The total cost for one day of shipboard operations could be enough to fund an entire season of a terrestrial archaeological site excavation. But this example really does not represent a fair comparison. Such daily costs associated with doing research at sea are typically devoted to the study of natural history phenomena in the oceans and on the ocean floors. The following questions can be asked: Is the study of human history beneath the sea just as important as the study of natural history beneath the sea? Are cultural resources as significant as natural resources? Should federal dollars be equally spent to protect these resources? Should archaeology be federally funded to the same level as other oceanographic sciences? If the answer is yes to any of these questions, then we can justify the cost of conducting "archaeological oceanography." Many of the well-established geophysical tools and techniques that have been employed by archaeologists in shallow water can also be used on larger ships and in deeper water, thereby employing an oceanographic approach. Deep-water oceanographic techniques do not differ greatly from shallow-water techniques, but a focus here is to present methodologies for surveying that optimize time on board expensive scientific research vessels.

Archaeologists have used side-scan sonar, subbottom profilers, magnetometers, and visual imaging techniques, although not nearly as extensively as scuba techniques, to search for and map submerged sites, especially shipwrecks (Oxley and O'Regan 2001). For exploration and mapping of terrestrial sites, use of ground-penetrating radar (GPR) has become more widespread to acoustically image the subsurface details of sites. Collection of sediment cores to ground-truth the GPR data and to characterize the depositional context of terrestrial sites is also common in terrestrial site surveys. In a similar manner to the way these geological techniques have been employed to investigate terrestrial archaeological sites, oceanographic techniques are now being employed to investigate underwater archaeological sites. These techniques (discussed below) are all commonly used during oceanographic research and exploration cruises and represent important methodologies employed to characterize underwater archaeological sites and landscapes.

Established Archaeological Survey

Archaeological survey strategies and techniques, particularly for terrestrial sites, are well established (Banning 2002) and include different approaches for exploration, reconnaissance surveying, and intensive site surveying. Regional scale surveying techniques (Dunnell and Dancey 1983) and sampling strategies (Nance 1983) are also well established, but these are also mainly for terrestrial archaeology. Archaeological survey can involve different techniques and methodologies, depending on the site. From a theoretical standpoint, there should be almost no difference between surveying on land or under water, except for the obvious logistical differences. For example, on land aerial photographs can be used as a base map similar to the underwater use of side-scan sonar mosaics. From a practical standpoint, however, there are significant differences. Firstly, many underwater sites are in regions of very poor visibility, so surveyors must rely more on acoustic strategies than visual strategies. Secondly, survey techniques for shipwreck archaeology differ from the techniques for surveying terrestrial (including inundated) sites. Ancient shipping trade routes or more modern naval battle locations—regions where shipwrecks would be expected, for example—would have well-defined boundaries that would bias the survey strategy. Thirdly, and perhaps most importantly, underwater surveys are much more difficult logistically, and the rigid limits set by cost, time, and weather for work at sea could significantly influence survey strategies.

To complete a well-planned archaeological survey, whether on land or under water (shipwrecks or inundated terrestrial sites), the entire region of interest should be mapped and investigated, even if there are no suspected sites in parts of the survey region. The absence of sites in particular locations provides scientific data and evidence to support the regional archaeological interpretation. For example, to search for shipwrecks along suspected trade routes, surveyors must also search away from the suspected trade routes to verify working hypotheses about delineation of the suspected routes.

Established guidelines for underwater survey exist, primarily, for the purposes of cultural resource management. Several federal agencies in the United States, such as the Army Corps of Engineers, the Minerals Management Service, the National Park Service, and the National Oceanic and Atmospheric Administration, either suggest or require that survey operations follow their guidelines. These guidelines vary depending on the particular archaeological sites and the scope of work. For the most part, the survey guidelines were established to protect submerged cultural resources from being damaged by activities that involve disturbance of the seabed, such as dredging, construction projects, and oil well drilling. For these activities in U.S. waters, compliance with the National Historic Preservation Act of 1966 is a requirement, and a complete site survey and characterization is necessary and must be approved prior to further site activity. Many individual coastal states have rules and regulations in addition to the requirements by federal law. Organizations such as UNESCO (United Nations Educational, Scientific and Cultural Organization) have worked to develop international guidelines and codes of ethics for conducting archaeology under water.

A variety of geophysical methods have been used in land-based archaeological exploration and surveying, including, but not limited to, satellite remote sensing, airborne imaging, ground-penetrating radar, and magnetic techniques (Renfrew and Bahn 2000). These are primarily tools for prospecting. Other terrestrial archaeological methods that involve geophysical techniques include archaeomagnetism, radioisotope studies, dendrochronology, palynology, paleontology, and provenance studies. These are primarily analytical methods and are useful in absolute dating and understanding past environmental conditions and archaeological associations. For buried terrestrial archaeological sites, a regional sampling strategy can be employed to test for potential sites (Nance 1983). This could include coring or excavating test pits situated in high-probability locations.

For the marine environment, similar sets of prospecting and analytical techniques exist. The primary focus here will be on marine geophysical exploration and surveying techniques, but techniques analogous to those used on land can be used once an archaeological site is identified for higher-resolution investigations. Intrusive techniques have been employed as part of the survey phase of underwater archaeology (Oxley and O'Regan 2001). This primarily involves limited sampling of material from the site to better characterize and understand its nature, such as the collection of organic material for radiocarbon dating. Excavation, such as trial trenching on land to test whether a site exists, is intrusive and can be very destructive, and this is not very practical for investigating underwater sites (Oxley and O'Regan 2001). Techniques for underwater site excavation are well established (e.g., Green 1990), and typically involve intensive surveying to carefully map the site prior to excavation and subsequent site disturbance. New oceanographic methodologies that employ remotely operated vehicle systems for high-precision site surveys can now be utilized in both shallow- and deep-water settings (Foley and Mindell 2002).

Archaeological Oceanographic Surveys

The nature of the survey strategy is dependent on whether archaeological sites are known to exist within the region to be surveyed. Exploratory and reconnaissance surveys can take many forms, but for targeting inundated archaeological sites, certain methodologies work better than others. A full range of geophysical methods can be applied to archaeological oceanography, and these methods help to define this new field. Interpretation of the survey data will help to delineate sites for further exploration and detailed investigation. These oceanographic methods include bathymetric mapping, side-scan sonar surveying, high-resolution reflection surveying (including subbottom profiling and lower-frequency seismic methods), magnetometer surveys, and visual imaging surveys using remotely operated vehicle (ROV) systems (Oxley and O'Regan 2001). Other geophysical methods, including electrical resistivity and marine gravimetry methods, can be used to explore for and characterize underwater archaeological sites.

For the survey of submerged terrestrial sites, the strategy must be different from shipwreck mapping surveys because prehistoric sites are typically buried in the shelf sediment. The process of coastal inundation due to rising sea level is generally destructive to archaeological sites. If sites are rapidly buried, there is a greater chance for preservation of delicate materials. But for the most part, what survives are the nondelicate cultural and human remains—lithic artifacts (stone tools, points), kitchen middens (mammal and fish bones, shells), gravesites (human bones and associated artifacts), stone foundations of dwellings, pottery, hearths, postholes, and other cultural features. Remote-sensing methods would not typically be able to distinguish these cultural remains from natural features on and in the sediments. Visual methods must be used to identify specific anthropogenic features from natural features.

A new methodological approach is presented here that involves both remote-sensing and visual inspection techniques. The remote-sensing strategy is used to identify potential archaeological environments on the seabed. For shipwreck exploration this involves using geophysical prospecting techniques to locate man-made targets on the seabed. For submerged terrestrial sites the remote sensing strategy is used to identify paleoshorelines, ancient river channels, tidal inlets, lagoons, and embayments. Once these features are located through mapping and geomorphologic analysis, a systematic approach to develop the visual survey is used to target regions where submerged archaeological sites are predicted based on the environmental setting. One aspect of the archaeological oceanographic survey methods presented here that is common to all techniques is accurate and precise navigation. By employing the Global Positioning System (GPS), navigation can be accurate to within a couple of meters using differentially corrected signals. For the shipwreck and submerged landscape studies presented in later chapters, the geophysical techniques utilized are bathymetric mapping, side-scan sonar imaging, subbottom profiling, video and still camera photography, and geological sampling. These are described below.

Side-scan Sonar Surveying

Side-scan sonar is commonly used to acoustically map the seafloor. A side-scan sonar towfish is typically deployed off a survey vessel and towed behind the ship through the water at a given altitude above the seafloor (figure 1.1). Echo is an example of a side-scan sonar towfish (see figure 2.7). For use of this particular system, which is capable of operating in deep water, the towfish is tethered to a depressor weight that acts as a heave compensator, thereby allowing the towfish to be unaffected by the ship's vertical motion. The system emits acoustic pulses at set intervals that are focused with a defined beam pattern according to the design of the sonar transducer. The range, or imaging distance to either side of the centerline, can also be set, as can other data acquisition parameters. Because the instrument images to both sides, twice the range indicates the effective swath width that represents the width of seafloor that is mapped along track (figure 1.1). The acoustic sonar pulses are transmitted with a set frequency. Some dual-frequency side-scan sonar systems exist (Echo, for example), which enable high- and low-resolution data to be collected at the same time.

By keeping careful track of the layback, or the horizontal distance behind the ship that the instrument is towed, which is a function of the amount of cable that is spooled out, features on the seabed can be precisely located. The layback represents an offset that can be used to compute the GPS position of features on the seafloor. Acoustic targets stand out on the sonar record as anomalous features. These targets are either natural or man-made features that can be later inspected using visual surveying techniques, as discussed later. Large targets that have vertical relief will have an associated acoustic shadow that is clearly visible on the sonar record. The shadow results from a loss of acoustic information on the seabed because the target is essentially blocking the sonar pulse from reaching points inside the shadow. As with most acoustic data, the interpretation of targets relies on groundtruthing by visual inspection, to determine if the features

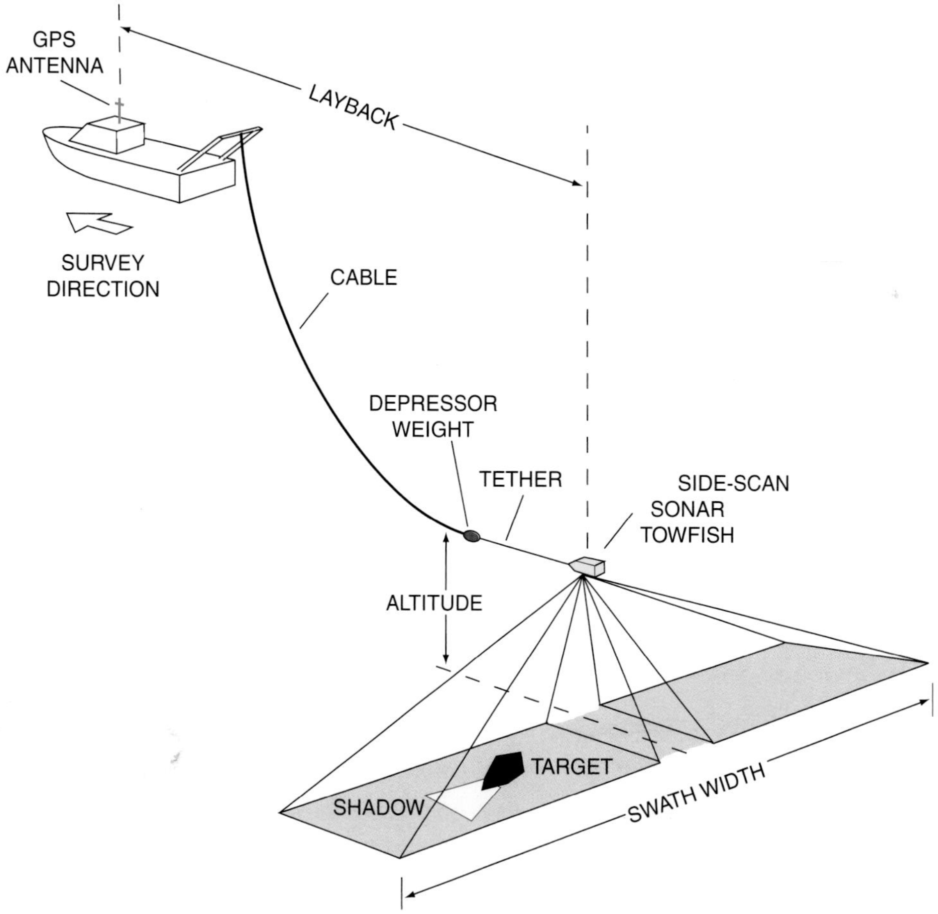

Figure 1.1. Use of side-scan sonar (such as Echo, shown in figure 2.7). Refer to text for explanation.

are natural or man-made. An exception to this is modern shipwreck targets that typically can be recognized solely by their acoustic character. To fully characterize shipwreck targets, visual inspection is still required, however. For more information on side-scan sonar theory and operation, refer to Fish and Carr (1991).

Bathymetric Mapping

The initiation of a terrestrial archaeological survey typically involves examination of topographic maps, aerial photographs, and satellite remote-sensing images (Banning 2002). Marine archaeological surveying must commence with a similar data set. Multibeam (swath) and single-beam (echo sounding) sonar mapping methods are used in reconnaissance surveying to initially interpret the seafloor topography to give a first-order depiction of the submerged landscape. Multibeam bathymetric sonar systems use hull-mounted acoustic transducers and receivers, with signals that sweep through a swath beneath the ship (figure 1.2). Typically, a hull-mounted multibeam sonar system can resolve features

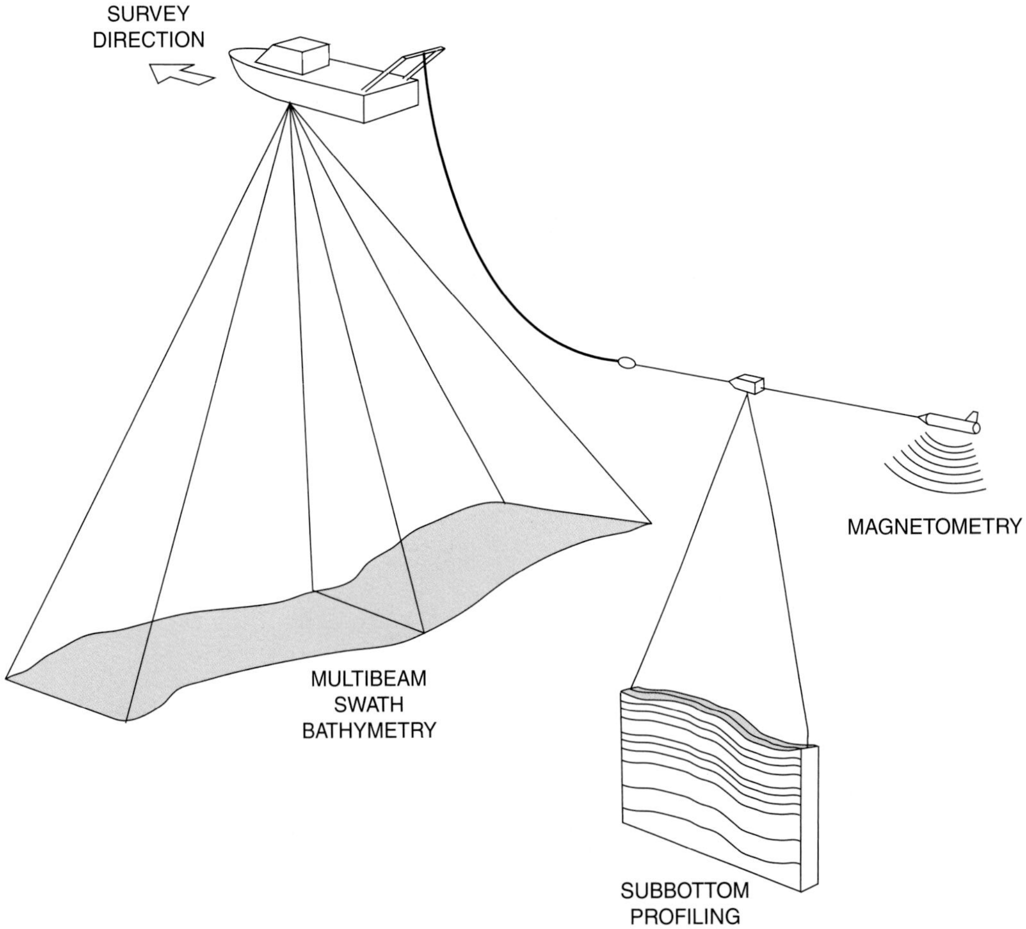

Figure 1.2. Geophysical surveying and prospecting equipment, including multibeam bathymetric mapping, subbottom profiling, and towing a magnetometer. Refer to text for explanation.

on the seafloor on the order of tens of meters in size, depending on the water depth, acquisition parameters, and characteristics of the sonar transducers. Advances in this technology, particularly with deep-towed and robotic systems, have resulted in much more detailed bathymetric maps, with centimeter-scale spatial resolution (Singh et al. 2000). In addition to collecting new data, preexisting bathymetric data from older surveys are useful and there are excellent resources for large data sets, although at much coarser resolution. The NOAA National Geophysical Data Center is an excellent resource for processed bathymetric data sets. Global bathymetric grids based on satellite gravity data exist, but with a resolution of only a few kilometers (Smith and Sandwell 1997). Higher resolution grids from shipboard surveys exist for most U.S. waters. Once the general bathymetric features are determined, more detailed geophysical work can commence.

Subbottom Profiling

High-frequency seismic reflection methods, also called subbottom profiling (figure 1.2) typically use lower-frequency acoustic signals than side-scan sonar to map features below the seafloor. Very low-frequency seismic reflection methods are used to image deep within the earth's sedimentary sequences and crust, but this will not resolve shallow buried features. Oil companies employ this technology for hydrocarbon exploration. Echo (figure 2.7) is equipped with a high-resolution (Chirp) subbottom profiler. Typically these systems are towed behind a survey vessel. A sound pulse is transmitted vertically through the water column, and any density changes within the seafloor sediments produce reflections that are recorded to produce seismic-stratigraphic profiles. High-resolution subbottom profiling data can be used to map anomalous features buried below the modern sedimentary cover and to interpret the recent sedimentary history. Data collected in this manner typically need to be postprocessed to correct for ship and towfish navigational parameters, and to remove artificial noise. The processed data, combined with seafloor mapping data (bathymetry and side-scan sonar), can be used to produce a complete three-dimensional picture of the geological and archaeological landscape. These systems do not have high enough resolution to image sites that are buried in the shallow sediments. However, a recently developed ROV-mounted subbottom profiler has been successfully employed to investigate shipwrecks (Mindell and Bingham 2001). As with side-scan sonar targets that must be visually verified, seismic data can also be verified by groundtruthing. This typically involves the collection of marine geological samples (usually sediment cores) along a seismic-stratigraphic profile that correlate to the subbottom imagery. For more information on the application of subbottom profiling to the investigation of submerged archaeological sites refer to Stright (1986). For theory and operational details of high-resolution (Chirp) subbottom profiling, refer to Schock et al. (1989).

A marine magnetometer can be towed in tandem with a side-scan sonar or subbottom profiler (figure 1.3), or towed by itself. The magnetometer measures the strength of the earth's magnetic field. Magnetic objects on the seafloor will locally disturb the earth's magnetic field and create an anomaly measured by the

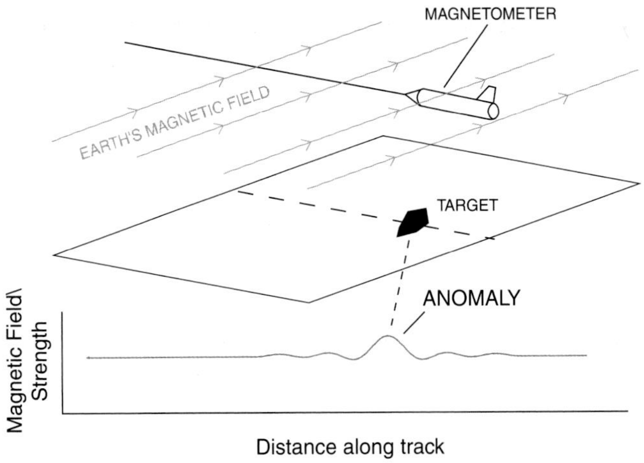

Figure 1.3. Magnetic field anomaly measured by a magnetometer, created by a target on the seabed. Refer to text for explanation.

magnetometer (figure 1.3). The anomaly is displayed on the shipboard acquisition system, and the location of the object or feature on the seafloor represents the magnetic target, similar to a side-scan sonar target. Again, ground-truthing is important for target identification and characterization. For more information on the application of marine magnetometers to shipwreck archaeology, refer to Clausen and Arnold (1976).

Visual Imaging Techniques

Visual imaging of acoustic and magnetic targets on the seafloor is a necessary component of archaeological oceanographic surveys. The visual identification of objects first identified by other means (such as side-scan sonar or magnetometer) determines whether the object is natural or man-made and enables archaeologists to evaluate the object's archaeological significance. A number of diverse methods can be used to accomplish this, including scuba diver inspection, inspection by remotely operated vehicle systems, and inspection by towed video camera systems (Oxley and O'Regan 2001). In addition, human-operated submersibles can be used for visual imaging. ROVs are particularly useful, especially when deployed from a research vessel equipped with a dynamic positioning (DP) system. This system of thrusters provides enhanced maneuverability and enables the ship to hold to a fixed location above the seafloor while the ROV is driven directly to the target (figure 1.4). Advances in subsea navigation, lighting, video imaging, and manipulation have improved the quality and general capabilities of ROV systems, especially for archaeological survey.

Some ROV systems can be operated with two vehicles in tandem to enhance their lighting and imaging capabilities (figure 1.4). Argus (see figure 2.2), is an imaging towsled, lighting platform, and depressor for the ROV Hercules (see figure 2.1) and ROV Little Hercules (see figure 2.6), which can be coupled to Argus via a fiber-optic tether (Coleman et al. 2000). The Argus–Hercules tandem ROV system was modeled after the Jason–Medea ROV system developed by the Deep Submergence Laboratory at Woods Hole Oceanographic Institution (Ballard 1993). Improvements were made to the vehicles' lighting and video systems, and the equipment was customized for the investigation of submerged archaeological sites, not general-purpose oceanography (Coleman et al. 2000). A sector-scanning sonar instrument mounted to an ROV is very useful for locating targets on the seafloor. The location from a side-scan sonar survey, for example, can be erroneous by up to several tens of meters due to layback and other navigational errors. In deep water or in areas of poor visibility the sonar device can pick up targets acoustically more than 100 meters away, then those fixed locations are used to help navigate the ROV to the targets. Geophysical instrumentation and oceanographic sensors can be mounted on ROVs, and, in conjunction with precision navigation, used to generate very high-resolution maps and images of the seafloor and archaeological sites such as shipwrecks (Singh et al. 2000; Foley and Mindell 2002). Geological and archaeological sampling equipment can also be mounted on ROVs to facilitate the collection of seafloor samples and other data that support the archaeological survey and visual identification.

Subsea tracking and navigation of the ROV systems are critical for locating targets on the bed. An ultrashort baseline (USBL) subsea navigation system can be employed to track one ROV or multiple ROVs at the same time (figure 1.5). A specialized transducer is mounted below the hull of the ship that transmits an acoustic signal that gets repeated by a transponder mounted to the ROV. The resulting time delay and depth and direction of the repeated

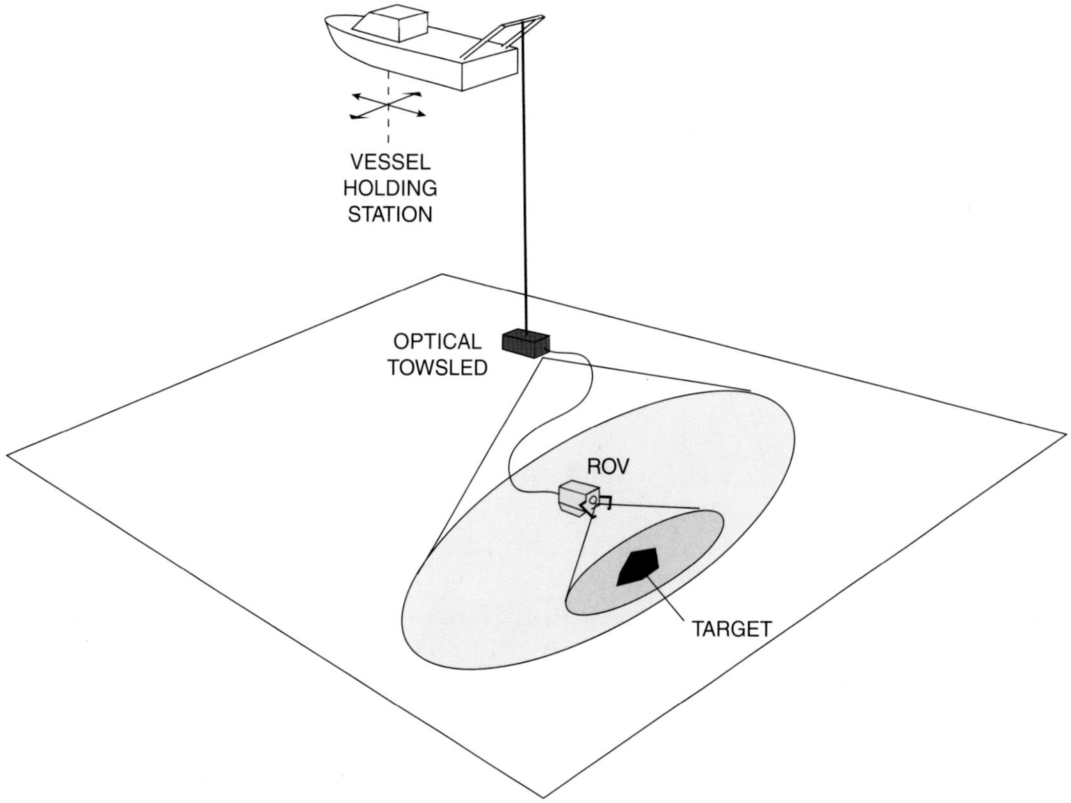

Figure 1.4. Use of remotely operated vehicles (such as the two vehicle system Argus/Little Hercules, shown in figures 2.2 and 2.6). Refer to text for explanation.

signal that is received by the ship translates into horizontal and verticle offsets. These offsets are then used to compute the position of the ROV in plan view relative to the ship, giving GPS positions for the ROVs. Sophisticated navigational software can be used to plot the ship and ROV positions accurately on a computer screen in real time. Precise subsea navigation is a critical component to conducting visual surveys and detailed mapping of submerged archaeological sites.

Human-operated submersibles (figure 1.6), which have limited capability compared to ROVs because they can stay on site only for short periods of time, can be used for archaeological site investigations, but not typically for reconnaissance surveys. Like ROVs, they can be equipped with scanning sonar, high-quality video cameras, lights, and sophisticated sampling equipment. An advantage to using a submersible is that it can operate independent of the surface ship, which may not be equipped with a dynamical positioning system. The submersible can use sector-scanning sonar to find the target of interest, then drive directly to it and operate with fine precision around the site. Submersibles can also typically lift heavier objects and carry more equipment and/or samples than ROVs, so for some tasks they have advantages.

For visual surveys of underwater archaeological sites, it is advantageous to have a number of different systems and to utilize a number of different techniques. Because of the high cost of shipboard operations, archaeological oceanographers need to avoid down time due to equipment malfunctioning and repair. That is another advantage of the tandem ROV system. If one of the vehicles needs repair, the other can be used independently.

In summary, the oceanographic tools and methodologies mentioned and described briefly in this chapter have been commonplace in marine geological

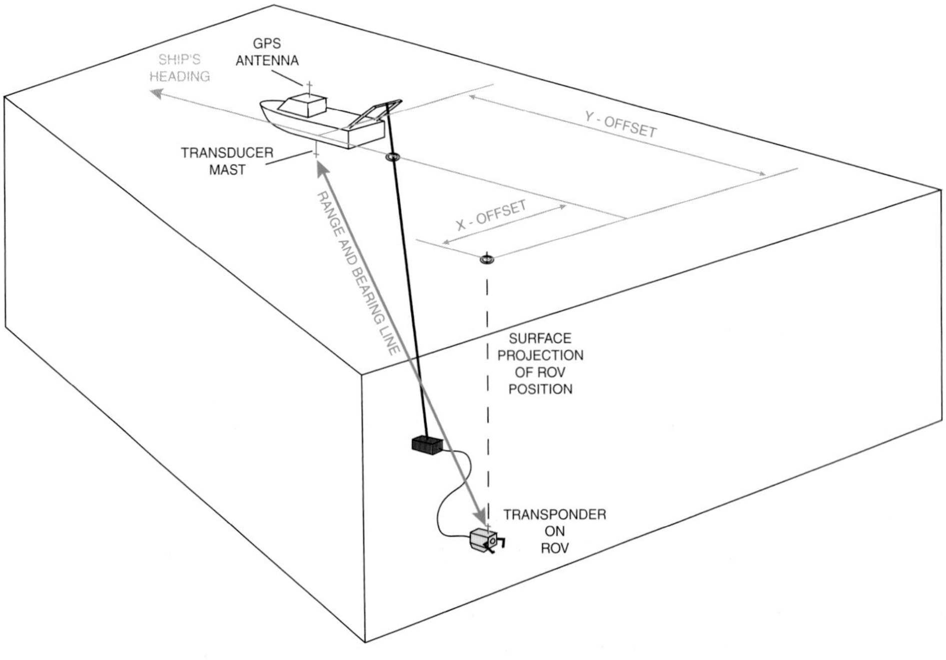

Figure 1.5. Use of precision subsea navigation for tracking an ROV. Refer to text for explanation.

Figure 1.6. Human-operated submersible PC-8B. The Bulgarian Academy of Sciences Institute of Oceanology submersible, sitting on the afterdeck of the research vessel *Akademik*, is a 3-person submersible equipped with scanning sonar, lights, still and video cameras, thrusters, and a manipulator. (Photo courtesy of Delcho Solakov, Bulgarian Academy of Sciences—Institute of Oceanology)

surveys, and also for marine biological transects and habitat characterization surveys. These same tools and methodologies are now being applied to deep-water archaeological studies, in the context of the new science of archaeological oceanography. For shallow-water archaeological studies, many of these tools and methodologies have been employed for several years, but not typically for long cruises operating 24 hours per day, and involving expensive scientific resources. With growing interest in this nascent discipline, we hope expensive assets like ocean-class ships and deep-diving ROVs and submersibles become available to archaeologists for investigating sites of historical significance.

References

Ballard, R. D. (1993). The MEDEA/JASON remotely operated vehicle system. *Deep Sea Research Part I* 40:1673–87.

Banning, E. B. (2002). *Archaeological Survey*. New York: Kluwer Academic/Plenum.

Clausen, C. J., and J. B. Arnold (1976). The magnetometer and underwater archaeology: magnetic delineation of individual shipwreck sites, a new control technique. *International Journal of Nautical Archaeology* 5:159–69.

Coleman, D. F., J. B. Newman, and R. D. Ballard (2000). Design and implementation of advanced underwater imaging systems for deep sea marine archaeological surveys. In *Oceans 2000 Conference Proceedings*. Columbia, MD: Marine Technology Society.

Dunnel, R. C., and W. S. Dancey (1983). The siteless survey: a regional scale data collection strategy. In *Advances in Archaeological Method and Theory*, Vol. 6, ed. M. B. Schiffer, 267–87. New York: Academic Press.

Fish, J. P., and H. A. Carr (1991). *Sound Underwater Images: A Guide to the Generation and Interpretation of Side Scan Sonar Data*. Falmouth, MA: Institute of Marine Acoustics.

Foley, B. P., and D. A. Mindell (2002). Precision survey and archaeological methodology in deep water. *ENALIA: The Journal of the Hellenic Institute of Marine Archaeology* 6:49–56.

Green, J. (1990). *Maritime Archaeology: A Technical Handbook.* London: Academic Press.

Mindell, D. A., and B. Bingham (2001). A high-frequency narrow beam sub bottom profiler for archaeological applications. *IEEE Oceans Engineering Conference Proceedings.*

Nance, J. D. (1983). Regional sampling in archaeological survey: the statistical perspective. In *Advances in Archaeological Method and Theory,* Vol. 6, ed. M. B. Schiffer, 289–356. New York: Academic Press.

Oxley, I., and D. O'Regan (2001). *The Marine Archaeological Resource.* IFA Paper No. 4. Reading, UK: Institute of Field Archaeologists.

Renfrew, C., and P. Bahn (2001). *Archaeology—Theories, Methods, and Practice.* London: Thames & Hudson.

Schock S. G., L. R. LeBlanc, and L. A. Mayer (1989). Chirp subbottom profiler for quantitative sediment analysis. *Geophysics* 54:445–50.

Singh, H., L. L. Whitcomb, D. Yoerger, and O. Pizarro (2000). Microbathymetric mapping from underwater vehicles in the deep ocean. *Computer Vision and Image Understanding* 79:143–61.

Smith, W.H.F., and D. T. Sandwell (1997). Global seafloor topography from satellite altimetry and ship depth soundings. *Science* 277:1957–62.

Stright, M. J. (1986). Evaluation of archaeological site potential on the outer continental shelf using high resolution seismic data. *Geophysics* 51:605–22.

The Development of Towed Optical and Acoustical Vehicle Systems and Remotely Operated Vehicles in Support of Archaeological Oceanography

2

James B. Newman, Todd S. Gregory, and Jonathan Howland

ROVs

A remotely operated vehicle, or ROV, is an underwater vehicle controlled by an operator who is not physically on board the vehicle. In practice, this almost always means that a cable connects the vehicle to a support vessel, with the operator on board the vessel. These vehicles use video cameras and other sensors to provide feedback to their remotely located operators and observers. ROVs have broad application in a variety of underwater activities, including search and rescue, survey and support for underwater cables, pipelines and oil recovery, scientific exploration and experimentation, ship hull inspection, and salvage work. And, of course, they are used in underwater archaeology. This section describes the major components of these vehicles, with an emphasis on deep-diving vehicles, and will treat the Hercules ROV developed by the authors (figure 2.1) as a primary example.

Most ROVs are close to neutrally buoyant, so that small thrusters (motor-driven propellers) can drive them up and down in the water column, and they have additional thrusters to move them about horizontally. Thrusters are propellers driven by electric or hydraulic motors.

ROVs are characterized primarily by size, usually measured in horsepower, thrust, or in-air weight, and by depth capability. In general, smaller vehicles (under 100 kg air weight) are inappropriate for work beyond a few hundred meters. As depth capabilities increase, vehicle weights inevitably grow, as heavier components are required to counter high ambient pressures, and thrust requirements are thus increased as well. Increased depths also mean higher horizontal forces from currents and horizontal motion imposing significant drag on the vehicles' umbilical cables.

Figure 2.1. Hercules during recovery. (Copyright Institute for Exploration)

Shallow-water inspection vehicles are often thin-walled aluminum or plastic cylinders with an internal camera behind an acrylic dome, and thrusters stuck on the outside. The classic example of this kind of vehicle is the Benthos Mini-Rover. To handle pressures beyond a few hundred meters of depth, the weight of the cylinder requires additional flotation material to achieve neutral buoyancy, and the vehicles quickly get larger and more complicated.

The classic *open-frame* vehicle is a more complex and more capable system. As the name implies, it consists of a framework, usually aluminum, supporting the various components of the vehicle, with a buoyancy module on top to bring the complete package to neutral buoyancy. Electronic components, including video cameras, are generally mounted in metallic pressure housings to protect them from seawater and pressure. Larger open-frame vehicles with manipulators are often described as being "work class" vehicles. Hercules falls into this general category. The term *intervention* is used to refer to manipulation, jetting, suction, and other interaction with the environment.

Towsleds

A less common class of remotely operated vehicle is referred to as a towsled. This type of vehicle has no buoyancy module and has significant weight in water, so it is supported by a taut cable from the support ship. Its positioning is largely controlled by movements of the ship. Towsleds are generally less expensive to acquire and to operate than ROVs, but cannot be used for close-in inspection or intervention.

Combining a heavy towsled with an ROV, with a neutrally buoyant cable connecting them, allows the towsled to absorb ship-induced motion, and to

Figure 2.2. Argus towsled. (Copyright Institute for Exploration)

counter horizontal current forces with its weight, and relieves the ROV from those disturbances. This means that the ROV can perform more careful work and use less power than would be the case if it was connected directly to the support vessel. This is the concept behind the Institute for Exploration's (IFE's) Argus towsled (figure 2.2). In addition, Argus carries video cameras and lights, so it illuminates and provides views of the ROV and the worksite.

Figure 2.3 shows Argus, dangling from the stern of a ship on its steel-armored cable. A yellow neutrally buoyant tether connects Hercules to Argus. The figure also shows the light patch that Argus provides for Hercules.

Telemetry

Communications between the ROV and the support ship are generally referred to as telemetry. This includes commands being sent down to the vehicle to control thrusters and other functions, and signals up from the vehicle, including sensor data, status indications, and video. The development of fiber optic cables and opto-electronics to support high-speed data transfer over optical fiber has been the most significant technical advance enabling deep ROV technology since the development of the solid-state video camera. Over longer cables, fiber can carry orders of magnitude more data than coaxial or other electrical cables. Systems are commonly available to support multiple high-quality video channels over a single fiber, whereas the best coax-based technology would be hard-pressed to support a single video signal over longer cable lengths. Optical fiber also supports transmission of other high-bandwidth data, in particular, imaging sonars and digital still cameras.

Figure 2.3. Argus/Hercules vehicle/cable configuration. (Copyright Institute for Exploration)

Hercules and Argus use optical telemetry systems that multiplex (combine) various serial data channels and standard definition (NTSC) video onto a single optical wavelength in each direction. High definition video is transmitted on a dedicated wavelength by separate hardware. These optical signals are transmitted on a single fiber using passive *wave division multiplexors* that can handle four discrete wavelengths. Most sensors on Hercules and Argus have standard serial outputs, which are transmitted through the telemetry system from the sensor to a topside computer. A microprocessor on the vehicle handles a few analog sensors and accepts serial control commands from the topside control system.

Control Systems

ROV operator indications and controls are, by definition, remote from the vehicle. With a few trivial exceptions, it is not possible to provide sensor data to the operator without the use of computer technology. Similarly, the commands of the surface operator have to be transmitted over the telemetry link and turned into action by computing and control systems. For example, if a pilot pushes a joystick forward, that motion of the stick must be interpreted by the control system and translated into commands to a motor controller or hydraulic valve system to generate appropriate thruster actions.

In general, the computer control system of a deep-diving ROV has several primary functions:

- hotel (power monitoring and switching, general health and status monitoring);
- sensor reporting and control;
- intervention and intervention control (manipulators and other tools);
- vehicle control;
- operator interface.

Hotel function As explained in the section on Power Systems, one of the components of a typical vehicle power system is a bank of DC/DC converters and switches that control power to various sensors and actuators on the vehicle. These switches must be controlled by the hotel function, which operates between the operator or pilot interface and the devices themselves. In addition, safe and reliable operation of the vehicle requires monitoring of temperatures, voltages, leakage currents, and other possible vehicle fault conditions. These hotel systems can be fairly complex, particularly in the case of electrically actuated deep vehicles, which can require careful modulation of power demands on the system to avoid overstressing cabled power systems.

Sensor reporting and control Some instruments, such as depth and magnetic heading sensors, are required for the safe and effective operation of ROV systems, and virtually all ROVs provide control and reporting services using these types of sensors. Other sensors, such as north-seeking gyrocompasses and Doppler velocity logs, allow precise vehicle control and mapping, and any vehicle suited for precise archaeological mapping will be fitted with a suite of these sensors and software to control them and process and record their data stream. Other instruments, such as dissolved oxygen or conductivity–temperature sensors, add value to archaeological surveys, and effective ROVs can be expected to include their use.

Intervention control This refers to the ability to adapt to a variety of sampling tools, usually including the use of one or more manipulator arms, typically actuated by a hydraulic system and controlled by a dedicated portion of the vehicle's control system. An example of this type of tool is the suction system developed for Hercules (see chapter 4).

Vehicle control Modern ROV systems intended for survey and intervention utilize navigation sensors and closed-loop control systems to enable precise, computer-controlled maneuvering of the vehicle in 4 or more degrees of freedom (Whitcomb and Kinsey 2003). Automated station keeping in position, depth, altitude, and heading greatly enhances the effectiveness of the ROV in most of its modes of operation. In two-body systems such as Hercules/Argus or Jason/Medea, automated heading control of the depressor vehicle is also highly desirable.

Operator interface The pilot of an ROV has only a computer and various peripherals with which to control the vehicle. Automated control can decrease the workload on a pilot, but typically, a pilot is overtasked and saturated with information. Expecting a pilot to use "traditional" mouse and keyboard systems for delicate and complicated vehicle control tasks is not practical, so ROV control systems typically use joysticks similar to (and in some cases identical to) those used in computer gaming, and other dedicated human interface devices intended to help the pilot achieve the goals of the mission. Use of touch screens, in particular, is valuable in allowing fairly conventional computer interfaces to support less precision-critical operator input requirements.

The Hercules system is an example of an effective use of computing resources in vehicle control. It was derived from software written for the Jason vehicle (Howland et al. 2008) and has been modified to suit the specific demands of Hercules. A subsea PC-104 runs a very low-level operating system, communicating with a Linux-based topside computer via a serial RS-232 connection. The PC-104 converts high-level commands sent by the Linux system into direct control of relays and analog converters that actuate hydraulic valves. A similar microprocessor-based system is held in the subsea Argus housings, also communicating with the topside computer.

The Hercules pilot interacts with the topside control system using a separate computer running a Windows-based graphical user interface (GUI), accessed through a touch-screen, and using a custom "joybox" containing joysticks and a variety of knobs and buttons. The Windows computer communicates with the Linux-based topside system via network commands. The engineer (Argus pilot) uses a separate copy of the same GUI on yet another computer, and can control virtually all of the vehicle capabilities while providing assistance to the pilot.

Power Systems

There are four major elements to the power systems for ROVs. The first element is the topside or shipboard system, which usually includes a step-up transformer to provide an elevated voltage to run over the cable to reduce resistive losses. The topside power system should include a ground fault interrupter that shuts off power automatically if leakage to ground is detected. Output voltages of 750 to 3000 VAC are common. Some vehicle systems have frequency converters that raise the frequency of the AC power going to the vehicle (400 Hz is most typical), which allows smaller and lighter transformers on the vehicles themselves. Argus and Hercules use three-phase 50- to 60-Hz shipboard power without frequency conversion, and raise the voltage in the cable to 2400 VAC.

The second element is the umbilical that carries the power to the vehicle or vehicles. This is a specialized product that includes optical fiber, coaxial, or other specialized conductors for signals, a steel or synthetic strength element and conductors (usually copper) to deliver electrical power to the vehicle. These cables are optimized to support high voltage (meaning carefully specified and constructed insulation) and moderate current (meaning copper conductors of modest gauge) while limiting overall diameter and weight (for lightweight tether cables). A *slip ring* is also required at the topside winch to support the transfer of current between the rotating inner end of the cable and the nonrotating outside world. Some slip ring assemblies support both electrical (power) and optical (communications) transfer.

The third element in the ROV power system is the step-down transformer on the vehicle that reduces the voltage coming down the cable to practical levels for use by high-power devices and for subsequent conversion to lower voltages.

The fourth element is the bank of converters that take the reduced-voltage power to conditioned low-voltage DC power to operate the vehicle's control electronics, cameras, and sensors.

Hydraulic Systems

Most deep ocean ROVs incorporate some form of hydraulic system to provide power to tooling, manipulators, and (unless they use electric propulsion) thrusters. The fundamental principle is to provide a convenient and reliable means of converting electrical power into fluid power, and to distribute this fluid power to end user components, which convert the potential energy into linear or rotary mechanical motion. Industrial hydraulic systems, ranging from power steering systems in cars, to heavy construction equipment, to aerospace actuators, provide a vast array of off-the-shelf components that are adaptable for use on an ROV. There are five primary components to any hydraulic system: the fluid, the prime mover, the end users, the valves, and the plumbing system.

Hydraulic fluid is usually a petroleum-based product, though some vehicles use a water/glycol mixture and others under stringent environmental regulations use biodegradable synthetic fluids. The key features are lubricity and a suitable viscosity grade. The fluid acts as a lubricant for the mechanical parts, so must provide good lubrication and have high film strength. Lubrication is related to the viscosity of the fluid, which must remain within suitable limits over a wide range of temperatures. Deep-ocean water temperature is as low as 2°C, but the vehicle must be able to operate in tropical surface water without an excessive decrease in fluid viscosity. Hercules uses fluid derived from mineral oil at a constant viscosity grade of 22.

The *prime mover* consists of an electric motor coupled to one or more hydraulic pumps, generally at the highest practical voltage available on the vehicle, as higher voltages improve efficiency. One or more hydraulic pumps are coupled to the motor's output shaft, selected from two general varieties: fixed displacement and variable displacement. A fixed displacement pump provides pressurized fluid at a constant flow rate. Excess fluid not used by the end users is "dumped" to a reservoir by means of a relief valve. In a variable displacement pump the output pressure is fixed and the flow rate of the pump adjusts automatically based on the demand of the end users. The final feature of a prime mover is the nominal horsepower, which is typically characterized by the output power of the electric motor. In general, the fluid's output pressure and flow rate are both directly proportional to horsepower. Vehicles with end users that require large amounts of flow at high system pressure will require a motor of substantial horsepower. Common ranges of horsepower for ROVs are from 1 to 200 hp with hydraulic pressure typically ranging from 250 to 5000 psi. The system on Hercules uses a single variable displacement pump that produces up to 11 gallons per minute of fluid at 3000 psi and is driven by a 2400-VAC three-phase motor rated at 20 hp.

End users in a hydraulic system are either linear actuators (extending/retracting rams) or rotary actuators (hydraulic motors). Historically, the most common use of hydraulic components on ROVs is in thrusters utilizing hydraulic motors. While some deep-ocean ROVs have adopted electric thrusters to realize better power efficiency and controllability, hydraulic thrusters remain a favored option based on simplicity and ease of integrating on a vehicle with a variety of other hydraulic needs. The second most common mechanism on an ROV, using a combination of linear and rotary actuators, is a manipulator arm. Commercially available manipulators range in size, weight, range of motion, strength, dexterity, and price. With very few exceptions, manipulators are hydraulically powered devices requiring between 0.5 and 7 hp with supply pressures between 1000 and 3000 psi. Other hydraulic end users on Hercules, specifically for archaeological applications, are hydraulically driven water-jet and suction/excavation pumps, actuated sample and tool boxes, and actuators for camera pan and tilt functions.

The *valves* in a hydraulic system control the flow and pressure of fluid to individual end users. There are two fundamental families of valves, each with its own list of varieties: static devices and those that are activated electrically. Static devices include check valves, safety relief valves, pressure reducing valves, lock valves, and static flow control valves in various shapes and sizes. Electrically activated valves use the principle of an electromagnet to open and close the fluid ports, thus controlling flow. Simple varieties, known as solenoid valves, act as on/off devices. More sophisticated valves, proportional or servo-valve controlled, enable a fine degree of metering of flow rate to the end user. For example, a hydraulic thruster is controlled with a proportional or servo-valve, which allows for a fine degree of throttling between slow and full speed. In the archaeological excavation system on Hercules, the water pumps are driven by hydraulic motors that are controlled with proportional valves, giving very fine control of the jetting and excavation operations (for additional information, see chapter 4).

Finally, the skeleton of a hydraulic system is the *plumbing*, consisting of a reservoir, tubing and hose, and a filter. For an ROV, the plumbing must be composed of corrosion-resistant materials, such as 316 stainless steel, to ensure that corrosive degradation of the fittings and tubing will be minimized. The reservoir is, essentially, the tank where excess oil resides before being recirculated through the pump and plumbing system. The filter is a critical component required to ensure that any particulate matter in the oil will not contaminate delicate seals or damage precisely machined parts in the valves, actuators, or pump.

Cables, Winches, and Handling Systems

ROV systems by definition rely on cables for their power and data transmission, and often for mechanical connection to a support ship. Cables are a complex technology with many facets, but a few points are relevant here. For deep-water use and good quality video, fiber optics are required. Copper conductors to carry power are combined with appropriately packaged optical fibers and

Figure 2.4. Winch with steel-armored wire. (Copyright Institute for Exploration)

a strength member to create a complete cable. The assembled package must survive mechanical loading, running over sheaves, twisting, snap loads, and occasional hockling. Strength members are generally either steel or synthetic. Steel cables are less expensive and much heavier than synthetic. The weight can be an advantage in many situations, as it helps keep the cable vertical during ship transits and in the presence of currents.

For most systems, a winch will be required on shipboard, to deploy and retrieve the cable (figure 2.4). The winch must be fitted with a slip ring to connect fiber optic signals and power from the cable on the rotating winch drum to a fixed point on the frame of the winch. An overboarding sheave that holds the cable out away from the ship (usually on an A-frame on the stern) is also required. Either the sheave or the winch is usually instrumented to provide readings of the length of cable deployed and instantaneous tension.

Navigation

Most ROV operations require the use of an acoustic positioning system to provide vehicle position information. This supports the documentation and reacquisition of sites of interest, and helps the vehicle operators orient themselves and operate safely.

Acoustic navigation systems can be based on deployed transponders sitting on the seafloor around a site of interest. These are usually referred to as long baseline or LBL systems, since the distance between reference points is long relative to the wavelength of the sound used (12 kHz is typical). The main alternative approach is to use multiple acoustic receivers on the ship. These can be separated a modest distance, limited by the geometry of the ship (short baseline), or more commonly the receive hydrophones are combined in a single package (ultrashort baseline or USBL). The USBL system provides vehicle range and direction relative to the ship, and computation of geo-referenced position not only is dependent on the ship's heading and position (generally measured with GPS), but also requires accurate instrumentation of the ship's attitude (pitch and roll), except in shallow deployments in calm water.

LBL systems are generally more accurate than USBL, but their transponders must be deployed and surveyed, which takes time, and they are useful only in a limited area on any one deployment. Accuracies of a few meters are realistically achievable with LBL systems. USBL system accuracies are generally proportional to the range from the ship to the vehicle (slant range), and an accuracy of 1% of slant range is realistically achievable. For systems like Argus and Hercules that are positioned more or less vertically under the support ship, slant range equates closely to depth, so the uncertainty in the USBL system information can be assumed to increase proportional to vehicle depth.

Software is used to combine computed vehicle position with ship position and to show ship and vehicle positions on map-like displays. These capabilities are commercially available, and are referred to as integrated navigation systems. Examples are WinFrog (Fugro-Pelagos) and Hypack (Hypack, Inc.).

In addition to the acoustic systems that provide geo-referenced vehicle position, some vehicles use vehicle-based navigation systems. These can be inertial reference systems based on double-integrating accelerations, or acoustic systems using a Doppler velocity log (DVL), which use sound reflections off the seafloor to compute velocities. While these systems suffer from drift, they are valuable tools for tracking short-term movements of the vehicle. Hercules uses a DVL and a dedicated software package ("DVLNAV") to supplement the USBL system with high-resolution tracking (Whitcomb and Kinsey 2005).

Video Systems

The importance of high-quality video systems to any ROV-based investigation cannot be overemphasized. The difference between typical commercial ROV cameras and more expensive "broadcast-quality" systems can be the difference between identifying an object of interest and missing it entirely. Piloting a vehicle is much easier with clean, high-resolution video.

In recent years, high definition (HD) video cameras with high-resolution imaging elements (typically 1080 lines) have been packaged for underwater use. In combination with high-quality zoom lenses with well-designed optics, the HD format has enabled ROV-based imagery to be more detailed than imagery received by a human's eyes present in an occupied submersible. Of course, these video capabilities also lend themselves well to documentation and broadcast of the footage that would never be considered with more typical ROV video.

Telepresence

One of the most frequently voiced arguments for using human occupied underwater vehicles for scientific research in the deep sea claims that a human presence is required to generate an understanding of the environment. Unfortunately, the disadvantages of supporting human beings underwater are significant, including very high expenses and very poor use of time at sea. But technology has enabled an answer to this argument, at least in part, in the form of telepresence—using video and other data to give a remote observer a sense of being present at the study site.

The ideal telepresence system would simulate the observer's immersion in the environment under study. Turning one's head would provide a view in that direction, seamlessly. IMAX film comes close to this experience, but cannot support real-time use and requires enormous cameras. Limitations on resolution of video cameras are going to preclude this ideal for many years still. But there are practical approaches that can provide great results with current technology.

Almost any ROV system can generate telepresence, but careful use of high-quality video from multiple cameras, and additional information like vehicle heading and position presented seamlessly to the observer, can enhance the effectiveness of the system enormously. In addition, the experience can be shared in real time over satellites and high-speed networks, and recorded to be experienced again at later times.

One of the most powerful features of the Institute for Exploration (IFE) vehicle system is the "eye in the sky" that the HD camera on Argus provides. Looking down from above, it can image Hercules working on a site, providing an invaluable supplement to the primary HD video from Hercules. Multiple SD camera views on Hercules further assist observers in developing a situational awareness by providing different perspectives on the study site and views outside the narrow view of the primary camera. Moving the additional cameras on Hercules to high definition would assist this effort. A future vehicle optimized for supporting telepresence might have multiple high-definition cameras set up for panoramic viewing.

Lighting

In the deep sea, where daylight has been absorbed in the first few hundred meters, vehicle systems must necessarily be equipped with underwater lighting. Relative geometry of the lights and cameras is more critical than when viewing in air, because the particulates in the water generate backscattered light that tends to obscure viewing, much like driving in rain or fog with high beams on. To avoid backscatter, illumination of subjects of interest should be from a very different angle, so most of the volume of water between camera and subject is not illuminated. Typically, ROVs position their lights as high and as far out to the sides as practical. Hercules carries up to four ballasted HMI lamps that are fairly efficient (compared to incandescent lamps) and that provide a color temperature close to daylight. These lamps are based on lamps built for movie making and videography, adapted for underwater use.

For the best quality underwater video, the ideal is to provide lighting from a separate vehicle. Toward this end, the Argus towsled vehicle carries a pair of 1200-W HMI lamps, trained down and forward. By positioning Argus over Hercules at a work site, its powerful lamps can be used to provide an ambient lighting effect much like daylight. This effect is shown in cartoon form in figure 2.3.

Control Vans

Operating the IFE vehicles, or any sophisticated ROV system, requires a significant suite of shipboard electronics equipment. Most large systems that are moved between ships use 20-ft ISO containers (or "vans") that are configured as mobile control rooms. IFE uses two vans, with an wide passageway between them. The "control" van houses most of the electronics, and can be used by itself if necessary. The "imaging" van is primarily for science observers. Each van has a number of video monitors, and dedicated computer displays and intercom stations for each operator and observer (figure 2.5).

There are several specific watch-stander positions for Hercules operations. The ROV pilot is the person with primary responsibility for controlling Hercules thrusters, manipulators, and other functions. This person uses a dedicated joystick handbox and a touch screen to control the vehicle. The engineer (also know as Argus pilot) controls the winch that raises and lowers Argus, and all the functions aboard Argus. The engineer also plays a backup role for the ROV pilot, assists during manipulation, and generally helps maintain the pilot's situational awareness and ability to get the job done safely. The navigator is the only person authorized to give commands to the ship's bridge to change position or speed. The navigator operates the navigation computers, monitors navigation information, and defines ship movements for launch and recovery evolutions. The *video engineer* has responsibility for keeping the video and

Figure 2.5. Shipboard control van. (Copyright Institute for Exploration)

Figure 2.6. Little Hercules. (Copyright Institute for Exploration)

computer systems operating, for recording video and data as required, and for controlling zoom, focus and exposure settings on the HD cameras. In addition to these technical roles, a *data logger* and a *watch leader* are required on each watch. These roles are often filled by members of the science team.

Other Vehicles

Before Hercules was built, IFE developed a much smaller deep-ocean vehicle in 2000, strictly as a video-acquisition vehicle. Little Hercules (figure 2.6) is an open-frame ROV, approximately 600 lb in air, with four ½-hp electric thrusters and no hydraulics. It can carry the same high-definition camera and works in tandem with Argus. Little Hercules lacks manipulators and most of the sensor capabilities that are on Hercules.

The Echo vehicle (figure 2.7) is a towed, neutrally buoyant side-scan sonar platform capable of working as deep as 3000 m. It is towed behind a heavy depressor weight, to isolate the side-scan system from ship motions. Echo uses a dual-frequency system to simultaneously get higher resolution (400 kHz) and longer range (100 kHz) than would be possible with a single frequency. Typical height off the bottom for Echo is 50–100 m, and it can generate images as wide as 800 m. It also carries a subbottom profiler, a downward-looking sonar that can penetrate tens of meters below soft bottoms to reveal underlying structures.

Side-scan sonar uses sound, reflected at oblique angles off the bottom, to create an image of the seafloor made up of bright spots, representing good acoustic reflectors; and shadows representing lack of reflections (Fish and Carr 2001). These images, presented as a moving cascade over time as the system

Figure 2.7. Echo, with depressor weight at *left*. (Copyright Institute for Exploration)

moves over the bottom, paint a picture on which objects such as rock outcrops and shipwrecks appear clearly. As with most systems, compromises are made between area covered (range and speed) and resolution of the resulting image. Most important in this optimization is the frequency of the acoustic signals. IFE's most typical searches, for wrecks of ancient ships, requires finding fairly small targets often indistinguishable on sonar from modern wrecks, rock outcrops, and seafloor trash, so relatively high frequencies are required and ranges are fairly modest. Even so, visual confirmation of sonar targets is almost always required before confirming that a target is of archaeological interest (refer to chapter 1). An exception is illustrated in figure 2.8, which shows side-scan image of a modern shipwreck in Lake Huron, made using Echo's 400-kHz channel during a survey in 2001. The image clearly shows a ship upright on the bottom, with the stern broken off.

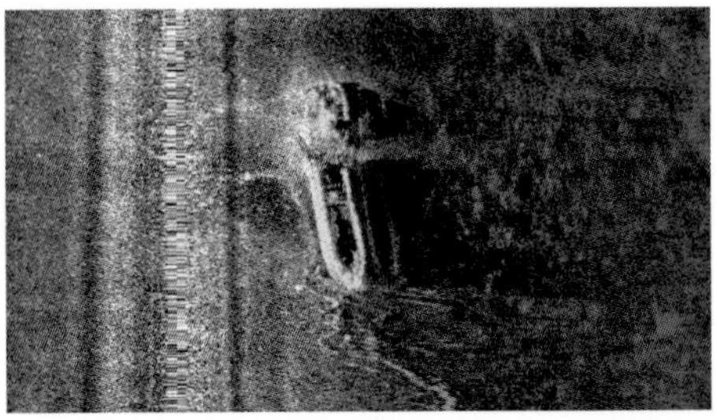

Figure 2.8. Echo image of a modern shipwreck in Lake Huron. (Copyright Institute for Exploration)

VEHICLE SPECIFICATIONS

Hercules Specifications (approximate)

Air weight 5200 lb
Maximum depth: 4000 m
Length: 9 ft
Width: 5.5 ft
Height: 6.5 ft
Payload: 250–350 lb
Tether cable voltage: 2400 nominal, 3-phase 50–60 Hz
Maximum power: 20 KVA

Argus Specifications (approximate)

Air weight: 4000 lb
Maximum depth: 6000 m
Length: 11 ft
Width: 3.5 ft
Height: 4 ft
Tether cable voltage: 2400 nominal, 3-phase 50–60 Hz
Maximum power: 7 KVA

References

Fish, J. P. and H. A. Carr (2001). *Sound Reflections: Advanced Applications of Side Scan Sonar.* Orleans, MA: Lower Cape Publishing.

Howland, J., L. Whitcomb, T. Thiel, and D. A. Smallwood (2008). A software system for control and operation of underwater vehicles. Woods Hole Oceanographic Institution technical report, in preparation.

Whitcomb, L. L. and J. C. Kinsey (2005). DVLNAV Homepage. Baltimore, MD: Johns Hopkins University (http://robotics.me.jhu.edu/~llw/dvlnav/ updated on 19 October 2005).

3 High-resolution Optical Imaging for Deep-water Archaeology

Hanumant Singh, Christopher Roman,
Oscar Pizarro, Brendan Foley,
Ryan Eustice, and Ali Can

The physical characteristics of the medium lead to fundamental constraints on high-resolution imaging underwater (Jaffe 1990, Singh et al. 2000a) for both optical and acoustic imaging. In addition, the platforms (Yoerger and Mindell 1992; Singh et al., 2004) from which these imaging sensors are deployed impose their own set of operating constraints as does the infrastructure associated with navigating these assets (Whitcomb et al. 1999). In this tutorial we look at these effects separately and in combination with a particular emphasis on the applications for deep-water archaeology.

If we consider optical imaging, the primary consideration is that of the rapid and nonlinear attenuation of the visible spectrum in seawater (McGlamery 1975). Individual images may be degraded in intensity due to nonuniform lighting and degraded in color due to the selective attenuation of color. Also, large objects cannot be framed within a video or other optical camera's field of view. Thus, obtaining a global perspective of an archaeological site of interest on the seafloor cannot be accomplished with a single image. Instead this involves running a carefully planned survey over the site, collecting a series of overlapping images, identifying common features in the overlapping imagery, and then merging these images into a larger composite photomosaic.

Several factors make this a hard problem. There may be constraints on the way underwater vehicles can perform surveys due to insufficient accuracy in small-area navigation and a lack of mechanisms to automatically control the vehicle. Physical constraints on the distance separating cameras and lights as well as constraints on the energy available for operating the lights also constitute major impediments. Finally, and most important, the unstructured nature of the underwater terrain introduces incremental distortions into the photomosaic as successive images are added into the composite mosaic.

In the rest of this chapter we look at some of the issues highlighted above. The following section looks at the issues associated with obtaining the best possible imagery in underwater environs. The next two sections look at how we register individual overlapping images to form a composite image or photomosaic, and utilize such imagery to construct three-dimensional image reconstructions or stereo photomosaics with a single moving camera. The final section concludes with a look at other modalities for high-resolution seafloor imaging and how they complement optical imagery.

Obtaining High-quality, High-resolution Imagery Underwater

Lighting, Backscatter, and Camera-to-Light Separation

For the purposes of high-resolution imaging underwater one of the primary issues is the amount of power available for lighting. The rapid attenuation of the visible spectrum underwater implies that we need considerable light energy to image objects that are far away from the imaging platform. Unfortunately, it is not enough to simply put more energy in the water. One other complication arises in underwater imagery due to the presence of suspended particles in the water column. As we increase the amount of light energy in the water, we also increase the amount backscattered by the suspended particles and obscuring the object of interest. Instead, one must make an effort to separate the camera and light source to minimize the common illuminated volume while ensuring uniform lighting over the area of interest. A good example of this phenomenon is illustrated in the image sequence shown in figure 3.1. The images vividly

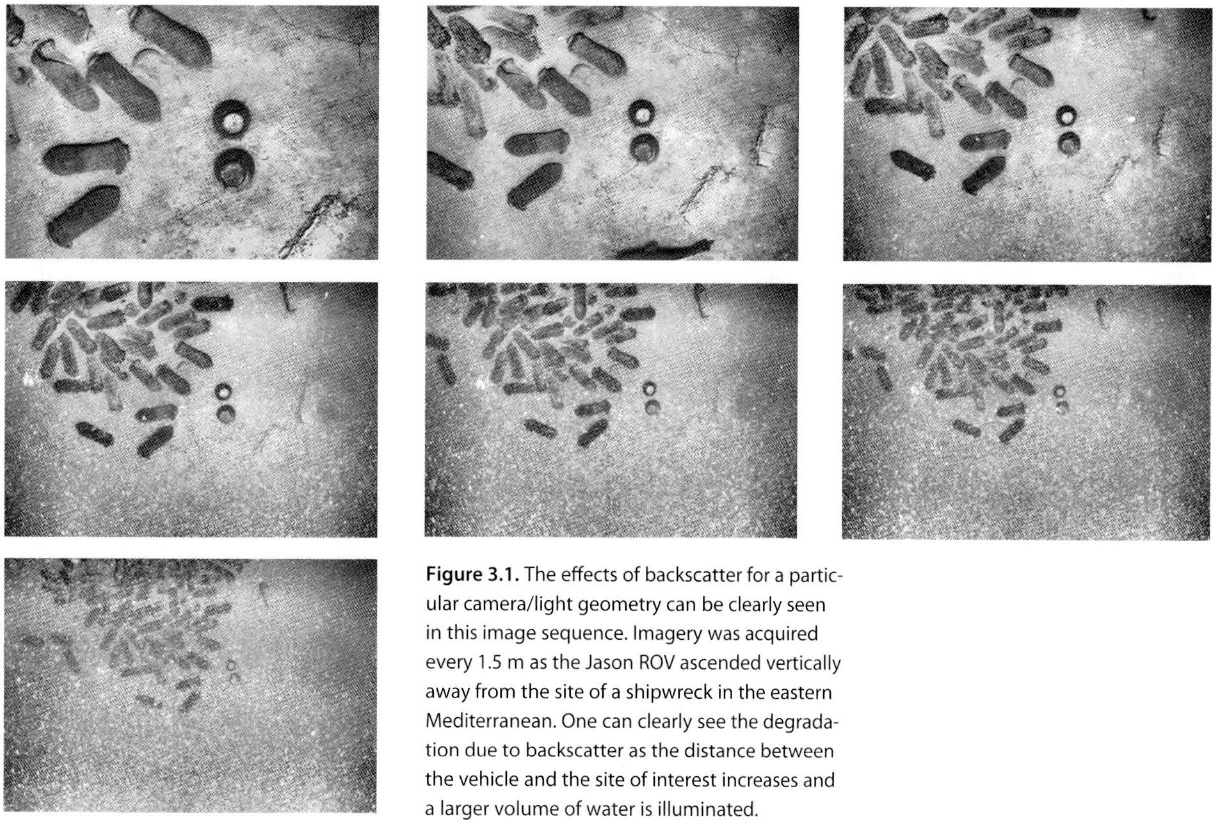

Figure 3.1. The effects of backscatter for a particular camera/light geometry can be clearly seen in this image sequence. Imagery was acquired every 1.5 m as the Jason ROV ascended vertically away from the site of a shipwreck in the eastern Mediterranean. One can clearly see the degradation due to backscatter as the distance between the vehicle and the site of interest increases and a larger volume of water is illuminated.

illustrate the trade-off in image quality versus image coverage for a given (in this case the Jason ROV) imaging platform.

Imaging platforms with larger baseline separation are thus preferable but ultimately the separation can become so large as to cast shadows due to inadequate light energy being reflected along the direct path from object to camera (Singh et al. 2000b).

Lighting Equalization

The nonlinear attenuation of the visible spectrum often yields nonuniform lighting across an image. Nonuniform lighting may arise in an image due to difference in the traveled path lengths between different parts of the image in relation to the light source. For example the center of an image may be more brightly lit than the edges where the light has traveled a greater distance to the camera. Standard methods exist in the image processing literature for dealing with such situations. They vary from simple bulk intensity adjustments to methodologies for adaptive histogram equalization based on different intensity distributions across an image (Eustice et al. 2002). Figure 3.2 shows a typical low-contrast image on the top as taken by the ABE AUV that manifests nonuniform lighting with a brighter center and a drop off in image intensity toward the corners. The bottom image is the result of adaptive histogram equalization applied to achieve a balanced intensity distribution across the image. While measures for image enhancement are subjective the bottom image is more uniformly lit and has far greater contrast.

Color Compensation

One other aspect of individual images under water is associated with color imagery. The color specific attenuation of the visible spectrum under water leads to imagery that, depending on the emitted light spectrum, color response of the camera, and distance to the object, can yield imagery that has a preponderance of green or blue. The red portion of the visible spectrum is highly attenuated and difficult to capture.

One way of overcoming this effect is by utilizing cameras with high dynamic range. The enhanced dynamic range allows us to separate out the components of color that are a function of attenuation and that vary significantly but slowly across the image as opposed to the higher frequency components that are associated with differences in color inherent to the objects being imaged. This process is illustrated in figure 3.3 where we can see the effects of our algorithm when applied to imagery collected on a 4th-century BCE shipwreck in the Aegean Sea.

Photomosaicking

Photomosaicking Planar Sites

As was alluded to earlier, a composite view of a large area under water can be obtained only by exploiting the redundancy in multiple overlapping images distributed over the scene through a process known as photomosaicking.

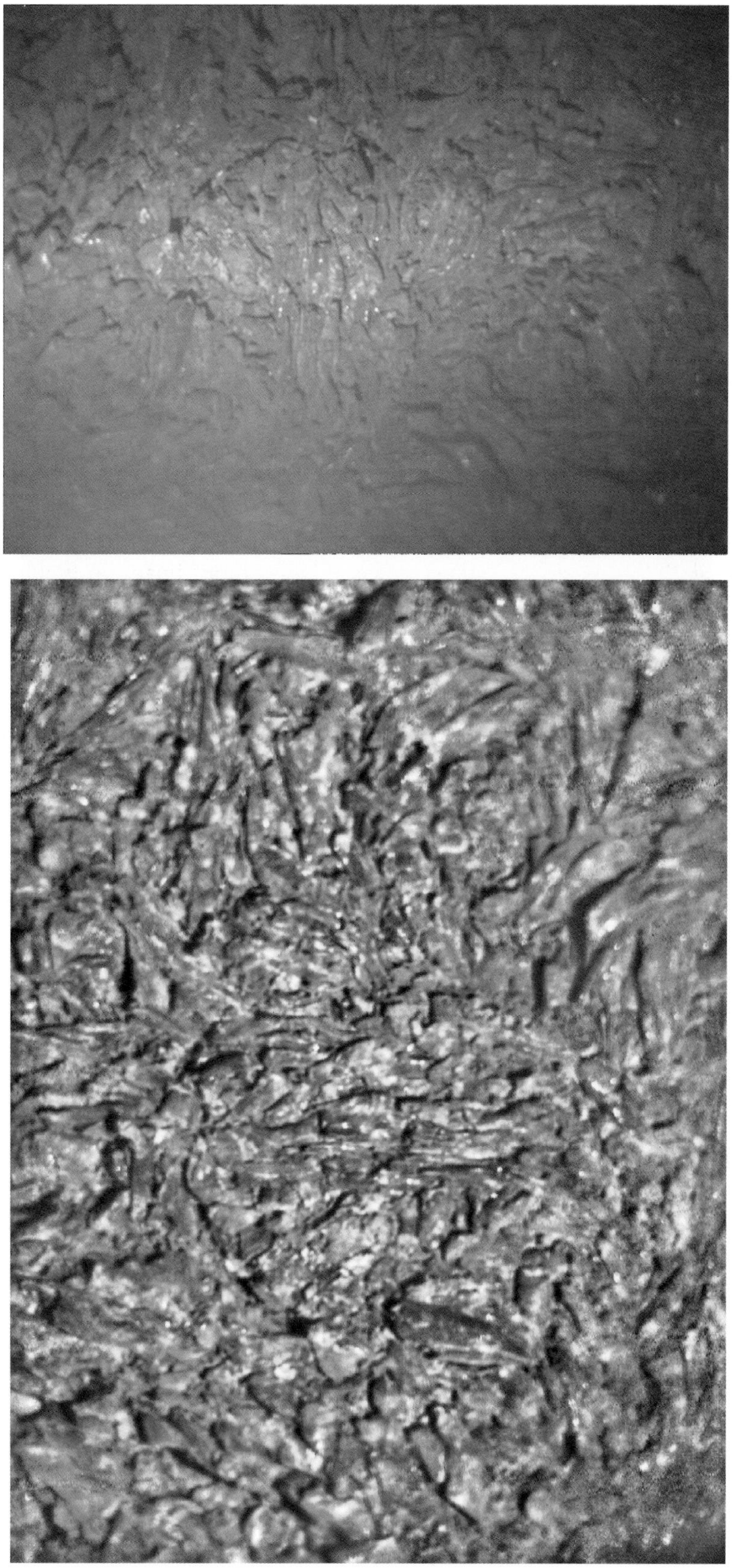

Figure 3.2. Original low-contrast imagery (*top*), and its adaptively histogram-equalized counterpart (*bottom*). The image on the right is more uniformly lit and has far greater contrast. (Copyright Institute for Exploration)

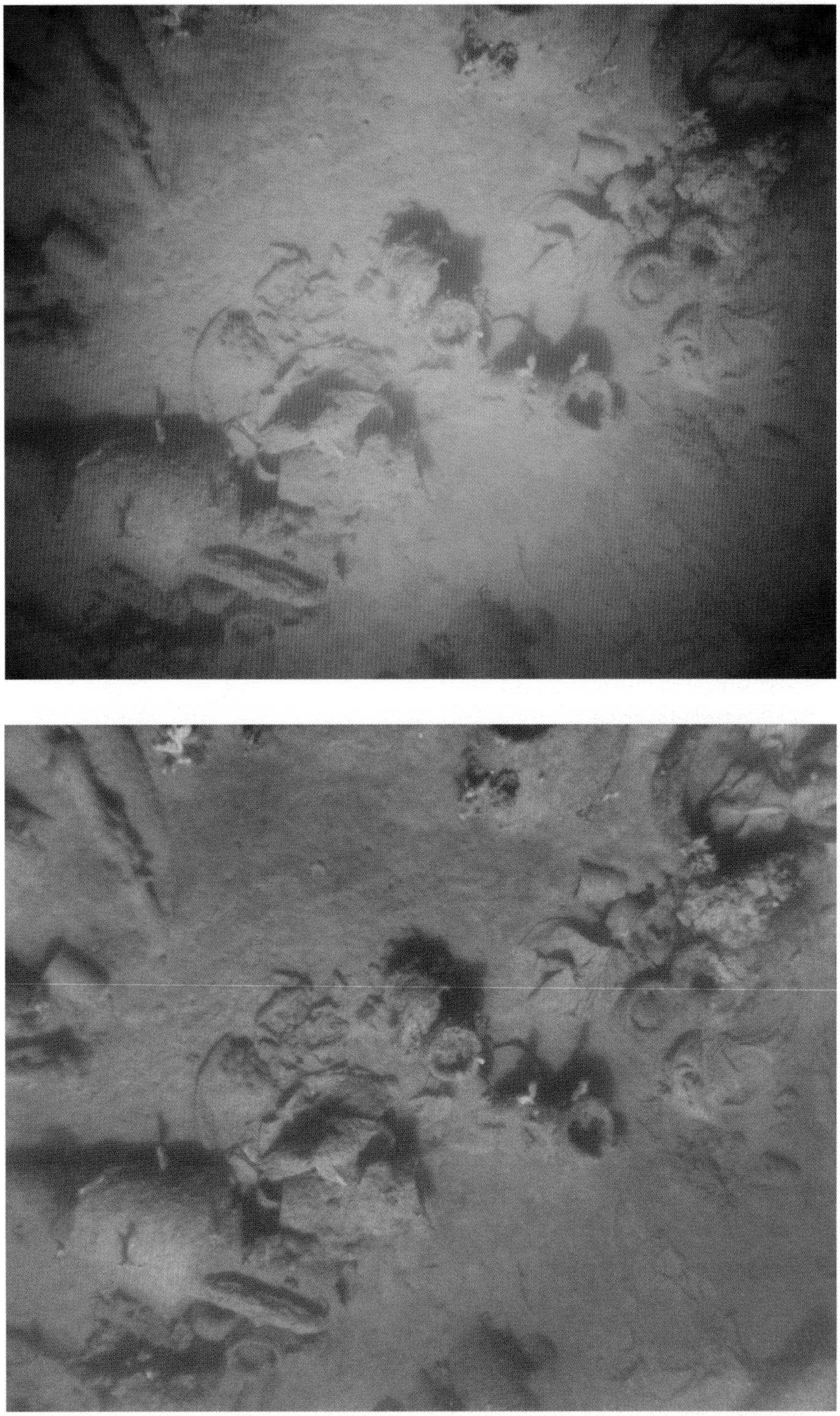

Figure 3.3. Original color imagery (*top*) and its color-compensated counterpart (*bottom*). The image on the right shows color fidelity that is independent of the camera, lighting system, and distance between camera and object. (Copyright Institute for Exploration)

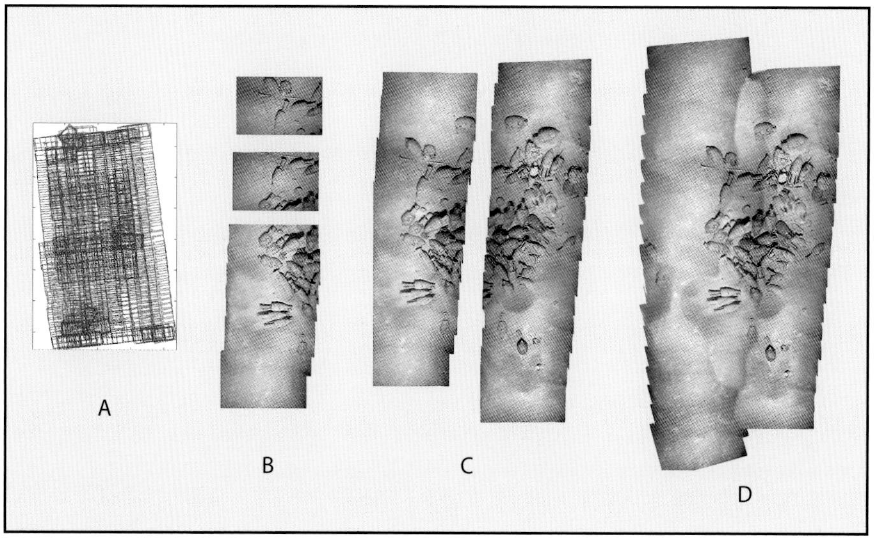

Figure 3.4. The Process of Photomosaicking
A. Step 1. Vehicle operators conduct a carefully planned survey over the area of interest to ensure sufficient coverage and overlap in the imagery. Image footprints are projected on the area of interest to allow operators to choose individual images for use in the mosaic.
B. Step 2. Individual images are processed to remove lighting and other artifacts. These images are then merged into single strip mosaics by identifying common features in successive images.
C. Step 3. Individual strips are then mosaicked together using a technique similar to that used in step 2.
D. Step 4. Ongoing work focuses on understanding and improving the quantitative nature of the mosaicking process.

Mosaicking assumes that images come from an ideal camera with known camera and lens geometry and that the scene is planar. Although there has been considerable effort in this regard for land-based applications, the constraints on imaging under water are far different. These assumptions often do not hold in underwater applications since light attenuation and backscatter rule out the traditional land-based approach of acquiring distant, nearly orthographic imagery. Underwater mosaics of scenes exhibiting 3D structure usually contain significant distortions caused by the requirement that the camera must be close to the scene being imaged. However, for planar sites one can build a reasonable qualitative representation of the objects being imaged. Typically, vehicle operators conduct a planned survey over the site to ensure sufficient, redundant coverage to allow us to combine, either automatically or manually, individual images into strips that are then assembled into a composite large-area mosaic as illustrated in figure 3.4.

An example photomosaic of the so-called Skerki D shipwreck (Ballard et al. 2000) is shown in figure 3.5. Composed from more than 180 images collected from four individual track lines, this mosaic provides a high-level perspective of the entire site that is physically impossible to obtain any other way. Note the use of image equalization techniques has minimized the effects of different track lines while preserving the variations in shading attributable to the depressions in which some of the amphoras are situated.

Photomosaicking Nonplanar Sites

If we consider large sites with nonplanar geometries where there is significant relief as measured with respect to the field of view, mosaics can become highly distorted. A good example is seen in figure 3.6, where we have tried to mosaic together three different passes over the *Tanit* Phoenician shipwreck (Ballard et al. 2002) located off of the coast of Israel.

For this site physical oceanographic or possibly ship impact effects have scoured out a depression around the wreck so that the difference in height from the central portion of the wreck to the depression is greater than 1 m. This difference in altitude is comparable to the 4-m altitude that the Jason ROV was flying while acquiring this imagery. Thus, if we consider three overlapping passes over the wreck (figure 3.6) the pass over the center is relatively undistorted, as all the amphoras within the imagery lie at roughly the same depth. The two passes on either side, however, span the central mound as well as the depression around the site. As we would expect from the geometry illustrated in figure 3.7, each of these passes gets skewed away from the amphoras pile as successive images are added to the strip. Also, as expected, the amphorae that have spilled over into the depression appear much bigger than those in the central pile even though they are all the same size.

Essentially, the problem is of projecting a three-dimensional unstructured terrain on a two-dimensional planar surface. Several possible solutions sug-

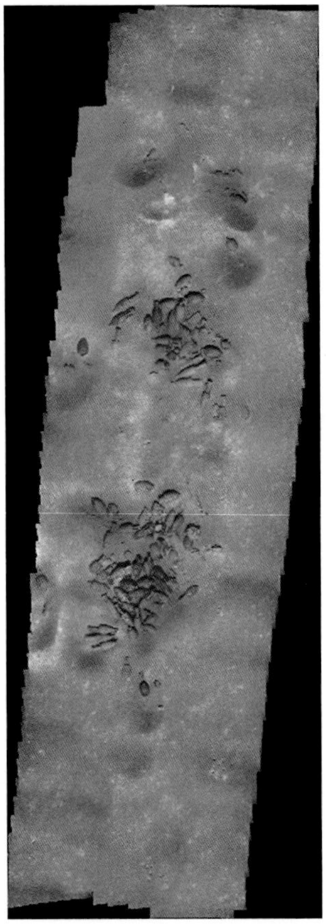

Figure 3.5. A 180-image mosaic of the Skerki D shipwreck constructed from high dynamic range imagery collected by the Jason ROV. This site is relatively flat so that the photomosaic provides a reasonable overall perspective of the site.

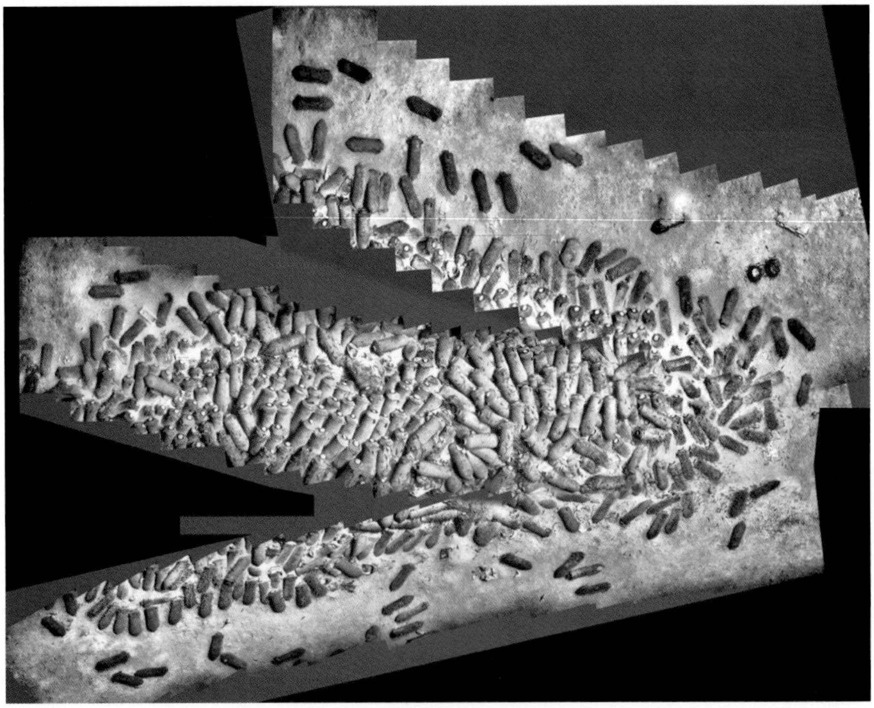

Figure 3.6. The effects of terrain on photomosaicking. Differences in depth of the order of a meter, compared to the 4-m altitude from which this imagery was obtained, result in distortions of the individual strips. The problem is essentially the fundamental difficulty of representing a three-dimensional unstructured surface on a two-dimensional plane.

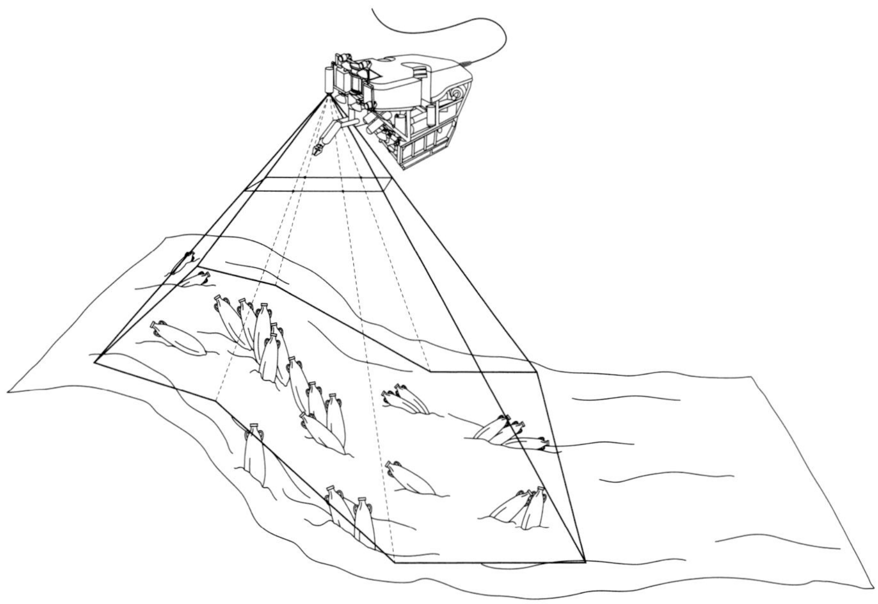

Figure 3.7. The geometry of image formation in an unstructured terrain. Equal sections in the image plane do not represent equal areas on the seafloor. Projective transformations do not accurately capture the image-to-image geometry. This can lead to significant distortions as shown in figure 3.6.

gest themselves. The overlapping imagery could be utilized to obtain a three-dimensional image reconstruction of the site under consideration as outlined in the next section. We could also utilize other sensor data to correct or compensate for the distortions.

In this case an alternative method was utilized that produced reasonable results. We surveyed the wreck from a much higher altitude to minimize the effects of perspective distortion. The imagery acquired was of quite poor resolution due to backscatter effects but was easy to mosaic as shown in figure 3.8 (top). The individual strips could then be overlaid on this mosaic, which served as an undistorted basemap for the individual strip mosaics distorted by the effects of terrain to form a consistent high-resolution mosaic of the site even in the face of significant terrain, as shown in figure 3.8 (bottom).

Three-dimensional Structure from Motion or Stereo Photomosaics

In contrast to mosaicking, the information from multiple underwater views can be used to extract structure and motion estimates using ideas from the structure from motion (SFM) problem and photogrammetry (figure 3.9). The basic idea behind these algorithms is to utilize geometric constraints between successive images to allow us to simultaneously estimate the three-dimensional structure of the underlying scene as well as the precise position of the camera for each image. By concatenating pairs of such relationships and then optimizing the results over the entire set of images one can obtain highly constrained three-dimensional structure and camera (vehicle) positions over large sites under water.

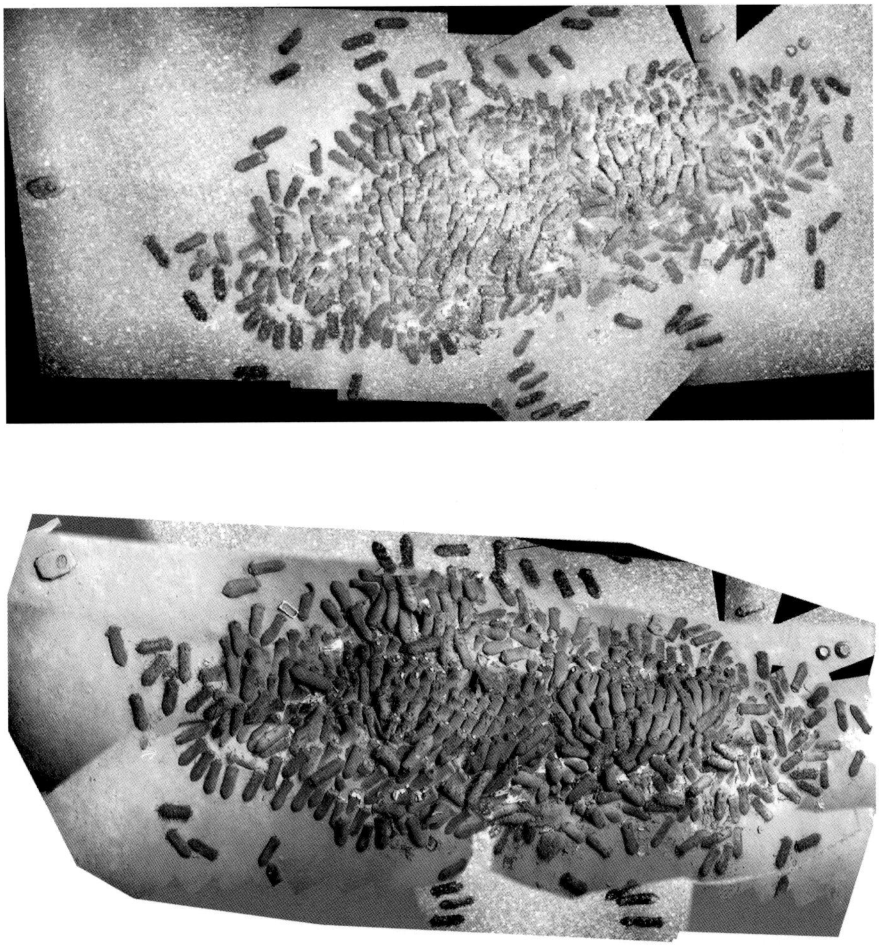

Figure 3.8. A high-altitude pass over the wreck results in imagery that is relatively distortion free and easy to mosaic, although it is marred by significant backscatter. This can, however, be used as a basemap to compensate for distortions in the low-altitude strips to build a high-resolution mosaic of the site.

In contrast to photomosaics, such reconstructions are quantitative and can thus be utilized for direct measurements of artifacts and for quantifying the spatial relationships between objects. The procedure however requires that the camera be calibrated to determine the precise relationship between the image pixels and ray angles of the incident light.

Conclusions

In this chapter we have tried to examine the issues associated with high-resolution imaging under water of large sites with optical sensors. While some of these technologies are relatively mature, a number of the techniques presented here are still the subject of active research efforts in the imaging community. In addition, it must be pointed out that optical imaging is often used in combina-

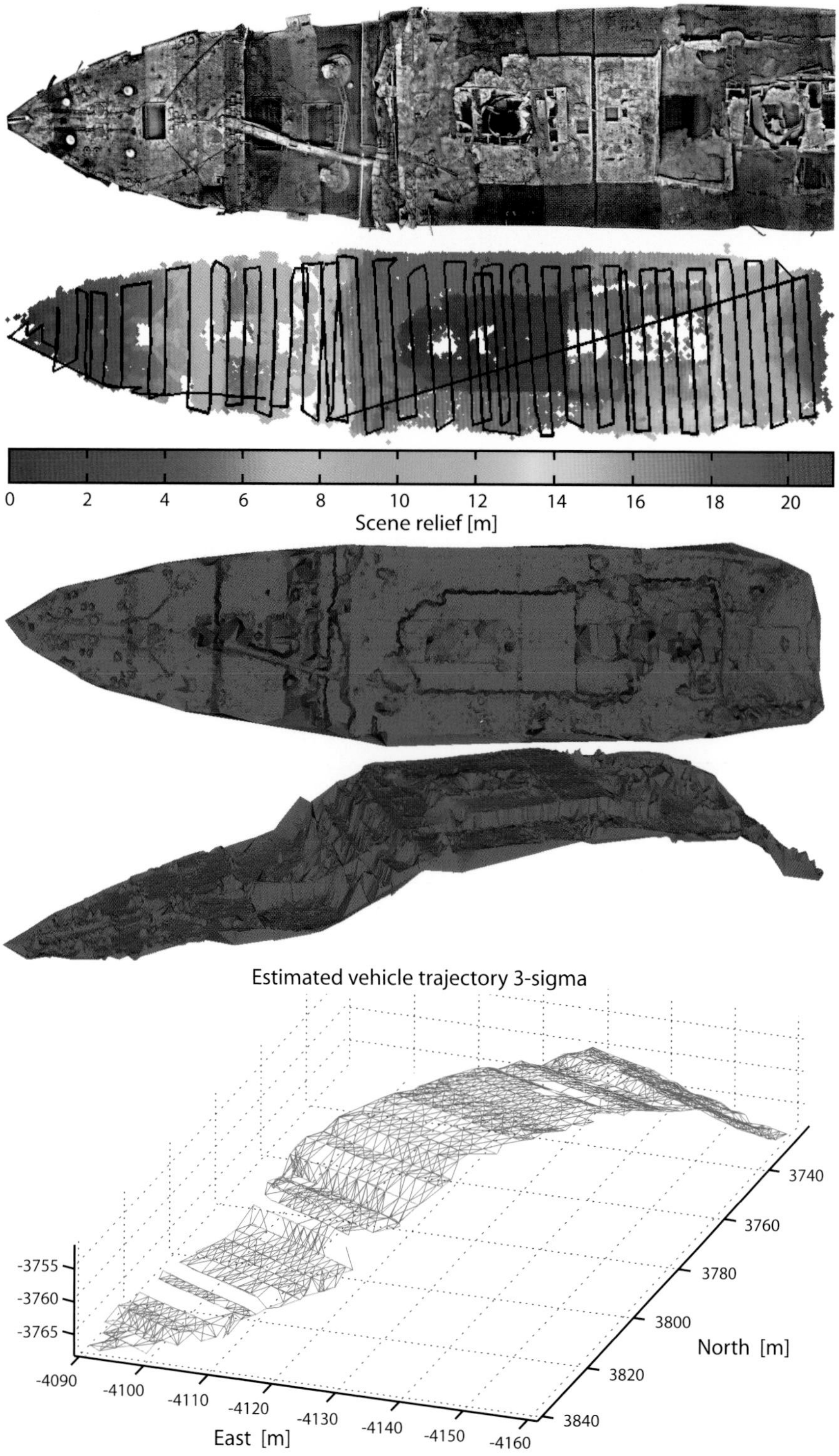

Figure 3.9. (*Top*) A photomosaic of part of the wreck of the RMS *Titanic*. (*Middle*) Different views of the 3D image reconstruction derived from the same set of imagery and (*bottom*) the camera state estimates simultaneously derived from the imagery. In comparison to the photomosaic we can make quantitative statements about the accuracy of our reconstruction and use it for measuration on the site of interest.

tion with other sensing modalities, including acoustics and lasers. There has been considerable work in related areas of GIS systems and computer visualization that may be germane to the archaeologist interested in high-resolution imaging of sites under water.

References

Ballard, R. D., A. M. McCann, D. Yoerger, L. Whitcomb, D. Mindell, J. Oleson, H. Singh, B. Foley, J. Adams, D. Piechota, and C. Giangrande (2000). The discovery of ancient history in the deep sea using advanced deep submergence technology. *Deep Sea Research, Part I* 47:1591–620.

Ballard, R. D., L. E. Stager, D. Master, D. Yoerger, D. Mindell, L. L. Whitcomb, H. Singh, and D. Piechota (2002). Iron Age shipwrecks in deep water off Ashkelon, Israel. *American Journal of Archaeology* 106:151–68.

Eustice, R., O. Pizarro, H. Singh, and J. Howland (2002). UWIT: underwater image toolbox for optical image processing and mosaicking in Matlab, in *Proceedings of the International Symposium on Underwater Technology*. Tokyo, Japan, 141–145.

Jaffe, J. S. (1990). Computer modeling and the design of optimal underwater imaging systems. *IEEE Journal of Oceanic Engineering* 15:101–111.

McGlamery, B. L. (1975). *Computer Analysis and Simulation of Underwater Camera System Performance*. San Diego, CA: Scripps Institution of Oceanography.

Singh, H, J. Adams, B. P. Foley, and D. A. Mindell (2000), Imaging for Underwater Archaeology. *American Journal of Field Archaeology* 27:319–328.

Singh, H, A. Can, R. Eustice, S. Lerner, N. McPhee, O. Pizarro, and C. Roman (2004). Seabed AUV Offers New Platform for High-Resolution Imaging. *EOS, Transactions of the AGU* 85:289, 294–295.

Whitcomb, L. L., D. R. Yoerger, and H. Singh (1999). Advances in Doppler-based Navigation of Underwater Robotic Vehicles. In *Proceedings of the IEEE International Conference on Robotics and Automation* 1:399–406.

Yoerger, D. R., and D. A. Mindell (1992). Precise Navigation and Control of an ROV at 2200 meters Depth. In *Proceedings of Intervention/ROV*, San Diego, CA.

4

The Development of Excavation Technology for Remotely Operated Vehicles

Sarah Webster

This chapter summarizes the tools and methodologies developed for an underwater excavation performed by a remotely operated vehicle (ROV), designed to meet the accepted professional archaeological standards of gentleness and controllability. The impetus behind this work was to enable careful excavation of archaeological sites that are out of the reach of human divers. These tools were developed at Woods Hole Oceanographic Institution (WHOI) and used during the 2003 partial excavation of a 6th-century CE Byzantine shipwreck off the coast of Turkey in the Black Sea (Webster 2004; also refer to chapters 7 and 8). The tools were attached to and used by the ROV Hercules, developed by Institute for Exploration (IFE), Mystic, Connecticut. Robert Ballard of the University of Rhode Island and the Institute for Exploration led the cruise along with co-principal investigators and chief archaeologists Fredrik Hiebert of the University of Pennsylvania and Cheryl Ward of Florida State University.

Background

Deep-water archaeological excavation using remotely operated vehicles is a nascent and quickly developing field. Diver-led excavations, which became common in the 1960s (Bass 1966), have utilized methodologies that are now well established in the field. The challenge for this project was to adapt both land- and diver-based methodologies to robotic vehicles and develop the technology to carry out tasks previously performed only by humans while maintaining the rigorous standards of current land and underwater excavations.

Because an excavation carried out with underwater vehicles requires collaboration between archaeologists and engineers, we enlisted the help of our chief conservator Dennis Piechota to facilitate discussions between the archaeologists and the engineers. Mr. Piechota helped create design specifications and excavation methodologies that met or exceeded the archaeologists' standards and were plausible given the technology available. For example, through the discussions it became apparent that while the engineers focused on making the digging tool powerful enough to be effective, the archaeologists' primary concern was making sure it was gentle enough to not cause harm. This resulted in a digging tool powerful enough to work under a multitude of conditions without sacrificing fine-tuned control at the low end of the power spectrum.

Through Mr. Piechota's moderated discussions we were able to determine a set of essential requirements for an archaeological excavation that served as guidelines for the tool design:

- fine control of the mechanics of the excavation process;
- documentation of the process in a way that would allow an accurate reconstruction of the areas disturbed by the excavation;
- safe retrieval of artifacts and samples composed of a variety of materials, in widely varying sizes and weights, and in varying states of preservation.

Several important principles were also brought to the engineers' attention concerning archaeological excavation. First, that excavation is a last resort. It is destructive in nature and is undertaken only and specifically to answer well-posed questions, disturbing as little as possible. And, second, that excavation is not about getting the "good stuff" (i.e., artifacts) by removing the "bad stuff" (i.e., sediment). Instead, archaeologists view the sediment as an integral part of the site to be carefully examined and either saved or removed with great care, and excavation tools should always be designed with this in mind.

Before getting into the details of the excavation plan and the tool design, a brief description of the site and the most relevant details about the platform on which these tools are implemented is necessary.

Archaeological Site

The first field season for most of the equipment and tools discussed here consisted of a cruise on the R/V *Knorr*, from Woods Hole Oceanographic Institution, off the coast of Turkey in the Black Sea in August 2003. The focus of the excavation was 6th century CE Byzantine shipwreck in 320 m of water that is exceptionally well preserved due to the anoxic conditions of the Black Sea (Ballard et al. 2001).

The top layer of sediment in the Black Sea, the layer in which we excavated, is coccolith ooze. It is typically around 30 cm thick and has a large clay fraction. We observed alternating dark and light layers in the sediment thought to correspond to a cycle in the amount of carbonate sediment deposition possibly due to yearly coccolith blooms (Degens and Ross 1974). The sediment was very easy to suspend, and did not appear to be as cohesive as typical pelagic oozes and clays.

ROV Hercules

The ROV Hercules (figure 4.1), was designed specifically for excavation. The vehicle design was led by Jim Newman of Woods Hole Marine Systems. Todd

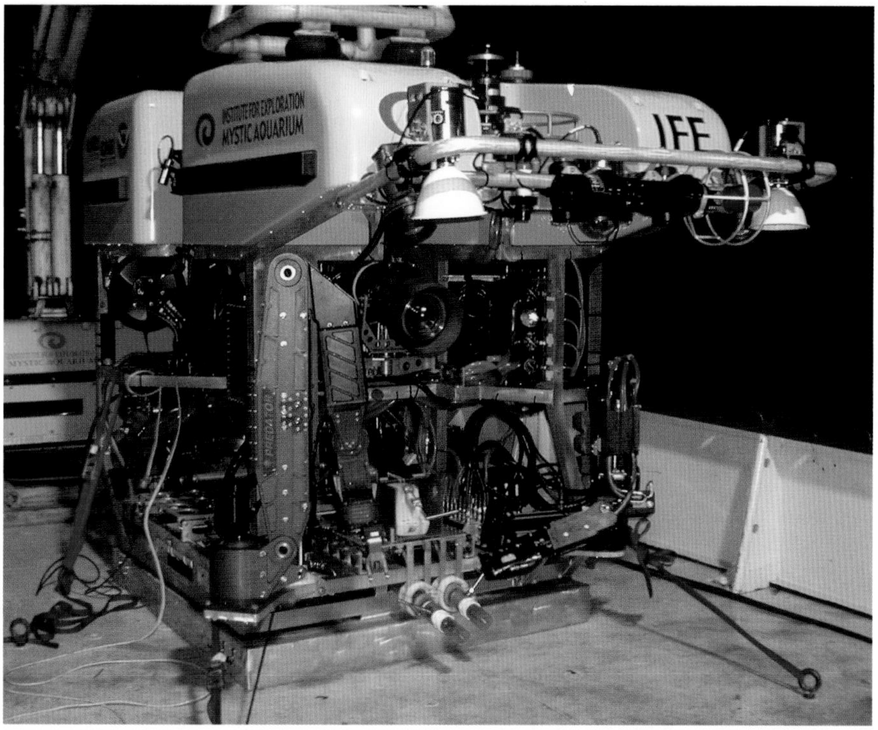

Figure 4.1. Hercules is a remotely operated vehicle (ROV) designed by a team of oceanographers, engineers, and archaeologists to excavate ancient shipwrecks in the deep sea. (Copyright Institute for Exploration)

Gregory of the Institute for Exploration was in charge of the mechanical design. Hercules is a tethered vehicle that connects to the ship by a cable that transmits power and data from the ship to the vehicle and video and data back to the ship. A 17.2-mm (0.68-in.) diameter steel cable with the Argus towsled on the end is used to get close to the seafloor. From there a 30-m-long neutrally buoyant tether leads to Hercules. The tether allows Hercules to work freely within this radius from Argus. Chapter 2 covers the design and specifications of this and other vehicles in more detail. The information presented here is only what is most relevant to the excavation tools.

Hercules has two manipulators: the Predator, designed by Kraft TeleRobotics is on the starboard side; the Magnum, designed by International Submarine Engineering (ISE) is on the port side. Both manipulators are hydraulically powered. The Kraft manipulator is position-controlled and the more dexterous of the two. It was used to do most of the tool manipulation for the excavation. A master–slave controller is used to operate this manipulator. As the ROV pilot moves a kinematically matched smaller version of the actual manipulator (the master), the robotic manipulator (the slave) moves in response. The Kraft manipulator is also equipped with force feedback capability. This allows appropriately scaled reactionary forces on the manipulator to be transmitted to the master controller and felt by the pilot. While this feature is extremely useful for a range of manipulation tasks, the forces encountered during excavation were generally too small to be transmitted by the force feedback.

The ISE manipulator is rate-controlled, stronger than the Kraft manipulator, but less dexterous. It was used mainly for lifting amphoras, though it could have been used for excavation if required.

The vehicle has two retractable drawers: the drawer on the front of the vehicle was configured to carry either tools or sediment samples and the side drawer was used to carry sediment samples or other archaeological samples.

Elevators

Elevators (figure 4.2) were used during the Black Sea project to carry samples and equipment to and from the seafloor independent of the vehicle (and much more quickly). Each elevator was outfitted with an acoustic release that drops its anchor weights enabling it to come to the surface for recovery by a small boat and raised onto the ship with the ship's crane. They were designed and built by Martin Bowen, WHOI, and Mark DeRoche, IFE, and described in detail in Bowen et al. (2000).

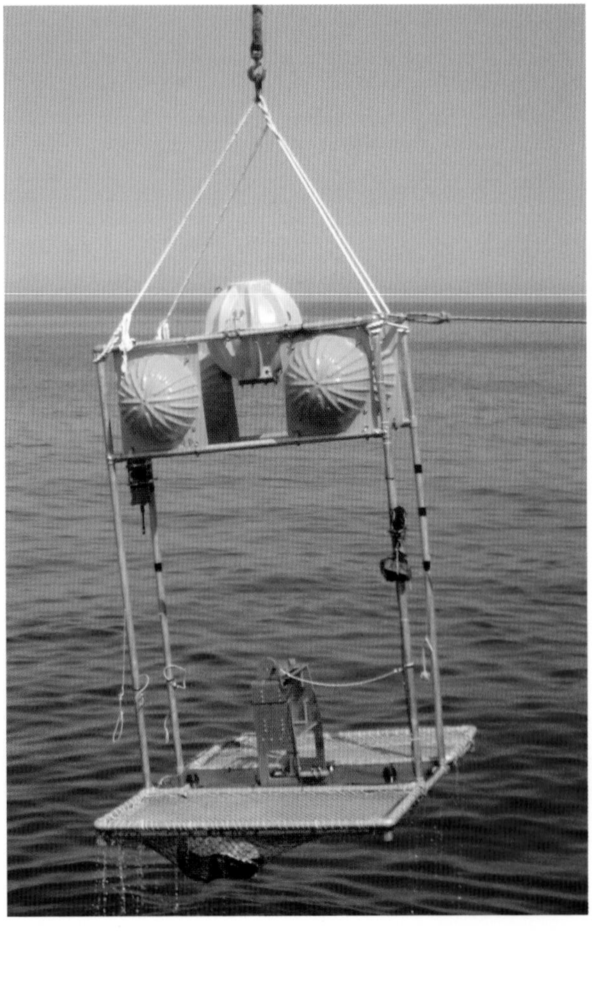

Figure 4.2. Elevator containing artifacts being recovered. (Copyright Institute for Exploration)

Excavation Plan

Before beginning the excavation we first mapped the site and tested the surrounding sediment matrix. Only after completing these preliminary tests did excavation begin under the direction of the chief archaeologist. To document the site we created composite photomosaics and three-dimensional microbathymetric maps of the wreck site using digital still cameras and high-frequency sonar. After mapping we tested the site by taking sediment cores around the perimeter of the site. These were used to determine sediment properties and to check for the presence of small artifacts in the mud matrix. After careful testing, we excavated the areas of the site specified by the chief archaeologist. Artifacts were recovered or documented and relocated under their direction. Had we encountered more oxygenated waters, we had a contingency plan to cover excavated portions of a site where we might have disrupted the anoxic environment through excavation. The plan consisted of draping a geotextile over the area in question and backfilling on top of the geotextile with mud. This method was neither tested nor used and therefore will not be discussed here. However, refer to chapter 5 for a detailed discussion.

Due to the sophisticated methods required for underwater archaeology and the early stage of development of the necessary technology, excavation tool development became a multiseason project. During this, the first season, we set certain critical goals, such as the ability to exhaustively map, gently collect artifacts, and carefully remove sediment. In subsequent seasons, as equipment was tested and proved (or disproved) in the field, the tools became increasingly sophisticated, resulting in a toolset that could meet the highest archaeological standards for excavation on land or under water.

Mapping

The first task of our excavation before touching the site was to map it extensively. Hercules has high-quality digital still cameras and high-frequency sonar for mapping and imaging purposes, respectively. Individual images were used to create a photomosaic of the entire site using the techniques described in detail in chapter 3. Sonar scans of the site provided microbathymetric data, which was used to create three-dimensional models of the site, as discussed in the article by Singh et al. (2000). The maps were used to document the site before, during, and after excavation. The goal was to eventually map continuously and constantly record the excavation surface as it was excavated.

Artifact and Site Measurement

In addition to creating maps, we needed a precise method for measuring artifacts and relative distances while we excavated. This was a more involved problem than creating maps through post-processing, because we were mapping in real time. As artifacts were uncovered or moved, their relative positions were recorded. This was accomplished by using the techniques used to create small area maps described above in order to create a model of the site from which actual measurements were taken. More accurate position data used in post-processing enabled the archaeologists to measure artifacts and their positions.

Another tool that could complement what is currently available, is a device to record the position of the manipulator with respect to the vehicle. Because the position of the vehicle with respect to the site is known with great precision, the manipulator could be used like a pointing device to record the location (*x*, *y*, *z* position) of various artifacts or points of interest, all referenced precisely to the vehicle's location. The positions recorded could then be included as a standard digital event log entry to provide a postexcavation aid for measurement, location, and interpretation of artifacts.

Site Testing

During our work in the Black Sea we used two different coring mechanisms: tube corers and box corers. Tube corers preserve stratigraphy well only when the sediment is fine grained. In areas where small artifacts may be encountered box corers were used to better preserve their position within the sediment strata. Rod Catanach of WHOI originally designed both the tube corers and the box corers for use with the manned submersible Alvin.

The corers were stored on the vehicle in either the front drawer or the side drawer. The tool tray on the front drawer (figure 4.2) held 4 tube corers. The side tray, shown in figure 4.3, held a maximum of 3 tube corers and 2 box corers simultaneously. Figures 4.4 and 4.5 show the tube corers and box corers, respectively, as they were stored on the elevator.

Figure 4.3. Front tray—tube corers.

Figure 4.4. Side tray—box corers.

Figure 4.5. Tube corers on the elevator.

Tube Corers

The tube corers are made of clear polycarbonate tube with a 6.35 cm (2.5 in.) inner diameter and a flapper valve on top. The valve lets water out as the mud is pushed into the corers, but prevents backflow as the corer is removed from the sediment matrix. This creates suction in the corer tube that holds the sediment inside until it can be put it its holster. The holsters, made from 7.62-cm (3-in.) polyvinyl chloride (PVC) pipe with one end flared, were modified for Hercules by cutting a slot into the side so that the corer did not have to be lifted high to be lifted in and out of the holster. At the bottom of the holster, a large rubber stopper, seals the corer and holds its contents in place. Once the vehicle came

Figure 4.6. Box corers on the elevator.

on deck, the stopper was unscrewed so that the tube corer with its stopper could be removed and handled without disturbing the sediment core sample.

The original corers had 29 cm of vertical core space. Several of the tubes were cut down to allow them to fit under the frame of the vehicle so Hercules could carry more core tubes. The polycarbonate tubes, after being cut, were 24 cm long, providing a working core space of about 21 cm. Corers were labeled as shown in figure 4.3 with both colored tape (on the handle, the tube itself, and the holster) as well as roman numerals written on the holster.

Box Corers

The box corers (figure 4.7) have a 15.24 × 15.24-cm (6 × 6-in.) opening. To collect the sample the corer is pushed into the sediment by the vehicle's manipulator. Two flexible "doors," originally in the up or open position, are then pulled shut to capture the sediment core and prevent it from falling out upon retrieval. These doors are actuated by twisting the handle of the box cores, which is connected to the doors with small metal wires.

The box corers were housed in custom-made holders. These boxes were designed to fit closely around the box corers to minimize excessive washing of the samples as they were brought up from the bottom. The holder was securely attached to the tool plate by just two screws. One side was low to facilitate getting the core in and out of the holder, which worked well. Instead of having to align the core up with a square hole below it, the pilots could bring the corer in over the low side, fitting it between the two high sides. Once against the back plate the corer was lowered into place. The presence of shells and coarse, non-cohesive sediment thwarted the recovery of intact stratigraphy from box core samples at the excavation site, but the samples were useful for characterizing the surface sediment.

Excavation

Many types of remotely operated sediment-removal tools exist for deep-sea work such as coastal and deep ocean dredging, offshore mining, and pipeline burial. However, the tools for these activities are all designed to remove sediment as quickly and efficiently as possible regardless of the specific properties of the soil and the debris it may contain. Archaeological excavation is a unique application of subsea sediment removal, requiring that the sediment be removed without disturbing surrounding areas or embedded artifacts and while maintaining visibility. This section explains the tools designed to address the various challenges presented by an archaeological excavation, how they are integrated onto the vehicle, and examples of their use.

Figure 4.7. Box corer, side view.

Excavation System

The primary component of the excavation system designed for the project was a combination water-jet/suction tool, nicknamed the "snuffler," that could gently clean around artifacts. The tool was designed to suspend sediment locally using a low-speed water jet while clearing the suspended sediment away with a suction hose to maintain visibility. The small water jet was held inside, concentric with the suction hose.

The water-jet/suction combination was inspired by the technique of "hand-fanning," where a diver holds a suction pipe with one hand while gently fanning the excavation surface with the other to suspend sediment and clear it away. The advantages of using a water jet versus a brush are that nothing solid ever touches the excavation surface and by controlling the flow as needed a water jet can be both gentler and more powerful than a brush.

Many tests were undertaken to find the optimal jet and suction configuration, discussed in the Excavation Tool Head section. In tank tests, the tool was found to be very effective at maintaining visibility and highly controllable and delicate enough for archaeological excavation. The tool was implemented on Hercules on a retractable hose that was housed on the underside of the vehicle, as illustrated in figure 4.8.

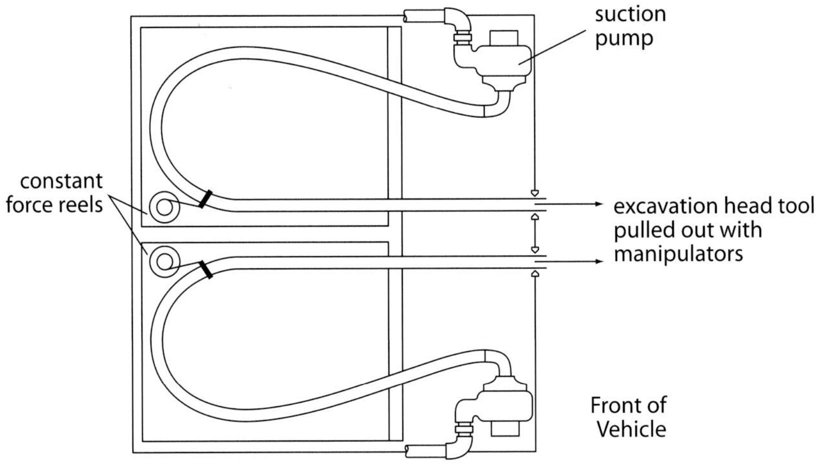

Figure 4.8. Snuffler configuration: top view of the bottom section of the Hercules frame.

The retractable hoses allowed the ROV's manipulators to use the tool to the extent of their reach while keeping the hoses out of the way when they were not needed. To use the tool the manipulator reached down in front of the vehicle, grabbed the tool, and pulled it out. When not in use the hose was retracted under the vehicle by a constant force reel. Two symmetrical systems were put on the vehicle, each able to reach six feet in front of the vehicle. Two identical systems provided backup in case one clogged, and allowed for two different configurations (usually a variation in jet nozzle size and position). The suction pump and water jet pumps were hydraulically powered and independently controlled to allow for precise tuning of the digging tool.

Excavation Tool Head

The working ends of the snuffler is shown in figure 4.9, a closeup of the tool protruding from the front of the vehicle with the hose in its retracted position. Two different configurations of the snuffler were used: one with the nozzle extending straight out from the suction hose, and one with a 45° bend downward as shown in figure 4.10. The first configuration, with the straight nozzle, caused the hose to kink as the manipulator maneuvered the tool to excavate. The 45° elbow alleviated that problem somewhat and a more flexible hose would correct it entirely. The 45° configuration was also a more natural position because the tool worked best when held perpendicular to the digging surface.

Figure 4.10 shows the tool with the concentric jet assembly that fits inside. For scale, the clear PVC pipe on the end has a 6-cm (2 3/8-in.) outer diameter. The function of the jet assembly was to mate the jet nozzle to the flexible hose that transports the jetting water up the suction hose (see Excavation Tool Hose section for more about the suction/jetting hoses). The water-jet cut Delrin collar holds the jet assembly centered inside the suction pipe. The jet tip could be adjusted longitudinally to protrude out from the suction pipe at different distances. For the work in the Black Sea, the jet nozzle was placed even with the end of the suction pipe. For harder sediments the jet could be placed deeper inside the suction pipe and a higher jetting force could be used.

Figure 4.9. Excavation tools, retracted position.

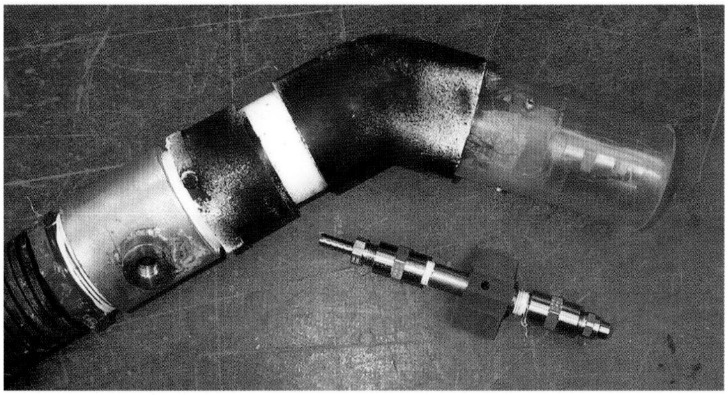

Figure 4.10. Working end with concentric jet piece shown.

Prior to and during excavation, the sediment was examined closely for small archaeological materials to be certain the sediment was clear for removal. In addition, to prevent small artifacts from being suctioned away, a stainless steel screen across the face of the tool prevented anything larger than 6 × 6 mm (¼ × ¼ in.) from passing through the pump. The screen did not interfere with the operation of the tool, and material caught by the screen was collected by the archaeologists after every vehicle recovery.

A significant amount of testing was carried out before the final tool was designed (Webster 2003). Figure 4.11 shows one set of tests used to determine the optimal relative position of the jet within the suction pipe by recording the maximum jet flow that could be used, while maintaining a constant suction flow. For this test, a 60° full cone jet nozzle was equipped with a 4.78-mm (0.188-in.) orifice and a constant suction flow of 3.7 L/s (60 gpm) was applied. The jetted water was dyed with Rhotamine RT to facilitate visualization of the process. In the test shown on the left of figure 4.11, the jet nozzle protrudes from the suction pipe. In this test the suction was able to capture all jetted water only at very low flow rates. In contrast, in the test shown on the right, with the nozzle inside the suction pump, at a full flow rate (0.20 L/s, 2.5 gpm) the jetted water could barely reach the bottom of the tank.

I explored many different operating parameters during these tank tests and was able to determine important system characteristics and an optimal envelope

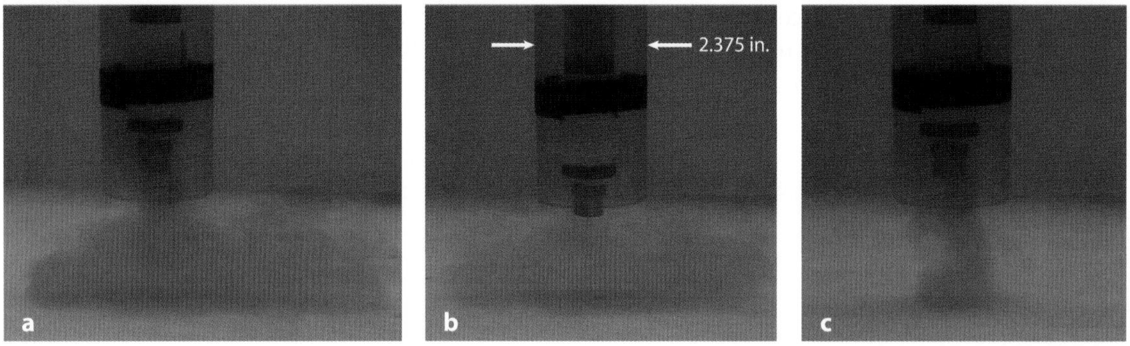

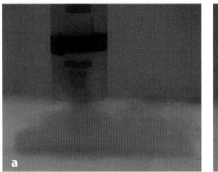

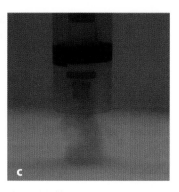

Figure 4.11. Test of maximum possible jet flow with varying jet positions and constant suction.

of operation. The tool was much more sensitive to the jet flow rate compared to the suction flow rate. The magnitude of the jet flow relative to the suction flow for this test setup was roughly 1:20. The jet flow required only 0.13 to 0.20 L/s (2 to 3.25 gpm), while the suction flow varied from 2.5 to 4.73 L/s (40 to 75 gpm). The height of the tool off the bottom for the most effective jet/suction combination was between 10 and 15 cm. Finally, the ability of the suction to capture surrounding fluid was evident. The suction could clear dye clouds from 0.3 to 0.5 m away by simply turning off the jet for 5 to 10 s and letting the suction flow continue. Larger dye clouds were easily cleared during the separate jet and suction tests by moving the suction head around and "vacuuming" them up. Also, as desired by the archaeologists, the tool suitably met the design requirement for fine control and delicate excavation forces at the lowest end of the power spectrum.

The jet nozzles used for the actual excavation were known as "0°" or "solid-stream" nozzles because, in air, the jet would exit the nozzle in a solid, round stream, as opposed to fanning out. Underwater in a static (no-flow) environment, a water jet will fan out into a cone shape with an approximate 22° spread due to entrainment within the surrounding water. Obviously, the less the spray spreads, the more powerful it can jet, which is why this tool could only be used at close range; the jet spreads too much and loses its digging power if it is too far away. The jet nozzles available for the excavation had orifice sizes ranging from 1.90 mm (0.075 in.) to 5.36 mm (0.211 in.) diameter. The smallest diameter was used in the Black Sea because the mud was very easy to suspend.

In the Black Sea we used the tool with the jet and suction running continuously, and tuned them independently as necessary. Another possible effective mode of operation could be operating the jet intermittently at higher flows than could be contained continuously by the suction. The jet would be turned on briefly at a high flow rate for a few seconds, then the suction could be allowed to clear the field of view. This method could allow for higher jetting forces while maintaining adequate visibility.

Excavation System Hose

The suction hose used originally for the excavation tool was a clear PVC reinforced hose. This hose had the tendency to kink just before the tool end because of the bending moment exerted by the manipulator on the tool. We attempted to correct this by tapering the transition from the snuffler to the flexible hose in order to spread out the bending moment. During the Black Sea expedition we verified that an optically clear hose was not critical to the operation of the tool, therefore a lighter, more flexible hose was suggested to alleviate this problem in the future. The use of the more flexible hose could eliminate the need for the 45° elbow at the snuffler.

The hose was stored under the vehicle when it was not in use, held up and protected from the bottom by 6-mm ultrahigh molecular weight polyethylene (UHMW-PE) sheets across the bottom of the vehicle. The hoses were spring loaded into the retracted position by retraction reels at the stern of the ROV. These constant force reels have a spring inside and a 3-m stainless steel wire, with a 10-ft total travel distance, the middle 6 ft of which have a roughly constant force (45 N (10 lb) for the reel we used), because the manipulators have slightly less than 6 ft of reach. We used 2 reels for each snuffler to give 90 N (20 lb) of total retraction force.

The design of the excavation tool required the hose from the jet pump to be plumbed inside the suction hose. To accomplish this, the jetting hose was connected to a 1.27-cm (½-in.) diameter steel tube, which was fed through a nylon stuffing tube to the inside of the suction hose. This connection was made at the very beginning of the hose, so the jetting hose could run the entire length of the suction hose. Due to the resulting loss of area in the suction hose, a 5.1-cm (2-in.) diameter hose was used instead of the 3.81-cm (1½-in.) diameter hose that the pump requires for its intake.

Backflushing System

The starboard excavation tool was designed to allow the suction hose to be backflushed in case it clogged, making the tool unusable. (Only the starboard side manipulator, the Kraft, has the range of motion to make this design possible, so it was not implemented on the port side snuffler.) The backflushing plumbing was designed so that the jet pump normally used to supply the water jet could alternately be used as the flushing source, controlled by a valve on the side of the vehicle. For an effective backflush, the suction line must be closed behind the place where the flushing water enters. This was accomplished using a ball valve inline with the exhaust of the suction pump. The jet water was then forced backward through the suction pump and out the snuffler to loosen debris that could clog the line. Fortunately, backflushing was never needed.

Both valves (the ball valve and the 3-way valve) were located on the bottom edge of the starboard side of the vehicle forward of the side sample drawer and held in place with custom brackets, as shown in figure 4.12.

Exhaust System

The exhaust from each suction pump was plumbed through a one-way check valve to prevent backflow through the pump and out the opposite snuffler. The two lines were connected at the aft end of the vehicle to a single exhaust hose that was coupled to the tether that connects the ROV to the towsled. Just

Figure 4.12. Backflushing valves on the starboard side of the vehicle.

Figure 4.13. Exhaust system hose. Note that the portion of the hose after the service loop floats above the vehicle when it is submerged.

before the exhaust hose was joined to the tether we installed a 0.6-m-diameter service loop, shown in figure 4.13. This loop prevented particles entrained in the remaining 6 to 9 m of exhaust hose from settling back down onto the one-way valves when the suction pump was turned off and flow stopped. In this configuration, the sediment remaining in the exhaust hose would settle into the bottom of the service loop and be carried away when the pump is restarted.

The sediment plume exhausted through the system was allowed to dissipate on its own with the aid of local currents. This method worked very well in the Black Sea and we never had a problem with visibility, even in fairly low current regimes. This was, of course, dependent on the particle size and current magnitude so different conditions would require a more refined plume management plan.

Excavation Process

Figure 4.14 shows a series of images taken during the excavation of part of the Byzantine shipwreck's cargo hold located amidships on the port side. The smallest diameter jet nozzle available, 1.9 mm (0.075 in.) diameter, was used because the sediment was fine-grained, water saturated, and flocculent. The excavation shown took approximately 7 hr of digging with the snuffler and 1 hr of brushing using the paintbrush tool to clean off the wooden frames. The estimated sediment removal rate during this work was 0.17 m^3/hr (6 ft^3/hr). The artifacts shown were recovered and presented to officials from the Sinop Museum in Sinop, Turkey.

Hand Tools

In addition to the water-jet/suction tool, I adapted some simple tools traditionally used by divers on shallow water excavations for use with the manipulator. Pictured in figure 4.15, from left to right, are a fabricated plastic scraper with a flexible plastic edge, a metal drywall scraper, a paintbrush with the bristles trimmed short for stiffness, and a plastic utility scoop. To enable the manipulator to hold the tools, they were outfitted with either a T-handle or a softball handle. The tools were painted black to minimize glare in the cameras.

Figures 4.16 and 4.17 show the brush and scoop during the excavation process. The brush was used to clean off wooden frames and other structural components of the ship. The scoop was used to pick up small artifacts and their surrounding matrix of sediment.

Figure 4.14. Excavation during an 8-hour time period. (Copyright Institute for Exploration)

Figure 4.15. Hand tools used by Hercules to conduct excavation activities.

Figure 4.16. Brushing one of the ship's beams. (Copyright Institute of Exploration)

Figure 4.17. Collecting a scoop sample. (Copyright Institute of Exploration)

On the vehicle the tools were stored in a holster made from a milk crate with thin plastic (PVC and UHMW-PE) dividers arranged to fit the different sized tools. The tools were usually secured upright with elastic bands to prevent loss during the ROV decent and to aid in the initial retrieval of the tool. We found that it was very important for the tool compartments to hold the tools upright to facilitate storage and retrieval.

A special holster for the scoop was created using one designed for the box corer. We mounted this device on its side with open cell foam hot glued into the bottom. We planned for the scoop to be used once, then held in this holster with the opening pressed against the foam to seal off the scoop and protect its contents. The enclosed space also helped reduce water flow around the edges of the scoop, further protecting the sample.

Pallet and Gantry Docking Stations

Two systems were designed and built by Todd Gregory, IFE, and Mike Purcell, WHOI, to allow the vehicle to anchor itself in a convenient and repeatable fashion. One was called a pallet, which could lie just outside of the excavation area, and the other was called a gantry, which would span the wreck site and allow the vehicle to sit suspended above the excavation area. While neither of these systems were used during the 2003 field season, they are included here because they were designed to solve several of the challenges associated with using ROVs to excavate sites in the deep sea.

The pallet is a simple T-shaped structure with a docking cone in the center. Each of the three ends of the crosspieces has a sand screw that the vehicle rotates to affix the pallet to the seafloor (figure 4.18). The vehicle has a device on its underside with a special hydraulic latch to secure the docking cone and vehicle. Had we needed to move heavy objects, which could upset the vehicle's balance or require excessive thruster use (potentially disturbing the excavation site and visibility), the pallet could securely hold the vehicle in place without

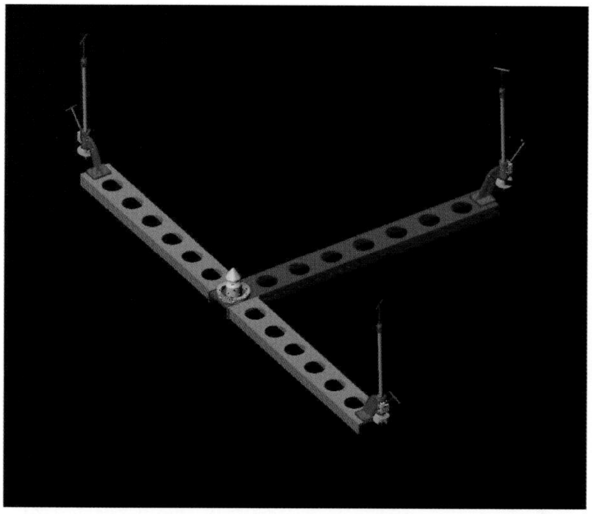

Figure 4.18. Design image of the pallet docking station with sand screws. (Image courtesy of Mike Purcell, WHOI)

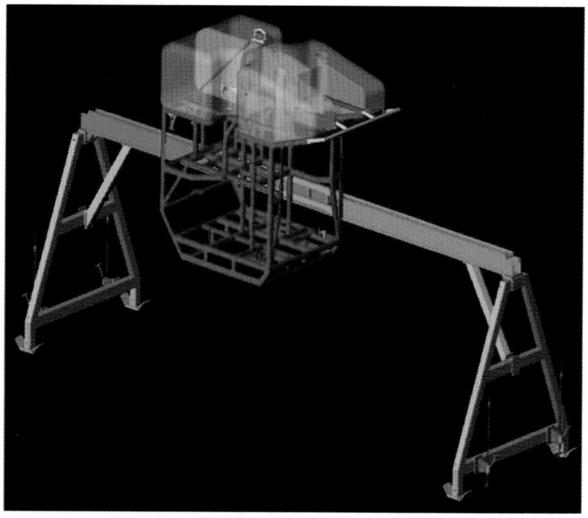

Figure 4.19. Design image of the gantry docking station. (Image courtesy of Todd Gregory, IFE)

the aid of thrusters. In addition, because the vehicle is hydraulically powered, any power that would have been diverted to the thrusters would then be available for the manipulators and excavation pumps.

The gantry consists of a crossbeam, supported on both ends, and held roughly 2 m above the wreck site, as shown in figure 4.19. The gantry is also designed to be secured to the seafloor with sand screws. The vehicle has a moveable (and removable) latching mechanism attached to its port side. The moveable latch could allow the vehicle to be positioned at varying heights above the wreck as necessary for excavation. The gantry could also provide the benefits noted above for the pallet, with the added benefit of providing access to the entire wreck, not just the edges. Without the gantry the center of a site would be unreachable.

Artifact Handling

Amphora Lifter

The amphora lifter (figures 4.20 and 4.21), designed by Todd Gregory, was used extensively to move and collect amphoras. The original design was dimensioned to fit a cylinder 60 to 75 cm in length with a diameter of 20 to 30 cm. We accommodated smaller amphora by covering the sides of the lifter with netting. The frame was made from ½-in.-diameter stainless steel hydraulic tube that was bent into a custom shape and covered with a closed cell foam tube to prevent any damage to the amphoras (figure 4.20). Using the amphora lifter required replacing the Kraft gripper with custom made gripper stubs to hold the stainless steel tubing, and therefore required a dedicated dive for amphora recovery because this tool could not be changed subsea. We also installed the amphora lifter on the ISE manipulator as shown in figure 4.21 with the netting in place. This amphora lifter was used successfully to move and recover amphoras from various archaeological sites in the Black and Mediterranean seas.*

Sample Boxes

Designed by Megan Carroll, WHOI, the sample storage boxes contained a latch assembly that allowed the vehicle to attach the boxes to either the elevator or the vehicle's side tray while on the bottom.Two different methods were used for opening and closing the lids. To operate the lid manually, the box was mounted with the opening facing the vehicle and had a loop with a knot for a handle. The lids were hinged such that the lid could open only through about 210° (instead of

**Editor's comment (Robert D. Ballard)*: A new amphora lifter was designed and employed in 2007 for a return trip to the Black Sea where we began excavation and artifact recovery activities at a Byzantine shipwreck site near the ancient Greek city of Chersonesos, off the Crimean peninsula in present-day Ukraine. This new system employed a water suction hose and cup attached to a tool that could be directly grabbed and used by the manipulator on the fly, without a hard modification to the grabber. The suction cup did not adversely affect the integrity of any artifacts it lifted. This lifter proved to be much more effective at retrieving artifacts from sometimes complex regions within the site, and was quicker and easier to implement. It also enabled the ROV to easily switch between tools to remove sediment and move objects. The amphora lifter as described here will probably no longer be used to lift artifacts during future expeditions.

Figure 4.20. Kraft manipulator equipped with the amphora lifter picking up an amphora. (Photo by David Mindell, MIT)

Figure 4.21. ISE manipulator equipped with the amphora lifter.

270°) to make it easier to close. Because the boxes and lids are made of polyethylene, they are slightly buoyant in seawater. The latching assembly provided enough weight to hold the box down, but with the lid insecure. To rectify this, the lids were modified to hold a small 0.5- to 1.0-kg dive weight held on by zip-ties. Figure 4.22 shows the box in its retracted position. The blue dive weight holding the lid closed and part of the yellow handle loop are visible, as well as the handle of the latching mechanism.

Figure 4.22. Sample box on Hercules. (Photo courtesy of Todd Gregory)

When the boxes did not need to be moved from the vehicle to the elevator, we used a configuration in which the lid would automatically open and close as the side drawer on which the box is mounted slides in and out. This was accomplished by mounting the sample box so the opening faced out with the lid attached to the vehicle frame with bungee cord.

Depending on the science mission and what we intended to collect, we put foam and/or divisions inside the boxes. The foam was attached with a glue gun, while the watertight divisions were thin UHMW-PE sheets cut to the right width and taped in place. The divisions worked especially well for maintaining a watertight enclosure.

Megan Carroll, WHOI, also designed the latching mechanism for use with the sample boxes and the tool trays. Figure 4.23 shows the latching assembly in one of the sample boxes. The intent of the latch was for the vehicle to be able to transfer sample boxes between the elevator and the vehicle drawers while on the bottom. Each sample box and tool tray has a latch assembly (figure 4.24). Any surface on which the item is to be attached has a latch lock (figure 4.25).

The latch handle is spring loaded upward. On the bottom of the handle is a small T piece positioned parallel to the handle (so that the pilot knows its orientation without seeing the actual latch). The latch lock, which is attached to the surface on which the box is to be locked to, has an opening that fits the T piece. With the T piece in line with the latch lock opening, the latch can be locked by pushing the handle down, fully compressing the spring, and turning 90° in either direction. Inside the latch lock, the T piece fits into a groove when it is turned 90° and the spring in the handle holds it in place.

The latch handle is also outfitted with a 6.35-mm (¼-in.) diameter pin welded onto the side of the latch rod parallel to the handle. This pin is designed

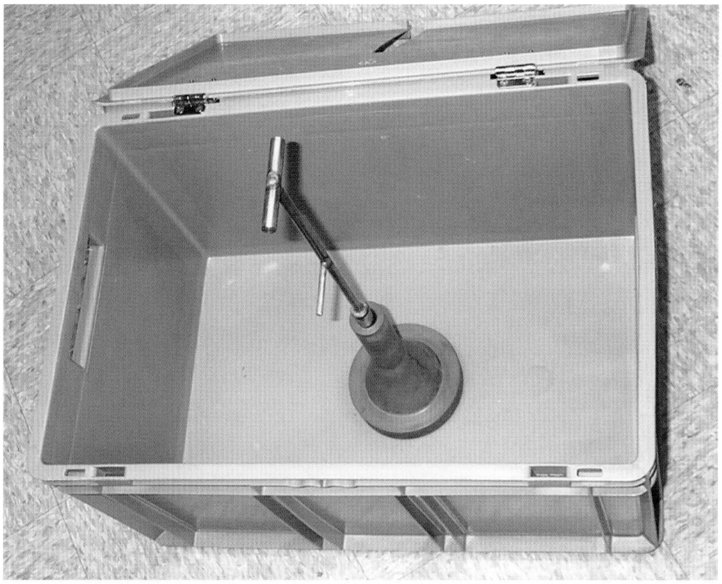

Figure 4.23. View of the latching mechanism inside a sample box.

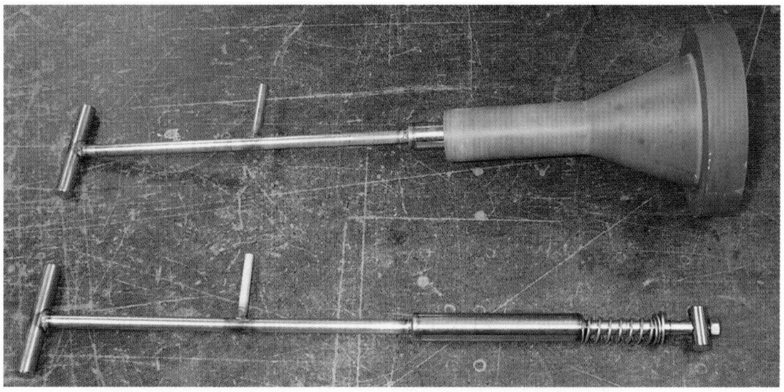

Figure 4.24. Latch assembly components.

Figure 4.25. Latch details.

to hold down the sample box lid while the box is being transported. A groove in the sample box lid allows the lid to move only when the pin faces the front of the box.

Clam Scoops

Clam scoops were originally designed for use with the manned submersible Alvin by Griffith Outlaw, WHOI. These were modified to fit several different applications for the excavation project and were fabricated out of aluminum instead of the originally specified titanium. The clam scoop is a unique scoop in that it can be actuated by the Kraft gripper without having to be permanently attached. The scoop is spring loaded open and must be housed in a box to hold it closed when not in use. Pins on the gripper fit into grooves on the scoop that allow it to be picked up and actuated; the scoop must be placed back in its box where it is held shut for the gripper to let go of it.

The three different kinds of modified clam scoops were made as shown in figure 4.26. The first kind was the standard clamp scoop with a 20.3-cm (8-in.) diameter bucket with sides, designed to pick up objects and their surrounding mud matrix. The second kind was the same size as the first but without a bucket; small crossbars were used instead to hold the tongs aligned. This version was designed to have netting stretched across the two inside faces of the scoop and be used to gently pick small objects off the bottom. The third kind was for picking up a particular style of amphora. This had a 25.4-cm (10-in.) diameter scoop without ends and the inside could be padded to protect the amphora.

The 20.3-cm (8-in.) diameter scoops were stored inside milk crates. Then thin UHMW-PE sheet was added to the side walls and built up with foam underneath to ensure that the scoop could slide in and out easily and stay closed when not in use. The larger scoop could be stored in a sample box.

Results and Conclusions

The ROV excavation system worked exceptionally well during the site work on the Byzantine ship and we encountered no major problems. We were able to

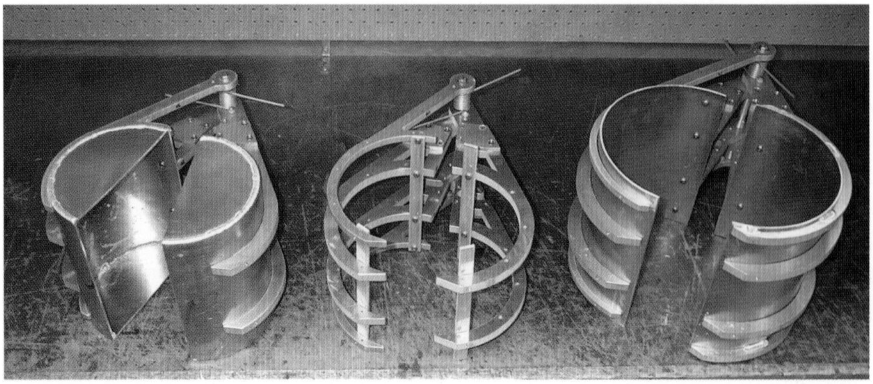

Figure 4.26. Clam scoops.

maintain visibility while excavating and the small amount of current in the area kept the exhaust plume of sediment from settling back onto the work area. The screen across the opening of the snuffler did trap some organic material but it did not affect the tool's performance. Once the vehicle was on deck, the material was collected by the archaeologists. The hose had a problem with kinking, but the concentric jetting piece kept it from collapsing completely and blocking all flow. This hose was later replaced with more flexible hose.

We were able to excavate in the sediment and around artifacts carefully and at an acceptable rate, 0.17 m^3/hr (6 ft^3/hr). The excavation tool met and exceeded design goals for low-end sensitivity in digging power, which is critical for archaeological excavation and one of the hallmarks that separates archaeological excavation from other types of underwater excavation. For future work, the next-generation excavation tools could benefit by creating the ability to allow the manipulator to automatically keep the jet nozzle a fixed distance away from the area being excavated. A hard-wired interface with a halt button and knobs to continuously control the jet and suction flows instead of the computer interface could also improve the system.

The Black Sea site proved to be ideal for excavation because of the excellent preservation of wood and artifacts and the soft, uncompacted sediment. The ability of the system to maintain visibility should hold true in sediment with similar particle sizes (i.e., oozes and clays). However, these sediments are typically more cohesive and may require higher jetting forces or alternative means to reduce the cohesion before digging. Sediments with high percentages of sand have not been tested with this tool and may require a different jetting to suction flow ratio or a different digging approach.

Finally, I would like to emphasis that these tools represent the results of our first field season and the first step of many in the development of excavation tools for ROVs. We designed them to accomplish a certain number of critical goals, which they did very successfully. Each successive generation will continue to improve upon the work described here, striving toward the ultimate goal of meeting and exceeding the highest archaeological standards for excavation on land or under water.

Acknowledgments

Thank you to Robert Ballard, expedition leader and chief scientist during the 2003 Black Sea project; Cheryl Ward, chief archaeologist for the shipwreck excavation; and Fredrik Hiebert, project archaeologist, for the cruise. Also Capt. A.D. Colburn and the crew of the R/V *Knorr* provided an exceptional platform from which to work. Chief conservator Dennis Piechota led the charge in development of the excavation methods described here. He was the liaison between the archaeological and engineering teams, promoting the understanding of each other's fields, which is critical for a multidisciplinary project. Jim Newman and the Hercules operations team worked tirelessly and cheerfully to keep the vehicle running and in the water. Todd Gregory was essential to the integration of the tools into the vehicle design. Matt Naiman helped a great deal with the testing setup and the jet pump. Albert Bradley gave an excellent crash course in amplifier design. The WHOI Media Relations Office provided top quality digital video cameras for indefinite periods during the testing phase. And thanks to the WHOI Machine Shop, who provided the highest quality work and top-speed turnaround times. This work was supported by ONR grant N00014-99-1-0109.

Disclaimer: Products and companies described herein do not imply endorsement by Woods Hole Oceanographic Institution.

References

Ballard, R. D., F. T. Hiebert, D. F. Coleman, C. Ward, J. S. Smith, K. Willis, B. Foley, K. Croff, C. Major, and F. Torre (2001). Deepwater archaeology of the Black Sea: the 2000 season at Sinop, Turkey. *American Journal of Archaeology* 105:607–623.

Bass, G. F. (1966). *Archaeology Under Water*. New York: Praeger.

Bowen, M. F., P. J. Bernard, D. E. Gleason, and L. L. Whitcomb (2000). Elevators—autonomous transporters for deepsea benthic sample recovery. *Woods Hole Oceanographic Technical Report*, 1–41. Woods Hole, MA: Woods Hole Oceanographic Institution.

Degens, E. T., and D. A. Ross (1974). Recent sediment of the Black Sea. In *The Black Sea*, E. T. Degens and D. A. Ross, eds. Tulsa, OK: American Association of Petroleum Geologists.

Singh, H., L. Whitcomb, D. Yoerger, and O. Pizarro (2000). Microbathymetric mapping from underwater vehicles in the deep ocean. *Computer Vision and Image Understanding*, 79:143–161.

Webster, S. (2004). Excavation tools for deep water archaeology. *Proceedings, Congress on the Application of Recent Advances in Underwater Detection and Survey Techniques to Underwater Archaeology*, Bodrum, Turkey, T. Akal, R. D. Ballard, and G. F. Bass, eds., pp. 271–278.

Webster, S. (2003). Testing of a jetting-suction tool for fine scale excavation. Unpublished manuscript, Woods Hole Oceanographic Institution.

5 Conservation of Archaeological Finds from Deep-water Wreck Sites

Dennis Piechota and Cathy Giangrande

> Let me tell you my favorite day-dream: You know that by a miracle a submarine television camera seeking the wreckage of a Comet aircraft in the Mediterranean came upon a deposit of amphorae at a depth divers cannot reach. But we have been able to make constant use of our latest invention, the "diving saucer," up to depths of 300 m. Imagine our feelings if we were able to uncover in our new domain an ancient wreck—completely untouched!
>
> —*Jacques-Yves Cousteau (quoted in du Plat Taylor 1965, p. 13)*

This chapter is an overview of the conservation activities common to four deep-water archaeological surveys undertaken from 1989 to 2003, which included wreck sites dating from the Iron Age to the Byzantine period, that were found in the Mediterranean and the Black Sea (table 5.1). It is intended that the practical information presented here will prove useful to others in planning the shipboard processing and final land-based conservation needs for future expeditions. Not discussed is general information on underwater conservation as this can be found elsewhere (Pearson 1980; Robinson 1998; Hamilton 1999), nor will the work of each expedition be described in detail, as that can be found in the monographs written for each site (Ballard et al. 2000; McCann and Freed 1994; McCann and Oleson 2004). Highlighted are selected problem areas that reflect the experience of the authors and that they feel deserve careful consideration.

The primary aim of archaeological conservation, whether of assemblages from deep-water sites, from land-based, or other marine or freshwater sites, is always the same: to stabilize the finds using minimally invasive treatments. The collection and care of finds and the documentation of evidence revealed during the disturbance of an archaeological site should be performed with care. The

TABLE 5.1.
DEEP-WATER WRECK SITES SURVEYED FROM 1989 TO 2003

Location & season	*Wreck*	*Date (century)*	*Depth (m)*
Skerki Bank			
1989, 1997	Isis	4th CE	820
1997	A	13th CE	766
1997	B	1st CE	770
1997, 2003	D	1st BCE	850
1997	F	1st CE	765
1997	G	1st CE	760
Ashkelon			
1999	A (*Tanit*)	8th BCE	400
1999	B (*Elissa*)	8th BCE	400
Black Sea			
2003	D	6th CE	400

highest standards must be set out and maintained during the process to retrieve the maximum amount of information from the items recovered, the samples taken, and the on-site analyses.

The physico-chemical dynamics affecting the preservation of underwater sites and their artifacts are complex. The site formation processes of sites accessible to divers have been extensively studied (MacLeod 1995; Murphy 1990; Pournou et al. 2001; Ward et al. 1999a, b).

The deep-water sites surveyed have the common trait of being little disturbed physically by post-depositional forces. At 400–800 m depth these sites experience low and near constant water temperatures year-round. Sunlight, which accelerates some forms of biodeterioration, does not penetrate below 200 m. At the seafloor the current speeds are often very low with minimal tidal effects. The low current and lack of wave action means that sediment transfer rates are typically low. At the Mediterranean sites the net effect of both the erosion and buildup of sediment is an overall accrual rate of only a few centimeters per millennium. The Black Sea site on the other hand experiences 30–100 cm of undisturbed sediment accrual per millenium. While the low temperature and current of the deep-water environment have a strong preserving effect, the large increase in water pressure that occurs with depth creates unknown consequences by altering the solubilities of gasses and minerals. In some cases, finds were recovered with excellent surface detail but with unexpected chemical alterations. Taken together, this environment defines the challenges and potential of deep-water archaeology and conservation. It clearly illustrates that the approach to conserving finds from deep-water sites needs to be reexamined.

On all four expeditions the conservation process was divided into two main halves: shipboard and land-based conservation. The shipboard phase concentrated on retrieval, documentation, sampling, and triage (initial cleaning, wet stabilization, and packing), while the land-based phase focused on dry stabilization. For the first Skerki Bank expedition in 1989, the shipboard conservation was carried out by M. L. Florian (Florian 1994). Since then the authors have worked together on all shipboard conservation. Once on land, the artifacts were treated by the author with his wife and fellow conservator, Jane D. Piechota, either at temporary conservation facilities at Woods Hole, Massachusetts or Groton, Connecticut, or at their conservation laboratory in Arlington, Massachusetts, with certain specialized treatments and analyses taking place at appropriate institutions.

Planning and Work Space Requirements

The first step in the process of shipboard conservation was pre-project planning for the recovery of all types of objects and for a range of pretreatments suited to these assemblage types. It was necessary not only to map out the work area given the restraints of space on board the ship but also to estimate supplies. Knowledge of the space allocated for conservation prior to the expedition is the ideal, but often this was not the case and flexibility and creative use of allocated space is paramount. Preparation times will vary, but 6 months prior to a 4-week season is a must as it provides time to discuss the goals of the expedition with the principals, to purchase or construct custom equipment, and to order all other supplies and have them shipped to the research vessel's loading port.

The following shipboard work areas are needed for stabilizing and packing finds for transit:

- a shaded or enclosed finds holding area, including one large desalination tank for amphora as well as several small tubs with access to both fresh and saline water;
- an adjacent shaded deck area for wet cleaning of larger finds and for examination of finds by groups of archaeologists, the film crews, etc.;
- dry area for database work and documentation, including sufficient worktop space for microscopes and photography;
- laboratory space with sink units for fine cleaning of smaller finds;
- refrigeration for wood, metals, and samples;
- sufficient space for packing of large objects to prepare for transport back to the land-based laboratory;
- secure dry storage space separate from the main work areas for storage of packed artifacts.

If the research vessel is docked in a nearby homeport prior to the expedition, as was the case during the 1997 expedition, the work spaces can be custom designed. Using information such as artifact sizes and types from the 1989 expedition helped in estimating the supplies needed for subsequent expeditions. A volume analysis of the artifact types and quantities was done for the 1997 and subsequent expeditions. This was coupled with a flowchart of likely shipboard treatments. By analyzing the conservation activities common to the past expeditions, a processing chart was generalized (see table 5.2).

As part of the preexpedition planning phase, one can use this information to run through the conservation processes for each class of expected artifact finds

Table 5.2.
Processing Chart used for 1997–2003 Cruises to Plan Space and Supplies Requirements

Step	*Location*	*Activity*
1	Elevator	Preliminary documentation A. Tag artifacts with field numbers B. Photograph in elevator to document received condition C. Protect from direct sunlight D. Begin misting surfaces with half-saline water
2	Deck	Move Artifacts to the processing area A. Place in wet padded carriers/trays and cover B. Clear walkways C. Transport in 2-person teams
3	Processing area	Place in holding tanks (ceramics, stone, glass, and wood) A. Attach polyurethane foam cushioning rings B. Isolate stone from ceramics in main tank C. Separate tubs for glass D. Separate tubs for wood
4	Processing area	Place metals in reducing baths A. Iron and concretions in sodium carbonate solution B. Cuprous and lead artifacts in salt solution
5	Processing area	Cleaning and sediment sampling A. Remove living coral, etc., and sample B. Remove sediment from interior of vessels and sample
6	Photography area	Photograph and document A. Measure and weigh artifacts B. Digitally image for "before" treatment documentation C. Enter information in database
7	Processing area	Artifact sampling A. Petrographic sampling of ceramics and stone B. C-14 and dendro sampling of wood
8	Processing area	Casting of iron concretion voids A. Sample and clean interiors of sediment B. Fill interiors with silicone rubber
9	Packing area	Pack for transit A. Wrap in wet cotton/polyester fabric and seal in polyethylene B. Cushion in bubble wrap C. Pack in heavy corrugated boxes D. Inventory containers and contents
10	Refrigerator	Refrigeration storage A. Refrigerate all wood and as much iron as space will allow B. Refrigerate all samples C. Inventory separately from bulk storage
11	Ship's holding areas	Bulk storage Lash corrugated containers in dry locations

to produce a list of supplies and equipment needed to accomplish each activity. This was done for the 1997 and subsequent expeditions and led the conservators to draw up a list of supplies from weighing scales to tweezers (see Appendix: Equipment and Supplies List at the end of this chapter).

The space and supplies required are also dependent on the number of conservation staff available and the rate of retrieval of the objects from the seafloor. Barring equipment malfunctions or poor sea conditions, the retrieval of artifacts can proceed on a 24/7 schedule. This means that conservators must be ready for anything to happen at anytime. Good communication with the archaeologists and director of the expedition meant that, although nothing was ever certain, the timing of object retrieval was approximated with good warning, as were the numbers of objects to be expected in each elevator or autonomous transporter (Bowen 2000). This gave the conservators time to make available trays for transporting the objects from the elevator to the holding tanks, as well as labels for tagging each object or sample.

No matter how much planning is undertaken, each expedition presents the conservators with varying facilities, depending on the size of the research vessel, the space needed by other team members and the equipment onboard. Often spaces need to have multiple uses, so understanding the possible types of work to be undertaken as well as estimates of the maximum number of objects to be treated are essential in deciding on even the most basic facilities.

It is also necessary to bring enough packing materials to securely pack the objects for their safe return journey. The importance of this step in the process has long been recognized (Carpenter 1987) and cannot be overstated; prepare incorrectly and once out at sea it becomes almost impossible to augment your supplies; and even when possible, it is more often than not costly. It is difficult to prepare for an exact number of artifacts or for particular material types. During the 1997 season, for example, 115 artifacts were recovered from 6 of the 8 wrecks that were surveyed. Eighty of these were ceramic, 19 metal with concretions attached, 9 wood fragments, 4 glass, and 2 stone objects. Among the ceramics were numerous amphoras, all requiring considerable amounts of packing materials to be safely transported. Other years few artifacts were collected either because of equipment malfunctions (Skerki Bank, 2003) or national export limitations (Black Sea, 2003) or because surface collection yielded a small number of diagnostic artifact types (Ashkelon, 1999) (table 5.3).

Conservator as Team Member

Deep-water archaeological surveys and excavations require the collaboration of a large group of specialists who work intensely for the period of the cruise. The composition of this group can vary greatly with each expedition. At times a member of the engineering or archaeological team may serve for a segment of the cruise only. Knowledge of such crew changes in advance will go a long way in assuring that the conservator has all data relevant to the shipwreck environment before the cruise ends. Important data sets for each site include seawater conductivity, temperature and depth (CTD data), current directions, and current speeds at the seafloor.

TABLE 5.3.
ARTIFACT BREAKDOWN BY SITE AND MATERIAL CLASS SHOWING THE WIDE RANGE OF RETRIEVAL VARIABILITY

Location & season	*Wreck*	*Ceramic*	*Wood*	*Ferrous*	*Stone*	*Cuprous*	*Lead*	*Glass*	*Site totals*
Skerki Bank									
1989, 1997	*Isis*	35	10	10	4	3	0	0	62
1997	A	6	1	0	1	2	0	4	14
1997	B	20	1	0	0	0	1	0	22
1997, 2003	D	29	1	2	2	2	3	0	39
1997	F	23	5	0	0	0	0	0	28
1997	G	6	1	0	0	0	0	0	7
Ashkelon									
1999	A (*Tanit*)	16	0	0	0	0	0	0	16
1999	B (*Elissa*)	17	0	0	1	0	0	0	18
Black Sea									
2003[a]	D	3	0	0	0	0	0	0	3
Artifact totals		155	19	12	8	7	4	4	209

[a]Black Sea 2003 artifacts from Wreck D and other shallower sites were given to Sinop Museum authorities, Sinop, Turkey.

Of course, for the conservator the most essential collaboration is with the project archaeologists. Their research goals and excavation plan should be discussed before the expedition to prepare the types and quantities of supplies and equipment. Decisions on the extent of sampling and excavation are made by the archaeologists and are dependent on the research questions they are attempting to address, resources, and other practical issues. If the research interests center on ship's technology, then wood and metals sampling and treatment supplies will be needed. If trade is the primary focus of study, then the retrieval of a large number of ceramics is likely and will require capacious soaking tanks and packing supplies.

The information technology (IT) specialist is a critical link between all workers. He or she will be involved in connecting all streamed data coming from the research ship, the ROV, and their sensors. It is important to let the IT specialist know your needs. CTD data, ROV location and heading, the archaeologist and engineer's log entries, and imaging logs may be available through this team member and these data will allow the conservator to reconstruct events after the expedition. It is very useful to postcruise conservation processing to define your needs and receive a set of these data before the end of the cruise. For example, an understanding of the erosion and sediment infilling of surface artifacts such as amphoras requires that their exact orientation be reconstructed with respect to the seafloor topography as well as the current directions and magnitudes of overlying waters. Specific questions should address what data will be collected routinely, whether an event logger will be maintained throughout the cruise, what data will be logged, and whether it will be accessible through various computer platforms and software packages.

The ROV engineers are responsible for designing and mounting the various survey and excavation tools on the vehicle. It is important to develop a close relationship with the engineer in charge of perfecting tools for each archaeological task (see chapter 4 on tool development). The conservator, with a working knowledge of the strengths and weaknesses of marine artifacts, is in a good position to act as a liaison in discussions between engineers and archaeologists on the design of robust yet sensitive tools. It is also important to understand how the presence of a particular tool limits the performance of other activities. For example, specialized artifact-lifting tools when mounted on the manipulator arm may prevent sediment data collection methods such as tube coring. Sampling activities should be scheduled for early ROV dives, e.g., when the site is being located and surveyed prior to excavation. Additional sediment sampling can usually be scheduled but it is important to know the limitations of the ROV in terms of how a particular science activity may affect the manipulator configuration and ROV stowage space.

The ROV pilots are a specialized group separate from the ROV engineers. Through the ROV they serve as the hands and arms of the archaeologists and conservators. It is important to understand their capabilities and perspective on the limitations of a particular ROV and for them to understand the types of archaeological materials they may be retrieving. For example, the fragility of an iron concretion is not apparent from its outward form. The conservator should tell the ROV pilot about its shell-and-void construction before the pilot attempts to retrieve one. Similarly, expected areas of cracking on ancient pottery should be described so that the pilot can avoid applying pressure on the wrong axes when lifting. And the conservator should make certain that the pilot has a sense of the spongy consistency of waterlogged wood.

If the expedition is organized and overseen by a chief scientist, the conservator should have a clear understanding of their goals and the schedule of activities. Having an oceangoing research vessel stationed in a far-flung location provides an opportunity for marine scientists to conduct investigations unrelated to archaeological activities. While these researchers are a valuable source for insights into the marine environment, they can affect day-to-day scheduling of the archaeological surveying and excavation. Their sample collection and processing needs may also place demands on certain limited shipboard resources such as laboratory space, refrigeration space, and the water supply. The chief scientist may also plan to incorporate educational activities such as filming. Deep-water archaeology has great popular appeal and filming presents the conservator with an opportunity to inform the public of the value and fragility of archaeological resources.

Finally, the captain and crew of the research vessel are essential players in determining the successful completion of shipboard conservation activities. The conservator should consult with the captain before bringing on board any flammable solvents and caustic reagents, as they may not be permitted or their use may be restricted to specific areas of the ship. On certain research vessels, such as those that comprise the UNOLS fleet, are specialized crew members who form the Shipboard Scientific Services Group. Through their knowledge of the ship's scientific support equipment the conservator can set up a temporary laboratory that achieves scientific and artifact processing goals. The crew is helpful in finding unused space to turn into storage areas, and when seas are rough, they can assist in securing all apparatus and desalination containers so that they escape harm.

The Care of Newly Recovered Finds

Once an elevator filled with artifacts is lifted from the sea the conservator's most pressing task is to prevent the artifact surfaces from drying by shielding them from direct sunlight and spraying the surfaces with half-saline water. Even a small amount of drying can cause precipitation of calcium carbonate and gypsum within the pores of an artifact (Lazar et al. 1983). This can lead to a dramatic loss of porosity, which will affect subsequent attempts at desalination.

The artifacts should be submerged in a holding tank of half-saline water immediately. Half-saline water is a mix of fresh ship's water and seawater tested to be roughly half the salinity of the originating seawater, 15 to 20 ppt for most marine waters. All other activities, such as archaeological study, photography and filming, documentation, and cleaning, should wait for at least one hour while the artifacts' surface pore salinity is allowed to lower in the bath. During periods of study the surfaces must be continually sprayed with water.

Lifting, Labeling, and Recording Artifacts

Artifacts should not be moved or retrieved until a precision survey of the site, including a photomosaic and microbathymetric map, is completed (Singh et al. 2000). When objects are disturbed the equilibrium they acquired during burial is lost. They should then be recovered if possible and, if simply moved, they must be recorded. A labeling system for these is needed so that they can be positively identified on the photomosaics. Electronic still camera (ESC) images and video captures of the object in situ are essential for recording purposes. During the survey, the decision to remove an object is made jointly by the archaeologists, the engineer working the ROV, the project director, and the conservator. The decision to select specific artifacts for recovery is based on certain prearranged criteria, such as their ability to assist in identifying the origin and age of the wreck site. The ROV and its lifting arm are carefully maneuvered to avoid damaging objects nearby or disturbing artifacts unnecessarily. At all times the opinion of the conservator on duty is taken into consideration before the object is removed. If an object appears too fragile or might damage adjacent artifacts, it is not lifted.

Lifting finds in their matrix is a common archaeological practice and can allow the retrieval of very fragile artifacts. Doing this remotely with an ROV requires great care by the ROV operator, as the electronically powered arm must be maneuvered into the sediment surrounding the object to secure it from beneath. Extra support for the matrix can be given by placing it in a tray secured to the elevator or the ROV. It is also important at this stage to record the orientation of the find on the seafloor, as this information assists in understanding the deterioration processes of the objects. Partially exposed ceramics deteriorate at various rates, depending on which part was exposed to seawater or submerged in the seabed as these provide different chemical and physical environments.

A necessary part of the conservator's work is the taking of sediment samples from within and outside the perimeter of the wreck. Sediment samples yield physico-chemical data on the relation of the artifacts with their surroundings and can be valuable in determining their deterioration mechanisms. These are retrieved using multiple coring tubes to collect sediment columns from several areas of the wreck site and are usually taken at the start of work on the site. Additionally, wood samples should be retrieved from ancient wrecks. These samples are taken for wood identification purposes, to verify details of hull construction, and to assess the physical condition of the wood.

A buoyancy-driven elevator is used for lifting the objects from the seabed through the seawater interface. It is customarily built on ship by engineers with consultation from conservators. Not only must it support the weight of very large amphora often brimming with sediment, but it needs to safely lift smaller objects or pieces of objects without losing them in the turbulence of the ascent. Particular care at this stage is of utmost importance, as finds are vulnerable because they are contending with pressure changes, their own weight, and that of the water that saturates them. Built with these concerns in mind, the elevator should be equipped with spacious net receptacles and a lid that can be closed prior to lifting, thus securing the objects.

Prior to sending down the elevator, artifacts to be lifted are chosen in consultation with archaeologists, conservators, and engineers. This allows engineers to reconfigure the netted compartments on the elevator to suit the type of objects to be lifted. During the Ashkelon expedition, for example, compartments were tailored to fit the amphoras, as they were very similar in size and shape (figure 5.1). Each netted compartment must be prelabeled to assist in identifying each object. This is particularly significant when retrieving amphoras that are similar in shape and size, as once on the surface it is essential to properly identify each find.

Figure 5.1. (*Left*) Amphoras from the Ashkelon deep-water wreck site A being retrieved in an elevator. (*Right*) Ashkelon amphora (AS99.A.005) and cooking pot (AS99.A.002) in the elevator on deck. (Copyright Institute for Exploration)

Once on board, the artifacts are put under the care of a conservator, who, besides keeping the objects wet using spray bottles filled with a mixture of salt and fresh water, must control access to the objects before their condition is assessed. Premade, numbered Tyvek (spun-bonded polyethylene fiber) labels are immediately tied to the objects upon recovery of the elevator to avoid mixing up similar looking objects and as an important aspect of registration of the objects. Water-repellent, these labels are strong and resistant to long periods of immersion in seawater or caustic solutions. Most important is the marking ink used; it must be waterproof and should not fade with time. Equally important is how the label is secured, as staples corrode and cotton twine will rot, so an inert twine, like polypropylene, must be used. The label should be firmly attached by tying it onto the handle of a pot. When this is not possible, the object should be placed inside a plastic tray on its own with the label and the tray should be labeled as well, or if the object is small enough, it should be put inside a plastic bag labeled with the number and with the numbered label placed inside. Wood samples can be tagged using stainless steel pins or Tyvek tags.

A computerized shipboard conservation treatment form is essential to describe the object as well as the conservation treatments undertaken on board. It allows for constant modification and the data files at the end of the expedition can be distributed on a CD for all those requiring the information. The files from these expeditions should be kept in a database. Certain computer software allows ease of access across platforms and is built to incorporate images as well as text.

A sample registration sheet is included (see figure 5.2). Besides spaces for recording obvious information, such as the registration number of the object and the material type, there are also places to record elevator number and dive number of Hercules (the ROV on the 2003 season at Skerki Bank). Space for inserting digital images of the object is essential, as are places for recording the artifact's gross dimensions and noting whether a sample of the artifact's matrix or content was taken for further investigation. Important also is a record of the coloring and condition of the submerged and exposed sides of an artifact, particularly of pottery and stone. The artifact's surrounding sediment should also be recorded and is essential for understanding the condition of the artifact and for determining the best method of conservation. Seafloor imaging conducted before and during a lift is also informative and space should be provided for them on the sheet. The sheets allow for the contact date and time to be recorded, which permit the archaeologists and conservators to refer back to the correct still images and video captures taken. The form is meant to methodically record all processes conducted on an object from the moment it is raised through its conservation on board and packing. It cannot be stressed enough how important it is to complete as much of this information as possible on board, as this documentation is crucial to the conservator on land when performing the final treatments and is an essential record of the artifact's transition from seabed to study.

Artifacts as Environmental Data

It is important to preserve the biological accretions and geochemical staining on deep-water artifacts as markers of the site chemistry and site development.

ASHKELON DEEPWATER PROJECT SHIPBOARD CONSERVATION

RETRIEVAL DATE/TIME: 6/12/1999 7:10 ZULU
ELEVATOR NO: 1 DIVE NO: 245
L x W x H: 29.6 cm 29.6 cm 27.0 cm

ID NUMBERS: AS99. A. 002
ARTIFACT: Cooking Pot
MATERIAL: Ceramic

DIGITOGRAPHY NIKON COOLPIX 950 SETTINGS: DATE:
☐ obv ☐ rev ☒ 4vue ☒ top ☒ bottom ☐ det 100/3.4/30 6/15/1999
☐ obv ☐ rev ☐ 4vue ☐ top ☐ bottom ☒ det 100/3.4/15 6/15/1999
photographed details of loss and crack

PHOTOGRAPH

PHOTOGRAPHY FILM: ISO: DATE:
B/W: ☐ obv ☐ rev ☐ 4vue ☐ top ☐ bottom
B/W: ☐ obv ☐ rev ☐ 4vue ☐ top ☐ bottom
COLOR: ☐ obv ☐ rev ☐ 4vue ☐ top ☐ bottom
COLOR: ☐ obv ☐ rev ☐ 4vue ☐ top ☐ bottom

NOTES: diameter of rim: 19.5 cm; wall thickness: 1.26 cm; rim thickness: 1.33 cm; inside diameter of mouth: 16.4 cm;

Small rim loss with a vertical crack extending about half way down sidewall. Unrelated sherd found in interior sediment cat #: AS99.A.002.1.

POTTERY/STONE pieces: 1 ☒ complete ☐ incomplete ☐ sampled
BURIED SIDE: ☐ friable ☐ abrasions ☐ cracked ☐ solubilized ☐ flaking ☐ spalled ☐ stained ☒ accretion
EXPOSED SIDE: ☐ friable ☐ spalled ☐ solubilized ☐ flaking ☒ cracked ☒ accretions ☐ abrasions ☒ stained ☐ crazed
Crack and small piece missing, exposed surface blackened

SEDIMENT
TEXTURE: ☐ fine ☒ medium ☐ coarse ☐ encrusted
COMPOSITION: ☒ clay ☐ pebble ☒ shell

WOOD/ORGANICS pieces: _______
☐ complete ☐ incomplete ☐ sampled
BURIED SIDE: ☐ spongy ☐ abrasions ☐ stained ☐ spalled
EXPOSED SIDE: ☐ spongy ☐ spalled ☐ stained ☐ crazed ☐ abrasions ☐ checked

METALS pieces: ______
☐ complete ☐ metal core ☐ incomplete ☐ stable ☐ mineralized ☐ active

CONTENTS
sample no/s AS99.A.002-1, 2, 3, 4
☒ sieved ☐ wood ☐ metal ☒ sampled ☒ pottery

TREATMENT
1ST TREATMENT: ☒ 50/50 soak ☒ wash ☐ fresh water soak ☐ acetone ☐ ETOH dewatering ☐ PEG400/3350 ☒ rinse ☐ NaHCO3
2ND TREATMENT: ☐ 50/50 soak ☐ wash ☐ fresh water soak ☐ acetone ☐ ETOH dewatering ☐ PEG400/3350 ☐ rinse ☐ NaHCO3

PACKING
☐ Nomex ☒ 50/50 polycotton ☒ bubbl
☐ Marvelseal ☒ triple bag HDPE
DATE: 6/22/1999 Box No.: 29

WEIGHT AND VOLUME
WEIGHT IN AIR: 3.62 Kg WEIGHT IN AIR W/ CONTENTS: _______ Kg WEIGHT IN WATER: 1.75 Kg VOLUME: 11.0 L

OBSERVATIONS: Several small scrape marks on ferromanganese discoloration from ROV gripper on opposite sides of pot. Completely filled with sediment. Ferromanganese staining indicates portion of pot that was exposed to open water.

Conserved by Dennis Piechota and Cathy Giangrande

Figure 5.2. Sample conservation database form used for all artifacts.

On the deep seafloor the environment that forms within the perimeter of a wooden shipwreck tends to be very different from the surrounding seabed. The worldwide oceanographic surveys and sediment coring programs have amassed an understanding of the seabed composition and chemistry that should be consulted before any expedition (Einsele 1967; Emelyanov 1972; Fabbri and Selli 1972; Degens and Ross 1974; Izdar and Murray 1989). But this knowledge, important as it is, can only be indirectly applied to the shipwreck sediments themselves. The large mass of decaying wood left by the hull alters the sediment chemistry profoundly by bringing anoxic conditions to within a few centimeters of the seafloor. Marine worms mine the original wood of the shipwreck and their succeeding generations remine the organic-rich sediments left behind. The bulk cargos, usually dominated by mounds of amphoras, provide habitats for marine life. Bacterial slime populates the voids of all porous objects, bivalves and solitary coral bodies attach to any hard surface, and shrimp benefit from the turbulence of currents flowing over exposed artifacts disturbing microscopic food particles (Giere 1993). Crabs and fish move in and out of the exposed amphoras seeking shelter and food. The sum of all these activities makes the wreck site an anomaly with respect to the chemistry of the surrounding sediments.

Organisms play a role in the development of the site and their residues on artifacts assist us in reconstructing the site. Surface amphoras can be partially covered with solitary corals and worm tubes especially along the sediment/water interface. It is important to note that preservation of these associated environmental accretions is essential, as they provide us with additional clues as to the orientation and condition of artifacts. Multiple linings of their calcareous skeletons can show that the artifact has moved or rolled after initial deposition. Geochemical staining can also indicate important environmental details. Black iron sulfide staining on the underside of an amphora indicates anoxic sediment conditions and this in turn may indicate the presence of preserved wooden ship structures. While calcareous accretions can be disfiguring and hide artifact details, these deposits should be considered environmental data and removed only when necessary (figure 5.3).

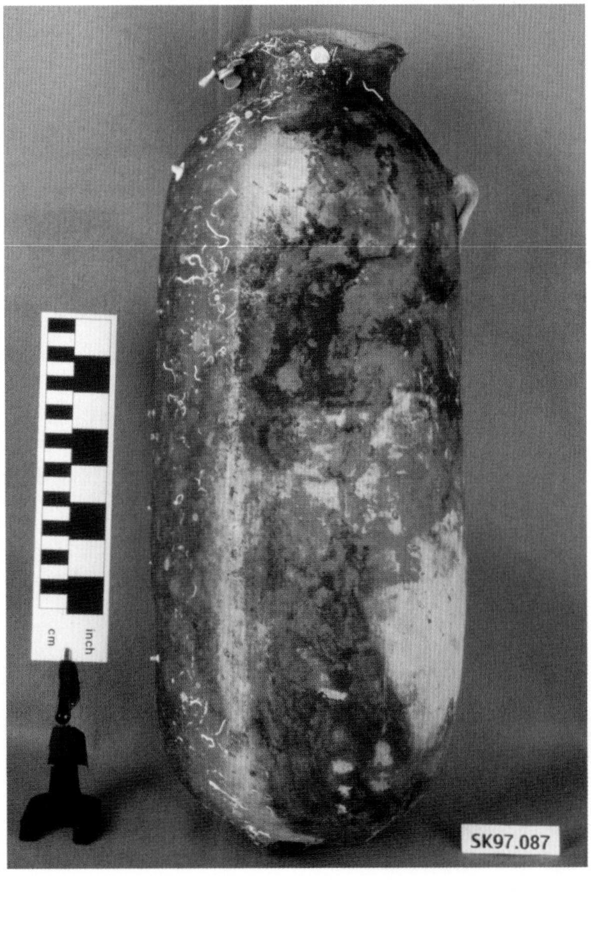

Figure 5.3. Neo-Punic amphora from Skerki Wreck F (SK97.089) showing the different microenvironments occupied by surface artifacts. Black sulfide and copper corrosion salts stain the submerged side shown on the right, while solitary corals and serpulid worms grow on the left side, which was exposed to seawater.

Iron Artifacts and Marine Concretions

The preservation of marine metals provides a great challenge to conservators. Over the period of their burial iron artifacts typically develop voluminous and disfiguring corrosion crusts that hide the original surfaces and character of the artifact. As this crust forms, the metallic core dissolves, leaving a void within the corrosion crust that often defines the shape of the original artifact (North 1976, 1982). Iron artifacts and fittings from ancient shipwrecks, though often completely mineralized from exposure to the saline and oxygenated waters on the seafloor, may still be unstable in air. While the concretion forms layers of minerals stable to the seafloor environment, the inner layers can become unstable upon exposure to air and reoxidize, causing loss of interior surface detail. This void can be destroyed easily by reoxidation of the interior surfaces during shipboard storage and by shock and vibration causing the concretion to collapse during transit to the land-based laboratory.

When a concretion is lifted from the seabed it should be transferred to an alkaline bath to limit subsequent reoxidation of the corrosion layers. It is essential to stabilize the void within the concretion immediately by filling it with a solid casting resin or molding rubber. This should be done the same day that the concretion is retrieved. Silicone rubber (GE RTV 11 with RTV 9811 as curing agent) is useful for this purpose because it is capable of setting or vulcanizing in the presence of seawater and caustic chemicals and can displace any water that remains in the voids. It does not adhere to epoxies or other resins that may be applied to harden the dried corrosion crust and its flexibility allows it to be teased out of the opened concretion reducing damage to the crust.

While there may be a strong desire to remove the corrosion layers, it should be kept in mind that the crust may contain preserved pseudomorphs: perishable organics whose shapes have been preserved by infusion with their corrosion products. Pseudomorphs of grains, seeds, lashings, and wood fibers from materials that were lying near iron objects have been found in the Skerki Bank concretions (figure 5.4). Lashing methods can give information on ancient technology, and grains can indicate possible cargoes or ship's stores. Wood fragments may be useful in identifying the woods used to construct the hull, decking, or other ship timbers. Therefore, the corrosion crust surrounding a marine iron artifact should be thought of as an artifact itself to be studied by micro-excavation and by micromorphological techniques.

The first step in addressing the preservation and analysis of a newly excavated iron object is to estimate the extent of mineralization by x-radiography. Marine iron corrodes to form many minerals, including magnetite, amorphous iron oxides, siderite, jarosite, and pyrite, and X-ray imaging allows one to see beyond these mineralized layers to determine if any of the artifact's uncorroded metallic core is still present.

Removal of the corrosion crust should start from the thinnest side. That side may need to be cut into several pieces to expose the silicone cast. If this is done carefully then the thicker half of the concretion can be preserved intact or in a few pieces that can be rejoined. After removal of the silicone cast the crust can be micro-excavated and examined for micro-artifacts or other inclusions of cultural origin. As these silicone casts are flexible, a permanent replica of the

Figure 5.4. Artifact information retrieved from within the corrosion layers of a Skerki Bank iron concretion (SK97.029). (*Left*) Lashings of braided bast fibers preserved in place inside the iron corrosion crust. (*Right*) Silicone rubber used to cast the form of the original iron object also seeped into voids left by decomposed barley and wheat grains from the ship's original stores. (Copyright Institute for Exploration)

artifact should be made by remolding the silicone copy and casting it in a rigid resin (see figure 5.5).

Packing and Shipboard Stowage Requirements

New ROVs (remotely olperated vehicles) and manipulators are being constantly developed to safely retrieve sensitive and fragile artifacts. This makes it increasingly important to maintain high standards for the packing and stowage of finds. Also, the time it takes for a research vessel to return to its homeport can be lengthy and include periods of rough sea conditions. The packing method must protect against shock and vibration, while at the same time prevent any drying from occurring. Even slight drying can cause saturated and near-saturated marine salts, including calcium sulfate and calcium carbonate, to precipitate within the artifact, leading to surface damage.

The method developed for most artifacts is simple and well-tested. Before packing, each artifact spends at least 24 hours in a tank of water maintained at half the salinity of the originating body of water. The artifact is wrapped in 50/50 polyester/cotton double-knit fabric that has been soaked in the same half-saline water. The stretch of double-knit fabric allows the wrap to conform to a variety of surfaces. Care is taken to pad out any projecting isolated corals to cover their sharp edges.

To maintain the wrapped artifacts in a wet condition for the complete transit period, up to two months, they are placed in three layers of high-density polyethylene (HDPE) bagging and well-sealed with package sealing tape. The bags are checked for leakage before proceeding. The triple-bagged artifacts are

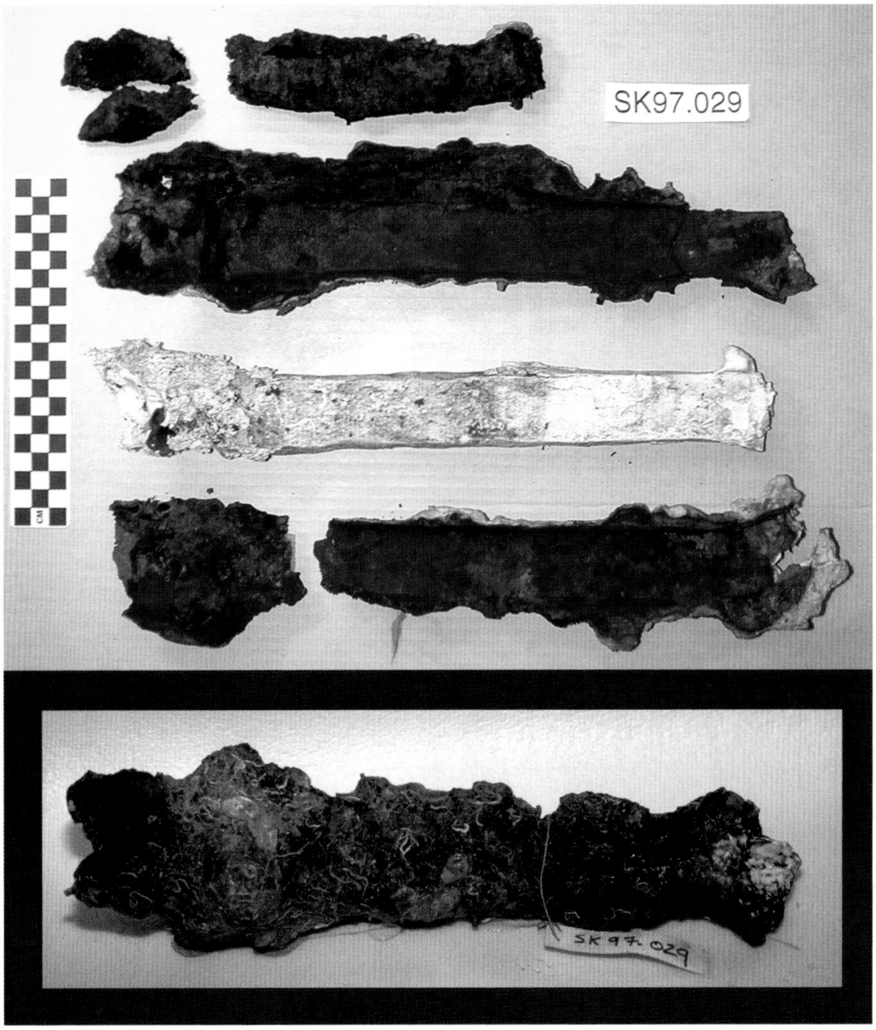

Figure 5.5. Skerki Bank iron concretion (SK97.029). (*Bottom*) Concretion before opening. (*Top*) After opening to preserve the concretion crust for later analysis and to remove the silicone rubber cast shown in the middle.

then cushioned by wrapping in two to four layers of bubble wrap and placed in heavy corrugated cardboard cartons. Bubble wrap maintains its cushioning performance even when wet and acts as a final barrier preventing water from leaking into the corrugated cardboard container.

The sealed boxes are inventoried and lashed in a single layer against the walls of available waterproof storage holds. They should not be stacked because humid conditions on ship can degrade the load-bearing capacity of the boxes. To prevent damage in the event of flooding they should be raised from the floor surface by a few inches on 2 × 4s or other waterproof stripping.

The above general instructions have worked well for stone and pottery, the most common material types encountered at the wreck sites we have surveyed. Other materials are treated similarly with added precautions. Metals, including completely mineralized concretions, are fully immersed in a reducing solution, 5% Glauber's salt for cuprous metals and 5–10% sodium carbonate for ferrous metals, and stored in waterproof plastic tubs. The cotton/polyester fabric is used here as cushioning and dunnage to prevent movement of the artifact with respect to the tub.

Organics are completely immersed in half-saline water, wrapped in cotton/polyester, sealed in waterproof tubs, and refrigerated. Refrigeration is essential for organics to slow the rate of biodeterioration from bacteria during transit. It is also strongly recommended for ferrous and cuprous metals to help the reducing solutions slow their corrosion rates. At no time should wet archaeological artifacts be allowed to freeze. The ship's scientific support staff should be alerted that the common practice of freezing oceanographic samples is not applicable to artifacts.

Transit and Storage of Artifacts

When a large quantity of artifacts is excavated far from a conservation facility the only practical shipping method may be by sea. This means that there may be long delays before the conservator can begin the land-based treatments. For example, at the Skerki Bank and Ashkelon sites there was a 2-month delay before unpacking and treatment could be started. The alterations experienced by artifacts when stored wet at room temperature in partially saline water are only poorly understood. Conservators assume that inorganic artifacts such as pottery are stable when stored wet for short periods of time because they are thought to be in an environment essentially similar to their seabed environment, but that assumption requires further investigation.

There is a large amount of decaying microfauna, predominantly bacteria, resident on the surfaces of all deep-water artifacts. This bacterial slime permeates the fabric of all porous artifacts. While sealing wet artifacts in waterproof bagging for storage and shipping works well to avoid damage due to salt crystallization it may also have deleterious effects. Once sealed, oxygen within the package is quickly consumed by aerobic bacteria allowing anaerobic and facultative bacteria to thrive. The incompletely desalinated artifacts contain sufficient sulfate ion concentrations to allow such sulfate-reducing bacteria to grow rapidly. Elevated temperatures accelerate bacterial growth and during summer transits, when expeditions tend to take place, ambient temperatures within ships' holds and storerooms can routinely rise above 30°C (86°F) due to the heat generated by the power plant. When the conservator opens these after 1 or 2 months a black bacterial mat covering all surfaces is usually encountered.The mat can be up to one-half centimeter thick and is thick enough to obscure surface details. While it quickly disappears when exposed to air the effects of its growth on the composition of some artifacts may be significant.

All the sites under consideration in this chapter were collected from above the carbonate compensation depth (CCD) and so calcium carbonate precipitation as an abiotic process may be expected to proceed slowly on or within artifacts and surrounding sediments (Morse and Mackenzie 1990). Anaerobic bacteria have been shown under natural and experimental conditions to rapidly accelerate this process by generating large amounts of dissolved carbon dioxide and by providing nucleation sites for precipitation (Baedecker and Black 1979; Paerl et al. 2001; Chafetz and Buczynski 1992). This process is more likely to be significant

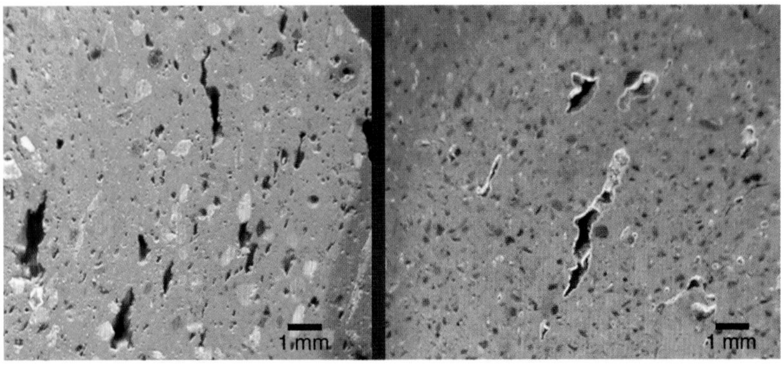

Figure 5.6. Macrophotographs of cross sections of two different Hazor amphoras dating from the 8th century BCE. (*Left*) Retrieved from the terrestrial site of Hazor (Stratum VI) showing a relatively unaltered appearance. (*Right*) Retrieved from the Mediterranean deep-water wreck site Ashkelon Wreck B, (amphora AS99.B.039) showing widespread void lining from the deposition of white marine carbonates that are suspected of reducing the porosity of the clay and inhibiting the desalination process.

for deep-water artifacts retrieved from above the CCD because carbonate solubility decreases with the lowered pressure and increased temperature (Leyendekkers 1975). Retrieving artifacts from the cold depths, especially from the anoxic sediments typical of shipwrecks, creates pore water within artifacts that one can expect to be more highly supersaturated with carbonates than artifacts collected from warmer and shallower sites (Berner et al. 1970). This condition could be responsible for the rapid postretrieval deposition of calcium carbonate within the pores, thus reducing the permeability of artifacts, especially ceramics, and lengthening the desalination process (figure 5.6) (Lavoie and Bryant 1993). While these so-called bacterially mediated carbonates are theorized as a significant problem only for artifacts excavated from deep sites, conservators should proceed carefully and take measures to limit the problem until it is investigated further.

Ideally, all ceramics should be stored for the shortest time possible and not on deck where containers are exposed to the heating effects of sunlight. If possible, ceramics should be refrigerated during the storage period; however, as refrigeration space is limited, this is not likely to be a practical solution for bulky artifacts such as amphora. Air-conditioned spaces should be used where possible. Biocidal pretreatments cannot be recommended at this point because they can be damaging to artifacts and because chemical fungicides are regulated by use and no archaeological uses have been defined within U.S. federal law (FIFRA 1996).

Shipboard Desalination

The desalination of artifacts begins on ship in a large (e.g., 4 [W] × 10 [L] × 2.5 [H]-ft) soaking tank constructed of plywood and lined with waterproof neoprene laid down on a ½-in. polyethylene foam cushion to protect the artifacts from the engine vibrations of the ship. The tank is fitted with a lid, turning it into a good work surface for photography, artifact examination, and packing.

Smaller objects are allocated individual soaking tubs to protect them from the larger objects, such as amphoras, which can roll in heavy seas. Even when fitted with polyurethane foam rings to prevent surface abrasion and contact with adjacent objects, smaller finds are best isolated in separate tubs.

The contents of the pottery objects should be carefully sieved for artifacts and biological remains. Multiple samples of the sediments should be taken for possible organic and inorganic remains, including charcoal fragments, fish bones, or olive pits, all substances commonly found in ancient cooking pots and amphora. Scrapings of the interior of amphora walls should also be collected to determine the presence of deteriorated resin linings. One should note vessel color differences, which can help in the identification of the vessel contents. It is necessary to be alert for fragile pseudomorphs preserved in the sediment. Inside one cooking vessel from the Ashkelon deep-water site the sediment clearly held the imprint of decaying timbers, indicating the incorporation of sediment into the decaying timbers.

All objects recovered should be desalinated first on board the ship using a mix of 50% seawater to freshwater. This procedure continues on land with tap water and demineralized water as discussed below.

Land-based Desalination of Low-fired Pottery

Low-fired pottery is the most common and voluminous artifactual component of ancient shipwrecks. As the primary cargo containers, amphora are large in numbers and dimensions. The main conservation activities for pottery are desalination and drying, with mending and resinous consolidation being required for a small number of objects. In 1989, when the first deep-water ceramics were treated, desalination was thought to be a benign and simple process. It was assumed that diffusion would remove most marine salts if the pottery were simply soaked in tubs of static freshwater followed by a short period of soaking in distilled water. Early on in the 1989 treatment, four observations showed that this process was not so simple. When transferred to freshwater some pottery developed cracks within the first hour of soaking. This early-stage cracking has since been observed in shipboard soaking tanks during the first hour after retrieval.

Early-stage cracking is theorized to be due to the swelling of the marine deteriorated clay fabric as salts are quickly pulled from within the partially fired clay domains. The removal of such ionic "glue" has the effect of swelling the spacing between clay particles (Bearat et al. 1992). This introduces stresses that at times express themselves in cracking. Secondly, it was observed that within twenty-four hours anaerobic bacteria were growing on all surfaces in the lower levels of the still water. Sampling this water for conductivity showed that it had a higher salinity than the top layers. Mixing destroyed the bacteria and homogenized the bath water's salt content. Thirdly, the still water inside the amphora and jugs had a much higher salinity than the surrounding water. Finally, the edges of some pottery fragments were cored with a drill and the resulting samples were stirred into a quantity of distilled water of equivalent weight to volume ratio as the general bath. When tested for salinity these samples revealed that while the

Figure 5.7. (*Left*) Overview of three neoprene-lined desalination tanks used for the 1999 Skerki Bank collections. (*Right*) Close up of tank A showing the tubing used to recirculate freshwater into the amphoras.

surfaces of low-fired pottery were desalinated sufficiently, the core clay between those surfaces still needed further desalination.

These observations and the concerns raised by bacterial growth during transit caused several changes in subsequent desalination procedures. The deep-water pottery retrieved during the 1997 and 1999 field seasons were sampled prior to treatment and placed in cascading tanks providing a constant low flow of cold freshwater from the tap and through each tank. A constant flow maintains temperatures below room temperature, slowing biological decomposition and increasing carbonate solubility. Within each tank this water is recirculated to homogenize salinity. Tubing from the recirculating pumps directs water into the interiors of amphoras and jugs (figure 5.7). The dynamic movement of water also reduces bacterial growth and the thin films of higher salinity that can develop at the surfaces of soaking pottery. This increases the efficiency of the diffusion process by maintaining a steep gradient between fresh and saline water. Marine salts of limited solubility, like gypsum and supersaturated concentrations of calcium carbonate, stand a better chance of diffusing into slowly moving bath water.

Even though these procedures solve many problems they do not achieve the ultimate goal of producing pottery specimens that are free of marine salts. While bath water tests showed the pottery surfaces to have low salinity, comparable core sampling showed that their cores did not. More study and experimentation is needed to evaluate the long-term effects of this disparity and to overcome this limitation in the soaking process. Rather than extending the soaking time, which can damage some low-fired pottery (Willey 1995), methods to increase the efficiency of desalination are needed as well as determinations of the ability of calcium carbonate to entrain and immobilize soluble salts (Warren et al. 2001).

Drying of Low-fired Pottery

Low-fired pottery retrieved after long exposure to a marine environment experiences cracking at two steps in the conservation treatment process: during its initial exposure to water of reduced salinity and during final drying. Swelling due to the loss of electrolytes in the clay causes the initial wet cracking. There is no lifting or misalignment along break lines in wet cracking; in fact, it can be difficult to detect. Cracking during drying, however, is particularly damaging in that the planar distortions to the curvature of the vessel can be large and irreversible. The clay of low-fired pottery is partially plastic when wet and loses most of this property with the loss of its free water. So as shrinkage stresses build up in the drying vessel the walls are capable of suddenly springing open; i.e., curved sidewalls tend to straighten as they shrink.

Conservators who have mended "sprung" pottery know that whole vessels are capable of accommodating the high internal stresses that are often introduced by the firing process. Until a drastic shock such as impact from a drop shatters the whole vessel the internal stress will not be expressed as a relaxation of the vessel's curvature (springing). While pottery degraded by marine exposure may not have sufficient strength to withstand any stress its ability to do so increases as it nears the dry state and a goal of the conservator should be to avoid incipient crack formation in the early phase of drying by distributing drying stresses within the vessel evenly.

A method of internal drying was used successfully on the Iron Age amphora from the Ashkelon deep-water site and will be the method for future work. Dry air is fed into the interiors of the vessels using aquarium air pumps. This produces a moisture gradient across the vessel wall; the exterior surface of the vessel remains damp while the interior surface dries. The shrinkage caused by such drying will be greater inside than out. This will tend to produce drying stresses that reinforce the curved character of the vessel walls, locking in the stress and avoiding crack formation. In conventional drying the outside surfaces shrink first, causing stresses that tend to "unpeal" the vessel and lead to unnecessary cracking. Exterior drying is prevented by enclosing the vessel in high-density polyethylene, where only the mouth of the vessel is exposed and through which fresh air is circulated to the interior by the air pumps. By limiting the surface area available for drying, this method also slows the overall drying process and that in itself should be beneficial to the avoidance of cracks.

Storage and Maintenance of Artifacts

After land-based treatment is completed with all artifacts dried, photographed, and catalogued each collection should be conditioned, i.e., each artifact examined individually for instability. This should be done annually for the first few years and then biennially for the next few years after that. Through time stresses within the dried pottery can express themselves as isolated cracking, and incompletely oxidized minerals within the crusts of iron concretions can express themselves as fresh corrosion and therefore require retreatment. If possible, the conservators who were involved in retrieving and treating the collections should also do the conditioning.

Proper storage conditions are important and ideally all remains should be stored in a climate-controlled environment that is continuously monitored and recorded using an electronic datalogger. Fluctuations in conditions can be harmful to all types of artifacts, particularly metals and organics. Costs for storage need to be factored in when considering the retrieval of artifacts.

Site Preservation

It is important that the conservator address the preservation of unexcavated portions of a wreck site. Since 1971 when George F. Bass reported on the complete excavation at the Yassi Ada wreck site this has been the ideal for all shipwreck archaeology (Bass and Doorninck 1971). Though some initial efforts have been made to lower the cost of deep-water archaeology (Soreide 2000), it is unlikely that it can be lowered enough to allow complete excavation of deep-water wreck sites. Instead, partial excavation to answer specific archaeological questions will prove the rule for the future and is supported by recent positivist trends in archaeological theory that cast the wreck site as a resource to be approached to test particular hypotheses (Adams 2001; Gibbins and Adams 2001). But partial excavation exposes the site to renewed biological attack by removing protective overburden. Shipworms, for example, may be able to access preserved hull remains through the excavation pit. An effort should be made to rebury or fill in excavations with sediment or sandbagging before leaving the site (Ferrari and Adams 1990; Stewart et al. 1994).

During the 2003 expedition an experimental procedure was developed to apply a geotextile barrier over the excavated portion of the Ashkelon wreck sites and to backfill the covering with local sediment. Due to geopolitical concerns the expedition was not able to enter the area of the site (off Egypt) that year so this method was not tested, but the planned procedure is described here to promote discussion of the problem of deep-water site preservation.

Terrestrial sites have been stabilized using geotextiles since the 1980s (Thorne 1988, 1989). A geotextile is a large-format, woven or nonwoven fabric engineered to be applied in or on soils or sediments to achieve specific goals such as sediment isolation, consolidation, or the prevention of erosion. It was chosen here to isolate the freshly excavated surface before backfilling with local off-site sediment. One can import sandbags or use local sediments as backfill. The length of time needed to safely drop sandbags near the site and then distribute them via an ROV led us to consider backfilling with local off-site sediments. Since local sediments could be confused with the wreck-site matrix, a geotextile barrier was planned.

Geotextiles are commonly made of either polypropylene or polyester polymers. While both are long-lasting, in a marine environment polypropylene is buoyant with a specific gravity of 0.9 and so unsuitable for submergence. Denser fabrics composed of polyester (S.G.: 1.4) can be expected to settle onto the seabed once spread over the excavation pit. Geolon HS400, a woven polyester by Mirafi Construction Products, was selected as isolation fabric and cut to a 5 × 15-m length.

Using an ROV to maneuver a large fabric underwater is the primary challenge for this method. In the 2003 season it was planned that the 5-m-wide fabric would be treated as a scroll with its two ends attached to lengths of perforated piping. One end would then be rolled to form the scroll. This rolled assembly would be delivered to the seafloor upcurrent of the excavation pit. The end holding the rolled fabric would then be sand screwed to the seafloor just off-site in such a way that the fabric could be unrolled by pulling on its free end. The ROV manipulators would grasp the piping of the free end and pull the geotextile up and over the excavation pit. Once the fabric was allowed to settle in place it was planned that the excavation pumps on the ROV would be used to backfill over the geotextile with local sediment.

It is clear that the application of a large fabric underwater will present a great challenge to the ROV pilot. Perhaps other methods for isolating backfilled sediment will prove more practical. It is hoped that raising the issue here will yield a fruitful discussion of the problem of site preservation of partially excavated deep-water sites.

Conclusion

The discovery, disturbance, or excavation of an underwater site automatically demands special obligations and courses of action, including proper recording and conservation, whether the site is one found in shallow or deep water. One becomes a guardian of the irreplaceable, internationally important cultural heritage that links the past to the present (UNESCO 2000; Oxley and O'Regan 2001). Artifacts can survive for thousands of years underwater and, although some objects must be lifted to positively identify a site, the use of advanced recording techniques can limit the number of objects one needs to retrieve.

During the course of these four expeditions, conservation procedures were refined and in some cases considerably improved upon, to ensure that all available evidence was retrieved and preserved to the highest standards. This provided the archaeologists with the maximum amount of information from the sites in question and their associated finds to formulate their theories and hypotheses. Iron concretions, for example, were shown to provide maximum information if cast immediately to avoid further deterioration as well as damage prior to shipping.

It is important that future conservation-related investigations focus on the biological and geological chemistry of deep-water wrecks. Data collection should be enhanced to better characterize the immediate environ of deep-water artifacts, as it has been shown that the chemical makeup of the site determines the nature and processes of deterioration of specific materials. With this information one can better formulate methods for mitigating deterioration. These and other advances in conservation procedures should be tried and tested to keep pace with the continuing improvements and developments being made by scientists, engineers, and archaeologists working in the underwater environment.

Acknowledgments

The authors thank Robert Ballard who, as chief scientist, organized and oversaw all the expeditions and made certain that conservation activities were always well-supported. A special thanks goes to Catherine Offinger who as director of operations made certain that whatever was promised was delivered. The conservators appreciate the opportunity to work with the chief archaeologists of each expedition: Anna McCann, John P. Oleson, Lawrence Stager, Fred Hiebert, and Cheryl Ward. There are many others we thank who in less official capacities helped with conservation activities, from visiting archaeologists such as Jon Adams and Shelley Wachsman to the many graduate students in archaeology and engineering. Dennis Piechota thanks Jim Newman, developer of the ROV Hercules, and Sarah Webster, the developer of archaeological toolset for that ROV, for the opportunity to engage in a valuable discussion on how ROV technologies might best be adapted to archaeological activities. Finally, the authors thank their spouses, Paul Hodgkinson for his support and, Jane Drake Piechota, who, as a conservator herself, critically reviewed drafts of this chapter.

Appendix: Equipment and Supplies List

This is a suggested list of supplies needed, which will be modified according to the type of site to be investigated. Guidance as to the kinds of objects to be expected will come from the project director and archaeologists. Quantities will depend on the length of the project and the anticipated types of finds. Generally, it is useful to err on the side of more, rather than less, as supplies are not easy to obtain once you are onsite.

All chemicals have to be clearly labeled and safely stored, preferably in a locked cupboard. Safety glasses, protective gloves, and waterproof aprons should be worn when handling these items.

Hygiene and Safety

waterproof aprons; nitrile gloves; safety goggles; eye wash bottle; roll of nonskid surfacing; dilute acetic acid for neutralization of mild alkali spills

Cleaning and Conservation

assorted brushes; cotton swabs; Orvus detergent paste; Acrysol WS 24 consolidant; Acryloid B-72 adhesive; assorted small plastic tubs; assorted plastic beakers and containers; garden hoses and nozzles; plywood soaking tank prefabricated in sections and reassembled after boarding; nuts and bolts for securing prefabricated sections of soaking tank; Neoprene liner for soaking tank; Pliobond rubber adhesive for repair of neoprene; spray bottles; submersible pump for draining soaking tank; containers of PEG 3350, PEG 400, PEG 200; salinometer

Chemicals and Rubber Casting

acetone in metal safety container; denatured alcohol in metal safety container; solution of 5% sodium carbonate; sodium bicarbonate, solid; sodium carbonate, solid; Glauber's salt; ivory black pigment to color silicone; Dow silicone RTV 11 rubber with 9811 hardener

Electrical and Tools

variable-speed electric drill with socket and drill sets; ratchet wrench with sockets; screw clamps and spring clamps; ring clamps; mat knife; scalpels; scissors; screwdriver, staplers with Monel staples; batteries; extension cords with GFI circuit; grounded multi-outlet

Science and Sampling

pH/redox meter with micro-electrodes; redox calibration solution; redox and pH storage solution; chloride titrator test strips; pH test strips; Ziplock LDPE poly bags; flexible shaft tool with cutoff wheels; micropipettes; sample jars and vials of various sizes from 2 to 32 oz; wash bottles; inspection microscope with lighting; extension tube for camera; microscope slides with cover slips; LED flashlight; 30-kg scale

Documentation

contractor's waterproof box for dry storage of paperwork; backdrop cloth for photography; metric measuring bar; weighing scales; digital camera and tripod; Tyvek tags; waterproof markers; roll Mylar film for 1:1 tracings

Packing

Bubble wrap for waterproof cushioning; polyester batting for delicate finds; gussetted low-density polyethylene (LDPE) bags; high-density polyethylene (HDPE) bags; cable ties; ratchet tiedowns; roll of aluminum foil, heavy weight; duct and package sealing tape; shock cords; tensioning tiedowns; polyurethane foam strips, 2 × 2 in., for making soaking tank cushioning rings; additional polyurethane foam, 2-in.-thick slabs; HDPE film, 48 in. wide; white double-knit cloth; corrugated boxes, heavy-weight, various sizes; roll of geo-textile, 15-ft width (for covering disturbed areas of site)

References

Adams, J. (2001). Ships and boats as archaeological source material. *World Archaeology* 32:292–310.

Baedecker, M. J., and W. Black (1979). Hydrogeological processes and chemical reactions at a landfill. *Ground Water* 17:429–37.

Ballard, R. D., A. M. McCann, L. Stager, D. Yoerger, L. Whitcomb, D. Mindell, J. Oleson, H. Singh, B. Foley, J. Adams, D. Piechota, and C. Giangrande (2000). The discovery of ancient history in the deep sea using advanced deep submergence technology. *Deep-Sea Research Part 1* 47:1591–620.

Bass, G. F., and F. H. van Doorninck (1971). A fourth-century shipwreck at Yassi Ada. *American Journal of Archaeology* 75:27–37.

Bearat, H., D. Dufournier, and Y. Nouet (1992). Alterations of ceramics due to contact with seawater. *Archaeologia Polona* 30:151–62.

Berner, R. A., et al. (1970). Carbonate alkalinity in the pore waters of anoxic marine sediments. *Limnology and Oceanography* 15:544–49.

Bowen, M. F., P. J. Bernard, D. E. Gleason, and L. L. Whitcomb (2000). Elevators—autonomous transporters for deepsea benthic sample recovery. *Woods Hole Oceanographic Technical Report,* 1–41. Woods Hole, MA: Woods Hole Oceanographic Institution.

Carpenter, J. (1987). The uses of soil stabilizing gel media in the conservation of large and small shipwreck artifacts. *International Journal of Nautical Archaeology* 16: 95–107.

Chafetz, H. S., and C. Buczynski (1992). Bacterially induced lithification of microbial mats. *Palaios* 7:277–93.

Chen, R., and K. A. Jakes (2001). Cellulytic biodegradation of cotton fibers from a deep-ocean environment. *Journal of the American Institute for Conservation* 40:91–103.

Degens, E. T., and D. A. Ross, eds. (1974). *The Black Sea—Geology, Chemistry, and Biology.* Tulsa, OK : The American Association of Petroleum Geologists.

du Plat Taylor, J., ed. (1965). *Marine Archaeology.* New York: T. Y. Crowell.

Einsele, G. (1967). Sedimentary processes and physical properties of cores from the Red Sea, Gulf of Aden, and off the Nile Delta. In *Marine Geotechnique,* ed. A. F. Richards. Urbana: University of Illinois Press.

Emelyanov, E. M. (1972). Principal types of recent bottom sediments in the Mediterranean Sea: their mineralogy and geochemistry. In *The Mediterranean Sea: A Natural Sedimentation Laboratory,* ed. D. J. Stanley. Stroudsburg, PA: Dowden, Hutchinson, & Ross.

Fabbri, A., and R. Selli (1972). The structure and stratigraphy of the Tyrrhenian Sea. In *The Mediterranean Sea: A Natural Sedimentation Laboratory,* ed. D. J. Stanley. Stroudsburg, PA: Dowden, Hutchinson, & Ross.

Ferrari, B., and J. Adams (1990). Biogenic modifications of marine sediments and their influence on archaeological material. *The International Journal of Nautical Archaeology and Underwater Exploration* 19(2).

FIFRA (1996). Federal Insecticide, Fungicide and Rodenticide Act. *136.*

Florian, M.-L.E. (1994). Appendix A: field conservation. In *Deep Water Archaeology: A Late Roman Ship from Carthage and an Ancient Trade Route near Skerki Bank off Northwest Sicily,* eds. A. McCann, and J. Freed. Portsmouth, RI.: *Journal of Roman Archaeology.*

Gibbins, D., and J. Adams (2001). Shipwrecks and Maritime Archaeology. *World Archaeology* 32:279–91.

Giere, O. (1993). *Meiobenthology: The Microscopic Fauna in Aquatic Sediments.* New York : Springer.

Hamilton, D. L. (1999). Methods of conserving archaeological material from underwater sites, http://nautarch.tamu.edu/class/ANTH605/File0.htm.

Izdar, E., and J. W. Murray, eds. (1989). *Black Sea Oceanography.* Boston: Kluwer Academic.

Lavoie, D. L., and W. R. Bryant (1993). Permeability characteristics of continental slope and deep-water carbonates from a microfabric perspective. In *Carbonate Microfabics,* ed. R. Rezak and D. L. Lavoie, 117–28. New York: Springer.

Lazar, B., S. Ben-Yaakov, et al. (1983). The carbonate system in hypersaline solutions: alkalinity and $CaCO_3$ solubility of evaporated seawater. *Limnology and Oceanography* 28:978–86.

Leyendekkers, J. V. (1975). The effect of pressure on the chemical potentials of seawater components, on ionic species in seawater, and on calcite saturation levels. *Marine Chemistry* 3:23–41.

MacLeod, I. D. (1995). In situ corrosion studies on the Duart Point wreck, 1994. *The International Journal of Nautical Archaeology and Underwater Exploration* 24:53–59.

McCann, A. M., and J. Freed (1994). *Deep Water Archaeology: A Late Roman Ship from Carthage and an Ancient Trade Route near Skerki Bank off Northwest Sicily*. Portsmouth, RI: *Journal of Roman Archaeology*.

McCann, A. M., and J. P. Oleson (2004). *Deep-water Shipwrecks off Skerki Bank: The 1997 Survey*. Portsmouth, RI: *Journal of Roman Archaeology*.

Morse, J. W., and F. T. Mackenzie (1990). *Geochemistry of Sedimentary Carbonates*. New York: Elsevier.

Murphy, L. E. (1990). *8SL17: Natural Site Formation Processes of a Multi-Component Underwater Site in Florida*. Sante Fe, NM: U.S. Dept. of the Interior, National Park Service, Southwest Region, Southwest Cultural Resources Center, Submerged Cultural Resources Unit.

North, N. A. (1976). Formation of coral concretions on marine iron. *International Journal of Nautical Archaeology* 5:253–8.

North, N. A. (1982). Corrosion products on marine iron. *Studies in Conservation* 27: 75–83.

Oxley, I., and D. O'Regan (2001). *The Marine Archaeological Resource*. Reading, PA: Institute of Field Archaeologists.

Paerl, H. W., T. F. Steppe, and R. P. Reid (2001). Bacterially mediated precipitation on marine stromatolites. *Environmental Microbiology* 3:123–30.

Pearson, C. (1980). Conservation and maritime archaeology. *The International Journal of Nautical Archaeology and Underwater Exploration* 9:147–50.

Pournou, A., A. M. Jones, and S. T. Moss (2001). Biodeterioration dynamics of marine wreck-sites determine the need for their in situ protection. *The International Journal of Nautical Archaeology and Underwater Exploration* 30:299–305.

Robinson, W. (1998). *First Aid for Underwater Finds*. London: Archetype Publications.

Singh, H., J. Adams, D. Mindell, and B. Foley (2000). Imaging underwater for archaeology. *Journal of Field Archaeology* 27:319–28.

Soreide, F. (2000). Cost-effective deep water archaeology: preliminary investigations in Trondheim Harbour. *The International Journal of Nautical Archaeology and Underwater Exploration* 29:284–93.

Stewart, J., L. D. Murdock, and P. Waddell (1994). Reburial of the Red Bay wreck as a form of preservation and protection of the historic resource. *Materials Issues in Art and Archaeology* IV:791–805.

Thorne, R. M. (1988). *Filter Fabric: A Technique for Short-term Site Stabilization*. Washington, DC: DOI Departmental Consulting Archaeologist/NPS Archaeology and Ethnography Program, National Park Service.

Thorne, R. M. (1989). *Intentional Site Burial: A Technique to Protect Against Natural or Mechanical Loss*. Washington, DC: DOI Departmental Consulting Archaeologist/ NPS Archaeology and Ethnography Program, National Park Service.

UNESCO (2000). *Background Materials on the Protection of the Underwater Cultural Heritage*. Paris: UNESCO.

Ward, I.A.K., P. Larcombe, and P. Veth (1999a). A new process-based model for wreck site formation. *Journal of Archaeological Science* 26:561–70.

Ward, I.A.K., P. Larcombe, R., Brinkman, and R. M. Carter (1999b). Sedimentary processes and the Pandora wreck, Great Barrier Reef, Australia. *Journal of Field Archaeology* 26:41–53.

Warren, L. A., P. A. Maurice, N. Parmar, and F. G. Ferris (2001). Microbially mediated calcium carbonate precipitation: implications for interpreting calcite precipitation and for solid-phase capture of inorganic contaminants. *Geomicrobiology Journal* 18:93–115.

Willey, J. (1995). The effects of desalination on archaeological ceramics from the Casas Grandes Region in Northern Mexico. In *Materials Issues in Art and Archaeology* IV, ed. P. B. Vandiver, J. R. Druzik, J. L. G. Madrid, I. C. Freestone, and G. S. Wheeler. Pittsburgh, PA: Materials Research Society.

PART TWO

Contemporary Shipwrecks in the Deep Sea

6

The Search for Contemporary Shipwrecks in the Deep Sea: Lessons Learned

Robert D. Ballard

The search for important contemporary shipwrecks in the deep sea is a relatively straightforward process since the history of significant shipwrecks like the *Titanic* or *Bismarck* is well documented. As a result, organizing an expedition to find a ship of such importance begins with a thorough search of the literature of the sinking in an effort to determine the size of the search area and the nature of the bottom terrain in which it was lost.

While this process is underway, the process of finding an excellent ocean-class research vessel to conduct the search and subsequent documentation is begun. Chartering such a ship on average costs between $12 and 20K per day (in 2005 dollars) whether private or federally owned. Outfitting the ship with the necessary deep-submergence vehicles, including various search sonars, imaging sleds, and remotely operated vehicles, as well as a professional team of engineers and technicians to operate those systems averages an additional $10K per day. Depending on the size of the search area and the nature of the documentation phase, the time needed to complete such an effort can take 4–6 weeks at sea, assuming a total search area of 200–500 square nautical miles in relatively smooth bottom topography. As a result, expeditions can cost between $1.5 and $2 million. As you go through this process, you will quickly discover that there are very few organizations willing to fund the search and documentation of contemporary shipwrecks in the deep sea unless the ship contains a valuable cargo or has major historical significance with strong public appeal.

The U.S. Navy has sponsored expeditions to find one of their ships or submarines that have been lost at sea or, during the Cold War, lost by the former Soviet Navy. These efforts, however, remain classified and are far from the public view. The decision to find them was made not based on an economic calculation but to obtain intelligence information on the enemy or to insure intelligence is not lost to the enemy. The Navy has, on rare occasions, provided support to find a contemporary shipwreck if the search effort involved the use of a new piece of technology they helped develop and wanted to test. But, by and large, the only organizations willing to sponsor such search efforts have been those interested in making television programs about the lost ship, such as the National Geographic Society, Discovery Channel, or History Channel.

Except for the *Titanic*, the National Geographic Society has been the primary sponsor for the shipwreck searches mentioned in this article. But even then, their budget for the production of a one-hour television documentary was quite limited. In the 1990s, that reached a high of $1.5 million but that also included the cost of postproducing the program after the ship was discovered. More recently, the amount of funds available for such searches exclusive of production cost has dropped to less than $500K.

The first major step that needs to be taken in the initial phase of a project like this is an accurate estimate of the size of the search area. Next is determining the size of the object you are looking for. Did the ship, for example, break up during the sinking process or is it intact? Another major factor governing the search strategy to be used is the nature of the terrain in which the ship was lost, since deep-sea terrain can vary from featureless abyssal plains, to the rugged midocean ridge, to seamounts, to deeply incised submarine canyons on the continental margin.

When searching in a flat region, side-scan sonar has proven to be the best search tool. The size of the target determines which frequency of sonar transducer to use. If it is a large, intact object, such as the aircraft carrier USS *Yorktown*, a low-frequency sonar (12 kHz) with a large search swath (10–15 km) towed on the surface can be used. If it is a small target, such as *PT-109*, a high-frequency sonar towed near the bottom should be used, the most popular being one with a frequency of 100 kHz and a search width or swath width of 1000–1200 m.

In highly complex terrains, we have commonly used visual imaging vehicles to search for debris associated with the sinking of the ship, since such debris can cover a much larger area than the ship itself. This visual search strategy was a valuable lesson learned from the mapping of the *Thresher* and *Scorpion* submarine wreck sites in the early 1990s (Ballard 1995). In both those cases, the submarine imploded a short distance beneath the surface, at an estimated depth of 500 m. These implosive events fragmented the submarines, introducing a vast array of objects into the water column. Depending on the weight of the object in water and its cross-sectional shape, undersea currents carried each object varying horizontal distances as they sank. The denser the object, the quicker it sank and the less it was deflected from the vertical by the undersea currents. The result was a debris trail that approached one nautical mile in length (Ballard 1995).

In flat terrain, a major portion of a debris trail will be easily seen in the sonar record, but in rugged terrain the objects comprising the debris trail are commonly masked by bottom features. In the case of the *Thresher*, detecting the debris field was further complicated by the presence of long, linear trails of glacial erratics dropped by the drifting icebergs as they melted, objects that resembled on sonar, pieces of ship debris. Although difficult to detect with side-scan sonar, objects within a debris trail, however, could be clearly seen by video cameras towed near the bottom and for that reason imaging vehicles are commonly used instead of sonars.

The purpose of this chapter is not to repeat what has been previously published but to review the lessons learned from these past search efforts for contemporary shipwrecks after years of reflection in hopes of helping others in the planning and execution of future efforts that might be undertaken.

Titanic

The discovery of the *Titanic* in 1985 (Ballard 1987) set the stage for a number of subsequent expeditions that resulted in the location of a number of important contemporary shipwrecks in the deep sea. It was a groundbreaking expedition as it demonstrated that technology now existed to locate such shipwrecks and that the unique characteristics of the deep sea contributed significantly to their long-term preservation. It also opened the door to deep-sea salvage operations and the emergence of new laws to govern that activity.

Unlike the expeditions that would follow, the search for the *Titanic* was fully financed by the U.S. Navy. This was made possible for two primary reasons. First, the Navy sponsored the development of the Argo imaging vehicle that located *Titanic* as well as the small, remotely operated vehicle, Jason, Jr., which was used the following year to go inside *Titanic*, and they saw the expedition as an excellent way of testing those new systems. The second reason was due to the fact that the discovery expedition in 1985, the follow-up expedition in 1986, and an earlier expedition in 1984 had as their primary objective the mapping of the USS *Thresher* and USS *Scorpion*, two programs that were classified at the time. It was only after the Navy objectives were completed, for example, in the 1985 discovery expedition, that the remaining days could be used to search for *Titanic*.

Previous search efforts to locate *Titanic* had taken many weeks and all had failed. Since we had only an estimated 12 days left in the expedition after the *Scorpion* survey was completed, we decided to approach the French Government in hopes of them joining us in the search effort. We had previously worked with France's major national oceanographic agency, IFREMER, on a number of large research programs in the past involving the development and use of deep-submergence technology, both manned and unmanned.

One of their senior engineers was Jean-Louis Michel who had recently developed a new side-scan sonar system called SAR and the search for *Titanic* presented an ideal opportunity to test this new system. The leadership of IFREMER in Paris also strongly supported this combined effort. This was in marked contrast to the leadership at the Woods Hole Oceanographic Institution, who felt that looking for *Titanic* was something an oceanographic research organization should not do. Fortunately, its scientists were free to do what they wanted to do if a federal agency such as the Navy funded the expedition. They were also aware that the 1985 expedition's primary objective was to map the *Scorpion* wreck site.

Jean-Louis and I developed a joint plan whereby the French would go first, something they had done on previous joint expeditions. Their mission was to totally survey the search area with their new SAR sonar system, while our mission was to image any targets they might find with our new imaging system Argo.

Jean-Louis chose the traditional search strategy of "mowing the lawn," beginning at one corner of the search area and working the ship back and forth with overlying side-scan sonar runs until the entire area had been "insonified" or "mowed" (Ballard 1987). What troubled me about this strategy was the fact that previous search efforts by both the Scripps Institute of Oceanography and the Lamont-Doherty Geological Observatory had used their side-scan sonars Deep-Tow and Sea Marc I, respectively, the same way and both efforts had failed to find *Titanic*. Since all of the groups searching for *Titanic* had access to the same historical database, the previous groups had searched the same 100 square nautical mile area and come up empty.

There were several possible explanations why these previous groups had not found *Titanic*. First, *Titanic* may have lain outside the search box. Second, *Titanic* might lie inside the deep canyon cutting across the search area, placing the ship in an acoustic shadow zone. Third, the previous searches had concentrated on the central portion of the search area, but *Titanic* may have lain near the perimeter of the search area. Fourth, *Titanic* may have been completely broken up and not detectable on sonar. Fifth, a massive landslide, triggered by the Grand Banks earthquake of 1929, that had broken several transatlantic cables in the area, may have buried *Titanic*. Or perhaps failure to locate the ship was due to some combination of the factors mentioned above.

Jean-Louis and his team arrived at the site on 5 July aboard their research vessel *Le Suroit* and immediately deployed the SAR vehicle to begin their search. This initial search phase lasted 2 weeks but failed to locate any interesting targets. I joined the ship for the second phase of the search that went for an additional 17 days, but once again no significant sonar targets were found. Ironically, as our discovery would later show, the French came extremely close to the sinking site of *Titanic* during the first leg when I was not aboard. In fact, their first search line with SAR came within a few hundred meters of the ship, but, as fate would have it, they spent the remainder of that leg and the time when I was aboard moving farther and farther away.

The French also towed a magnetometer at the same time they were towing SAR, so when the position of any sonar target was detected, the magnetometer data was checked for the existence of a possible metallic object. It also failed to detect anything of importance, although I have heard in recent months that they, in fact, had detected a magnetic anomaly on the first sonar run that occurred on the first sonar line when they passed closest to the ship's remains. It is amazing to me that such a claim would be made 20 years later. Jean-Louis Michel never mentioned such an anomaly after I came aboard and I am sure he would not have passed up such an opportunity for France to make this historic discovery. Finally, on 12 August, the French search effort was called off. It was hard to believe *Titanic* had escaped detection—70% of the original 100 square nautical miles had been searched.

The burden of responsibility now fell to the American team and our imaging vehicle Argo. Argo's side-scan sonar had a frequency of 100 kHz but it was not ideally designed to "mow the lawn" like SAR, Deep-tow, and the Sea Marc I used on previous expeditions. Argo was primarily an imaging vehicle. It commonly "crabbed" when being towed and to see the bottom it needed to be flown at a low altitude of 10–15 m. This greatly lessened its value as an acoustic search system, which normally operates 100–200 m above the bottom. We always had the side-scan sonar turned on but our eyes were on its three low-light-level black and white video cameras searching for man-made debris.

When we arrived in the search area we used the first 2 days to search *Titanic* canyon (figure 6.1) for any debris that might have been vectored by bottom currents and other mass wasting processes down its tributaries and into the central axis of the canyon. We also inspected all of the small sonar and magnetic targets detected on the previous search efforts by the French and American teams.

When these initial efforts were completed only 9 days were left before we had to return to port for another science program. Clearly, the previous search strategy of "mowing the lawn" had failed every time it was used so we needed to come up with a new strategy, one suited to a visual imaging vehicle like Argo. The actual area seen by Argo's low-light-level video cameras was small, and overlapping visual paths was an impossible strategy to follow.

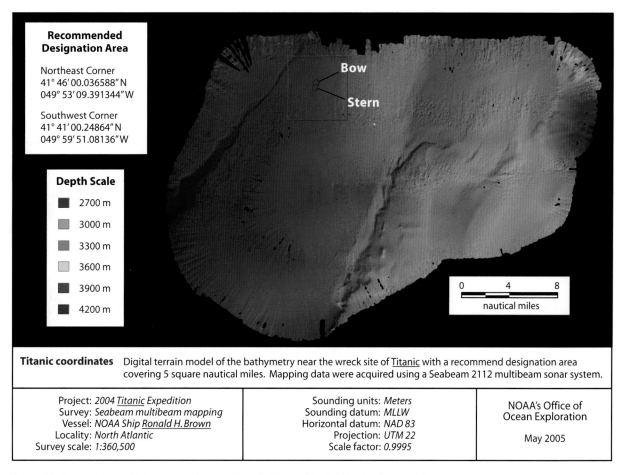

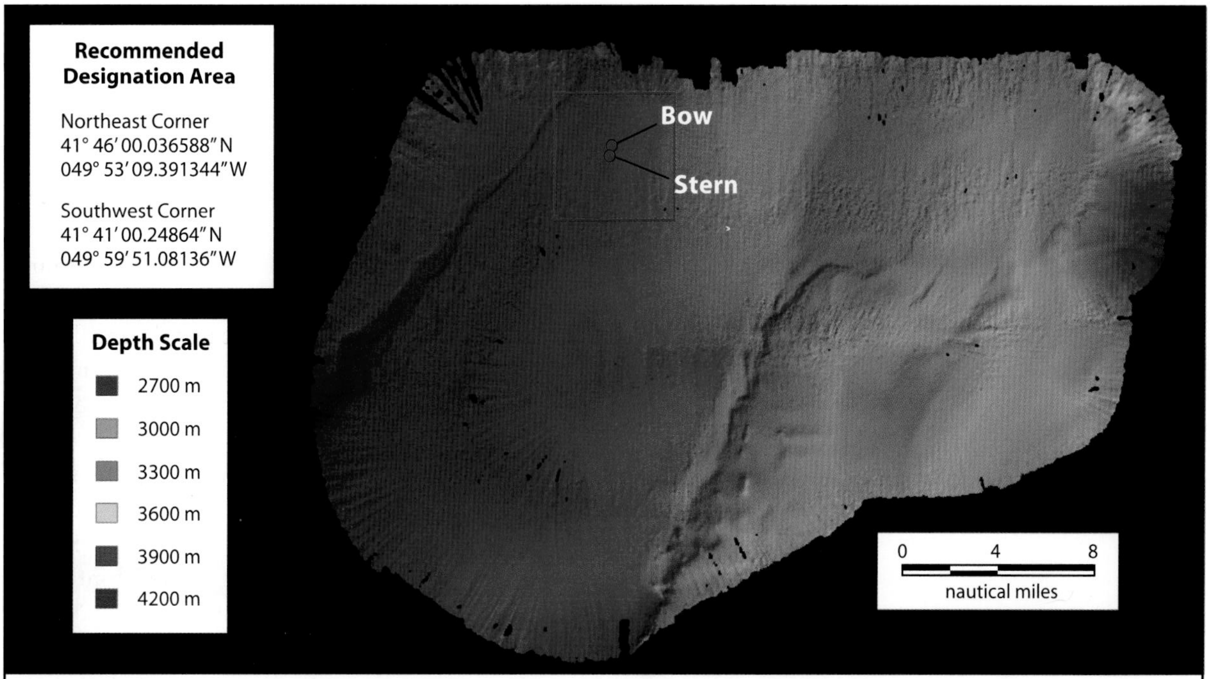

Titanic coordinates Digital terrain model of the bathymetry near the wreck site of Titanic with a recommend designation area covering 5 square nautical miles. Mapping data were acquired using a Seabeam 2112 multibeam sonar system.

Project: *2004 Titanic Expedition* Survey: *Seabeam multibeam mapping* Vessel: *NOAA Ship Ronald H. Brown* Locality: *North Atlantic* Survey scale: *1:360,500*	Sounding units: *Meters* Sounding datum: *MLLW* Horizontal datum: *NAD 83* Projection: *UTM 22* Scale factor: *0.9995*	NOAA's Office of Ocean Exploration May 2005

Figure 6.1. A recently compiled topographic map of Titanic Canyon (shaded blue) and surrounding area. This map was constructed based on Seabeam sonar data collected by the NOAA ship *Ronald H. Brown* during our 2004 expedition that revisited the site. The final resting place of *Titanic* is indicated on the map. (Copyright National Oceanographic & Atmospheric Administration [NOAA], graph created by Jeremy Weirich)

Reviewing the list of possible reasons others had not found the wreckage, we were left with the following possibilities. The first was that *Titanic* lay outside the original search box. With that in mind, the search area was enlarged to 150 square nautical miles, increasing it by 50%.

A visual run down the axis of *Titanic* Canyon had eliminated the possibility it laid there. Since all of the previous search efforts had concentrated on the central portion of the primary search area, we concluded that *Titanic* clearly lay outside that area since it was unlikely that *Titanic* had broken up to such a degree that SAR had missed seeing its major hull sections. And if she had broken up in the area not searched, with its three video cameras, Argo was the ideal tool to use.

SAR's sonar survey of the primary search area did, however, reveal that a massive landslide had occurred in the western portion of the search area, but Titanic Canyon (figure 6.1) blocked the landslide from affecting the eastern portion of the search area where *Titanic* unknowingly had come to rest (Uchupi et al. 1988). A review of these possibilities suggested that *Titanic* was to the east of the area SAR had already searched. Either *Titanic* lay in the eastern portion of the original search area or in the newly expanded search area farther to the east. SAR had, therefore, already searched 50% of the newly expanded search area, leaving 50% for Argo to cover in only 9 days. It would be impossible to visually search this entire area for a ship that was 90 ft wide with overlapping coverage. That would take months.

This is when the lessons learned from our previous, then classified, mapping efforts of the debris fields of the USS *Thresher* and USS *Scorpion* came into play. When the Navy agreed to support the development of the Jason, Jr. and Jason vehicle systems, they did so with the understanding that we would carry out a series of classified programs for the Navy using those vehicles. Although many of the programs remain classified, two of them have been declassified with the end of the Cold War—the mapping of the wreck sites of the USS *Thresher* and USS *Scorpion*, the only two U.S. nuclear submarines lost during the Cold War with the Soviet Union. During the late 1980s when the vehicle systems were under development, the U.S. Navy was contemplating the possibility of disposing of the nuclear containment vessels that house the reactors of aging nuclear submarines in the deep sea. This is not the reactor itself, but the compartment within the submarine in which the reactor was housed. Although these compartments had only a low level of radioactivity, the Navy needed to know what effect they might have upon the deep-sea environment in which they would be placed. When the *Thresher* and *Scorpion* were lost in the 1960s, they carried not only their nuclear containment vessels to the ocean floor, but also their reactors and, in the case of the *Scorpion*, two nuclear-tipped torpedoes. Clearly, the wreckage of these two submarines presented a worst-case scenario and the Navy wanted to know what effects the submarines and their associated debris were having on the bottom environments and the benthic animals living in and around those objects.

All that was known about the wreckage at the two sites in 1984 was that the large pieces of the hull appeared to be located in a series of impact craters in the soft muddy bottom of the North Atlantic. The *Thresher* was in 8500 ft of water near Corsair Canyon south of Georges Bank and the *Scorpion* was in 11,500 ft of water south of the Azores in the abyssal hill province of the Mid-Atlantic Ridge.

The *Thresher* wreckage was the first to be mapped, in the summer of 1984. This was the first time we had ever surveyed a piece of wreckage in the deep sea. All of our efforts prior to that had been geological mapping programs in the Mid-Ocean Ridge so we had no idea what we would encounter. We had experience with wreck sites in shallow water but the deep sea proved to be an entirely different environment for the subsequent shipwrecks we would discovery and explore.

Our plan was to initially bracket the wreckage site by running two lines, one south to north, the other east to west. Once the limits of the debris field had been determined, we would then concentrate on mapping every square inch of the area covered by wreckage as the Navy had requested.

Naively, we assumed the debris field would be circular in shape, with the major impact craters lying in the center of the wreckage field. On our first south-to-north run, that seemed to be the case. The lowering began on an undisturbed mud bottom with no indication of submarine wreckage. Once cameras began to see wreckage, its presence continued for a couple hundred meters before an undisturbed bottom terrain appeared to the north.

Turning to the east, Argo was then towed back toward the previous line but at a right angle, reentering the debris field and impact craters. On this run, however, the debris field extended much farther to the west. Instead of ending after a couple hundred meters, it continued for more than 1.5 km, with a length-to-width ratio of nearly 10 to 1.

Equally important was the nature of the debris. As Argo was towed to the west, the debris objects became lighter and lighter, creating a debris gradient with the lightest debris in the west and heaviest debris in the east near the impact craters that held the larger hull sections, including the nuclear reactor and its containment vessel, the object of our search effort.

Intermixed with the lighter debris were lines of glacial erratics, dropped over many thousands of years by melting icebergs. Some of the rock erratics were large. After the image run was completed, we ran a series of sonar lines parallel to the long axis of the debris trail. Many of the lighter debris fragments we not detectable by the sonar and it was impossible tell if some of the sonar targets were glacial erratics or fragments from the imploded submarine. It was clear that the heavy objects had fallen straight down while the undersea currents had carried the lighter ones west. A similar debris trail was mapped at the *Scorpion* site the following year. Based on this new insight, our strategy for finding the *Titanic* in only 9 days was to look not for the *Titanic* but for its debris field using Argo's low-light-level black and white video cameras.

Though popular accounts of *Titanic*'s final moments, including that contained in Walter Lord's book *A Night to Remember* (1955), had the *Titanic* sinking in one piece, some of the eyewitness accounts were at odds with that interpretation (Ballard 1987). Many survivors said the ship broke in half, particularly Jack Thayer, who made a sketch of the sinking sequence aboard the *Carpathia* as it took the *Titanic* survivors to New York. If Thayer's account proved correct, thousands of objects would have poured into the ocean at the surface, each following their own separate path to the bottom. It was clear from the logbook of the *California*, the steamship that was known to be to the north of *Titanic's* position, and possibly within its line of sight, that it had drifted to the south through the night after being stopped by the ice field that contained the iceberg *Titanic* struck. Based on that insight, and the southerly location of *Titanic's* lifeboats the next morning when the *Carpathia* found them drifting in the surface current, any debris that came out of the *Titanic* that lingered in the upper surface current before sinking to the bottom would have been carried south.

It was hard to know exactly how far the lightest material that did sink had drifted to the south during its 4000-m descent, but one nautical mile seemed to be a safe estimate based on our experience with the *Thresher* and *Scorpion*, a knowledge base I, unfortunately, could not share with Jean-Louis Michel as I explained my search strategy to him prior to the R/V *Knorr* arriving on scene. For that reason, the visual strategy for the Argo search effort was to conduct north–south search lines spaced one mile apart. Since the lifeboats had lingered in the surface current the longest, the plan called for beginning the search beneath where the *Carpathia* picked up the lifeboats. Clearly, the *Titanic* and its associated debris field should lie to the north.

When looking for a sunken ship using such a strategy, it is best to start away from your best estimate of where the ship and debris might lie. If you start at the highest probability location and nothing is there you are lost, as it is an equal probability in all directions. The first and southernmost east–west line was selected based on the *Carpathia's* reported position of the lifeboats to which was added an additional five nautical miles to the south to take into account the average error in navigation based on celestial navigation techniques being used at the time. The east limit of the start point was based on our earlier analysis that had created the larger 150 square nautical mile search area. The west limit would be the easternmost limit of the SAR survey.

As each line was completed, the ship would tow Argo one nautical mile to the north while still within the SAR coverage and then begin its next line to the east, repeating this pattern until the northern limit of the secondary search area was reached. If Argo detected nothing by the time the northern limit was reached, the ship would double back to the south, interlacing lines with a half-mile spacing. Based on our calculations, lines spaced a half-mile apart across the search area could be completed in the 9 days remaining in the expedition.

With this strategy in mind, we began our visual search with Argo on 28 August at 0145 in the morning. By late evening on 31 August, after overcoming a number of technical and personnel problems (Ballard 1987), we had completed our eighth east–west line across the search area and were turning west to begin line nine.

At 2348 on the night of 31 August, debris began to pass beneath Argo. Initially, it was hard to determine the origin of the debris since it was twisted pieces of rusting wreck. *Titanic* had sunk in the trans-Atlantic shipping lanes that had been the site of numerous Allied ship sinkings at the hands of German U-boats during World War II. But shortly after entering the debris field one of *Titanic's* round boilers came into view. *Titanic* had been found (figures 6.2–6.4)! The debris search strategy had worked.

After the discovery of *Titanic* in 1985, a return trip was made the following summer aboard the R/V *Atlantis II*, which supported the deep submersible Alvin.

Figure 6.2. Image of *Titanic*'s bow taken by the towed camerasled, Angus. (Copyright Woods Hole Oceanographic Institution)

Figure 6.3. Artist rendering by Ken Marschall of *Titanic*'s bow section made from an examination of the video images collected in 1985 by Argo and Angus and in 1986 by Alvin and the small ROV Jason, Jr. (Copyright Ken Marschall)

Figure 6.4. Jason, Jr. peering into the window of an officer's cabin on the boat deck on the starboard side of *Titanic*. The image was taken from submersible Alvin to which Jason, Jr. is attached. (Copyright Woods Hole Oceanographic Institution)

Bismarck

The sinking of the German battleship *Bismarck* was one of the great sea sagas of World War II. It came during the early years of the war before America had entered the conflict, at a time when Great Britain was most vulnerable. The German army had swept across Western Europe and was contemplating an invasion of England. Cut off from Europe, Great Britain was dependent for survival on the convoy routes from North America.

When the mighty Nazi battleship *Bismarck* set sail on its maiden voyage in May 1941 with the North Atlantic as its operating area, those very lifelines were now threatened. And when she sunk England's prized ship the HMS *Hood* with only three surviving from her crew of over 1400, many in Great Britain felt the war was lost. That was when Winston Churchill gave that memorable command to Great Britain's Navy to "Sink the *Bismarck*!" and one of the greatest sea chases of all time began.

The U.S. public's great interest in finding the *Bismarck*, like that for the *Titanic*, made it possible to obtain the funds needed to find her but even that proved difficult. A one-hour television documentary is only worth so much to an organization like the National Geographic Society (NGS). Even with some support from the Navy, notably the Office of Naval Research, which had a continuing interest in testing our new Argo search system, as well as funding from NGS, private funds were also needed to finance the expedition costs. That came from two businessmen, Don Koll and Marco Vitulli, who felt it was a good investment toward the profit they might receive from the release of the NGS television show as a home video.

We were also able to share the costs of the expedition's mobilization and demobilization and a portion of the transit costs with a new distant learning program we had created for middle school students and teachers who wanted to participate in "live" exploration, which Koll and Vitulli were also helping to fund. Later named the Jason Foundation for Education, this phase of the expedition had as it objective the discovery of ancient shipwrecks in the deep waters of Mediterranean Sea. Not only has this program continued to grow over the years, but this effort led to our new program in archaeological oceanography at the University of Rhode Island's Graduate School of Oceanography, which will be covered in other chapters of this book.

Unlike the *Titanic,* no previous attempts had been made to locate the *Bismarck* until our 1988 expedition (Ballard 1990). A search of the literature combined with British Naval records indicated that the *Bismarck* was sunk in approximately 16,000 ft of water west of Brest, France and southwest of Ireland. Based on three separate sinking positions obtained from the British battleships *King George V* and *Rodney* and the heavy cruiser *Dorsetshire,* the estimated area that needed to be searched was approximately 200 square nautical miles. Of these three ships, only *Dorsetshire* was present when the *Bismarck* actually sank since the other ships were running low on fuel and needed to return to Great Britain, resulting in us favoring *Dorsetshire's* estimated position over the other two; a decision that proved totally unfounded.

The western half of the *Bismarck* search area was a relatively flat, featureless abyssal plain, while the eastern half was mountainous (Ballard 1990). Our primary search vehicle would be Argo, the same towed sled outfitted with side-scan sonar of 100-kHz frequency and three low-light-level black and white video cameras that we had used to find *Titanic.* We decided on two separate search strategies. The sonar on Argo would be the primary search tool in the flat western region while, similar to the *Titanic* search effort, Argo's video cameras would be used to search for debris in the eastern mountainous area where numerous false sonar targets were expected. We had made modifications to the Argo sled in an effort to improve its tow behavior in the water in hopes of reducing its tendency to "crab."

Before sinking, the *Bismarck* was hit countless times by the British warships that surrounded her, fragmenting her superstructure. As she sank, she rolled over, dumping a large amount of debris into the water. Her four main guns were gravity seated and must have fallen out as well, introducing even more material into the sea. For these reasons, we were convinced that a visual search for a long debris field could assist us in locating the ship in the mountainous area in the eastern portion of the search area where rock outcrops and other geologic features could have produced numerous false targets to our sonar.

But other factors would ultimately dominate our 1988 search effort (Ballard 1990) using the chartered British ship *Starella* from Hull, England. The earlier Mediterranean phase of the 1988 expedition aboard *Starella* had been plagued by our troublesome winch, which our engineers were convinced could break down at any moment during the *Bismarck* search effort. With their concerns in mind, we made the decision to begin the search in the flat western region of the search area known as the Porcupine Abyssal Plain (figure 6.5). Once Argo had been lowered to its search altitude just above the ocean floor, there would

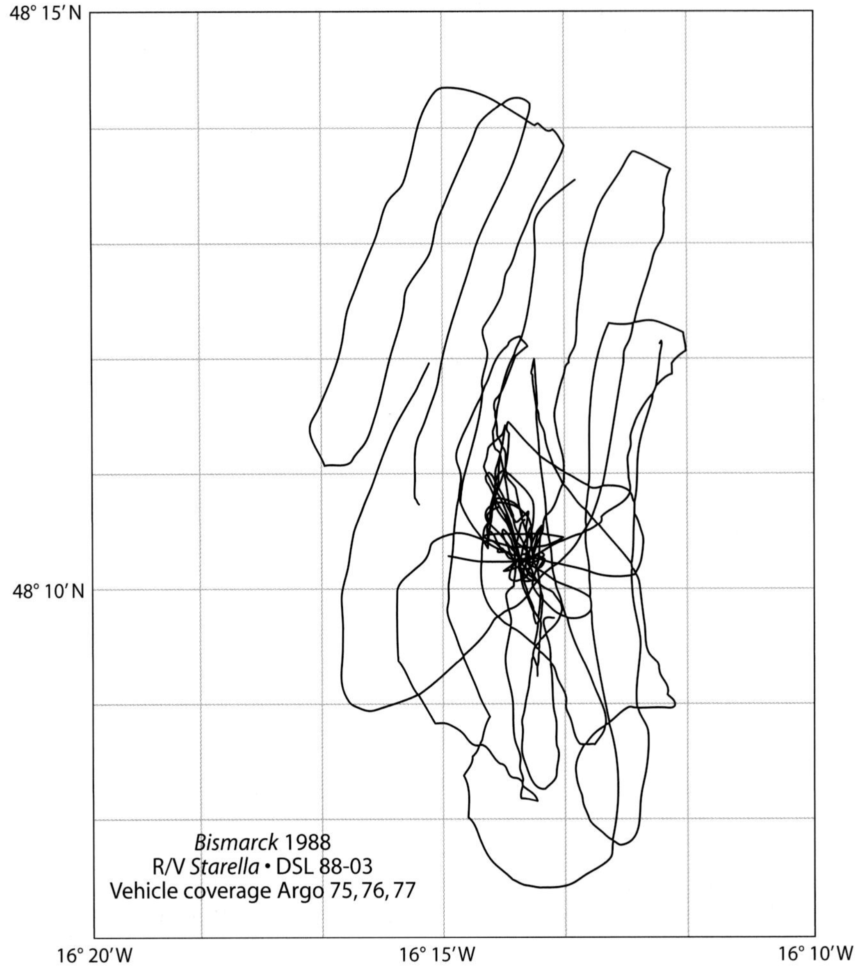

Figure 6.5. The area searched by Argo during the 1988 expedition aboard the R/V *Starella*. Central area of intense coverage was where a wooden sailing ship with its associated debris field was found and mapped. (Graphic created by Paul Oberlander)

be little reason to raise or lower the vehicle using the winch. Had we begun our search effort in the more mountainous region to the east, constant use of the winch would have been necessary.

The sea conditions in the North Atlantic where the *Bismarck* sank are seldom ideal and this expedition was no exception. Working in 5000 m of water is also time-consuming as it takes 3–4 h to lower Argo to the bottom and then the same amount of time to recover the vehicle.

But once the search began in the flat western region, we could settle in and let the tension subside. Argo leveled out at an altitude of 50 m, too high to see the bottom but ideal for a sonar search reaching out on average 500 m to either side. Although such high-altitude sonar searches are very boring, they cover a much larger area than a visual search and since the bottom was so flat and featureless, the sonar should detect even the smallest of metal debris. Equally important, this was not an area over which melting icebergs had dropped countless glacial erratics, creating unwanted false sonar targets.

Our first line ran from north to south along the base of the westward-facing mountain slope (figure 6.5). If the *Bismarck* had come down on the side of this slope, it would have generated a landslide and possibly slid downslope, an assumption that would prove correct during a second search effort in 1989. Once this initial line was completed, we reversed our course and headed back to the north with a line spacing of less than 1 km to insure that our sonar coverage was overlapping as we "mowed the lawn."

During the course of the 1988 expedition, only 30 square nautical miles were searched. More would have been covered but a considerable amount of time was lost when a debris field and impact crater were encountered. We spent several days trying to map its total extent in hopes that it might be associated with the running sea battle that occurred prior to *Bismarck's* sinking. In the end, unfortunately, the debris field led us to a large wooden sailing ship, perhaps an Inca class, four-masted schooner from the early 1900s and the expedition ended in failure.

Fortunately, the following year, 1989, we were able to convince the Navy, National Geographic, and our two California businessmen to sponsor a second search effort. As in 1988, it took place after the Jason Project was completed in the Mediterranean, making it possible to share the costs of mobilization, demobilization, and transit costs with a separately funded program.

Our support ship for the expedition was the *Star Hercules*, an excellent platform with dynamic positioning and plenty of deck space. Although our new remotely operated vehicle Jason had performed wonderfully on its maiden voyage for the Jason Project just days before, providing excellent "live" color images that were beamed back to 250,000 students and teachers across America, it would have to sit idly by for this year's *Bismarck* search effort. We had not yet received funding from the Navy to purchase a fiber-optic cable long enough to reach 16,000 ft. Instead, we had to take a step backward technologically to what we had used the year before, our old coaxial cable and Argo. But the *Star Hercules* had performed flawlessly the previous cruise and we had a brand new winch, so working in the eastern mountainous region should not pose a problem.

The 1989 search began where the 1988 had ended as we began a new series of west-to-east lines moving up the slope of the large extinct volcano that dominated the eastern region (figure 6.6). Traversing this terrain with Argo presented few challenges to our team, which had investigated an active volcano in the central portion of the Tyrrhenian Sea in the Mediterranean on the previous Jason Project expedition.

Throughout this phase of the search program we saw isolated pieces of man-made wreckage on the ocean floor but none of it was large or recognizable. We had to remind ourselves that the engagement between the *Bismarck* and the British fleet was a running sea battle. The *Bismarck* was hit countless times, and pieces of its superstructure were blown off into the sea as it steamed around trying to elude its pursuers. We also had no idea if the debris we were seeing even came from the battle.

Our search in 1989 had begun in the northeastern region of our search area and was slowly working to the south. We had covered the area around the reported sinking position given by the British battleship HMS *Rodney* during our 1988 search. The initial phase of the 1989 search had now covered the sinking site given by the heavy cruiser HMS *Dorsetshire*. This had been my favorite, since the *Dorsetshire* was the only ship present when *Bismarck* sank, but our

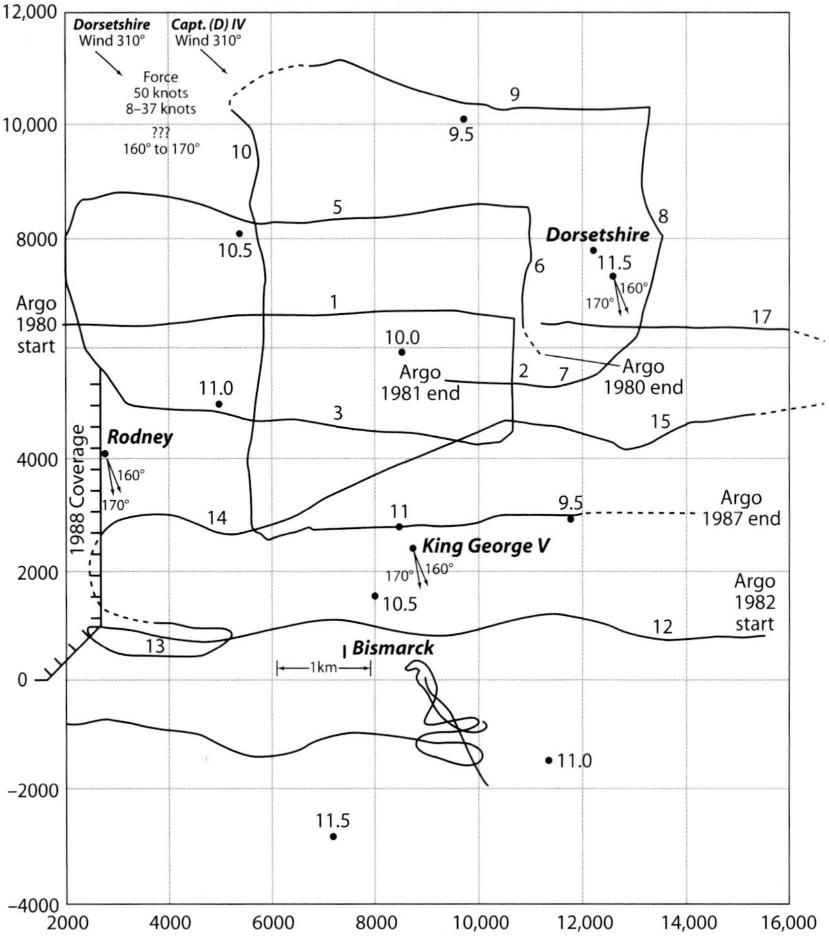

Figure 6.6. Track lines conducted in the 1989 search effort aboard the *Star Hercules*. Horizontal and vertical scale is distance in meters. Dots denote location and frequency of bottom transponders used to track the location of Argo. Estimated locations of the *Bismarck* by the British warships *Dorsetshire, Rodney*, and *King George V* are also shown, as is estimated wind direction and speed. (Graphic created by Paul Oberlander)

search in that area turned up empty. That left only the position given by the battleship HMS *King George V* to the south, the direction we were now headed toward with our east-to-west lines spaced one nautical mile apart; the same pattern we had used to find *Titanic.* The debris field we had encountered from what turned out to be a four-masted schooner was one nautical mile long. For that reason, we were confident that the debris field associated with the *Bismarck* sinking should be longer.

Although the 1988 search had been unsuccessful and the sailing ship had eaten up valuable search time, its north-to-south debris field showed us the current direction that must have also created a north-to-south debris field for *Bismarck's* debris. That is why we shifted our search lines from north-to-south during the 1988 search to east-to-west in 1989.

Six and half days into the search effort we had covered 80% of our original 200 square mile search area. We favored the *Dorsetshire* position to the north over the *King George V* position to south. Plus we had seen various pieces of random debris around the *Dorsetshire* position, which reinforced our premonition. All of this caused us to think that we should reposition our network of bottom transponders so we could expand our search to the north. But to complete our coverage within the original search area, we needed to conduct one final line to the south. Maintaining the discipline to stay with the original plan and resist temptation to modify it, proved critical, for on that final line to the south, halfway down, a continuous pattern of debris began to pass beneath Argo's cameras.

The debris was extensive, clearly not random debris thrown over the side of passing ships. As we continued east mapping the debris, the nature of the bottom changed abruptly, taking on an odd patchy and mottled appearance. Then it changed again into a bottom that looked highly disturbed as if a mixture of rock and sediment had been passed through a blender. Then we reentered a patchy and mottled bottom before finally returning to a typical flat and featureless mud bottom. After a while, we reversed direction and began running a parallel course to the west, farther south from our previous line. Once again the pattern repeated itself—a normal bottom followed by a patchy-mottled bottom, then across a highly disturbed bottom, through the patchy-mottled bottom again before passing over normal bottom with debris, and finally back to only normal bottom.

After much discussion, this pattern began to make sense. The highly disturbed bottom was caused by a giant landslide, triggered, we thought, by something large crashing into the bottom. The patchy and mottled terrain to either side was where the landslide had affected the surrounding terrain, causing the upper layer of mud to drain into the landslide, winnowing out the lighter material and leaving a coarse lag deposit behind. Since the bottom sloped up to the north, whatever caused the landslide in the first place must lie in that direction.

This conclusion was further reinforced by the nature of the debris itself. As we zigzagged back and forth, working our way up slope to the north, the debris pieces grew larger and heavier. Just as we had seen with the *Thresher*, *Scorpion*, and *Titanic*, undersea currents, in this case north to south, had carried the lighter material farther from the impact site, creating a long debris trail. The only new wrinkle was the existence of a large landslide but that was to be expected since the bottom slope was much steeper here than at the other sites, and the *Bismarck*, if that was what had caused the landslide, was much heavier, at 48,000 tons.

As we continued zigzagging to the north, we began to see that the debris trail was running north to south while the long axis of the landslide was running from southeast to northwest. Clearly, where the two lines intersected, we should find the impact crater, and that is exactly what we found. The only problem was that once the impact crater was found and mapped we had found only a piece of wreckage lying within the crater. It was a large piece, but only a fraction of the *Bismarck*. Could such a small piece have caused such a large landslide? Had the ship been buried on impact? This seemed unlikely. The *Star Hercules* was placed into dynamic positioning so we could hover over the impact crater and slowly move Argo back and forth, looking at every square inch of the crater.

Clearly, whatever caused this impact and associated landslide must have slid downslope to the south. We followed the landslide south for more than 2 km before it flowed out onto the abyssal plain that surrounded this undersea volcanic hill. Working our way back up the slope, we encountered one of *Bismarck's* main gun turrets resting upside down in the landslide region. Clearly, when the ship rolled over these large main guns, which were gravity seated inside barrettes,

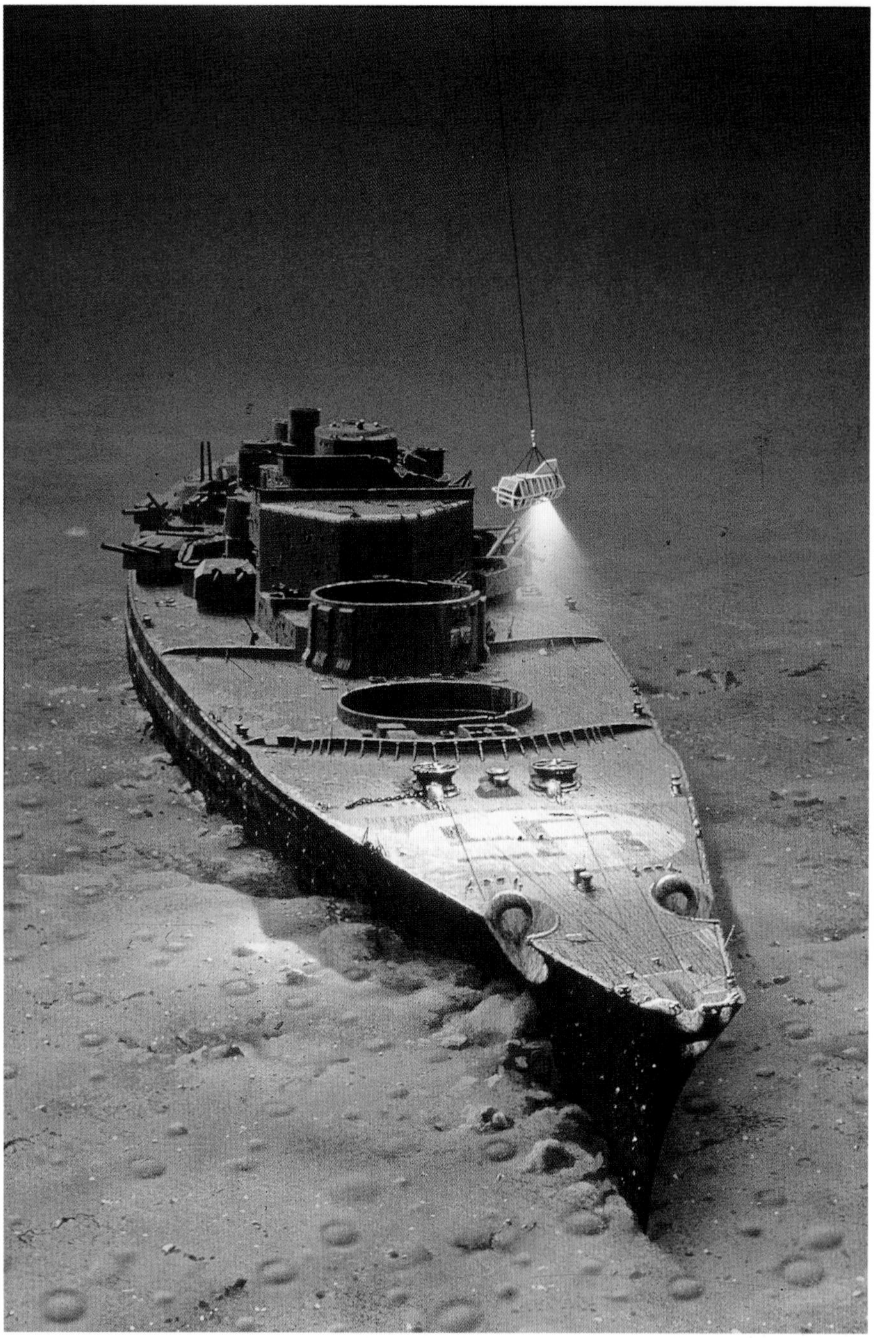

Figure 6.7. Artistic rendering by Ken Marschall of *Bismarck* during the discovery run by Argo. (Copyright Ken Marschall)

fell out and came to rest in an upside-down position. Clearly, the *Bismarck* had caused the landslide and associated debris. But where was the ship?

The only way to find it, now that we understood what we were looking at, was to conduct a systematic survey of the landslide area. Finally, on the tenth day of our search, two gun barrels attached to a turret loomed into Argo's camera frame; the *Bismarck* had been found, resting halfway down the landslide it had created, pointing west against the western edge of the slide (figure 6.7).

During the next two and half days, we used our remaining time to image the *Bismarck* as best we could. She was better preserved than *Titanic* but that was expected given her younger age. What was most amazing was that her teak deck was still in place along with two swastikas painted on her bow and stern.

Although the ship was last seen upside down, it had rolled back into its upright position on the way to the bottom. We also saw no damage to the hull from implosions as we had documented with the *Titanic's* stern section. This helped to support the statements made by the Germans that the crew scuttled the ship at the surface, permitting water to flood the entire length of the ship evenly, causing it to roll over on its side. The even flooding also "pressure compensated" the hull, avoiding subsequent implosive events as the outside pressure increased as she sank to greater and greater depths. This observation was not well received by the British press, since they preferred the sinking to have been caused by the torpedoes fired into the *Bismarck* by *Dorsetshire* and not by an act carried out by German hands. As far as I was concerned, the British had sunk the *Bismarck* regardless of who delivered the final blow.

Guadalcanal

Unlike our previous search programs, our two-year campaign off the island off Guadalcanal did not have a single ship as its objective but the exploration of an entire battlefield (Ballard 1993). The Battle for Guadalcanal involved numerous naval engagements carried out from June 1942 to February 1943. During this period of time, so many Japanese and Allied ships, nearly fifty in number, sank in the waters adjacent to the island that the waters were named Iron Bottom Sound.

We knew little about this body of water, due to the lack of previous studies in the area. Our efforts began in the fall of 1991 when we chartered a small coastal transport from Australia called the *Restless M* to conduct a preliminary survey of the Sound using a small team and a simple 100-kHz side-scan sonar.

The focus of our initial search area was the region south of Savo Island that lies at the northern entrance to the sound. It was around this island that a majority of the sea battles took place, beginning with one of America's greatest naval defeats, the Battle of Savo Island on 8 August 1942.

The side-scan sonar survey was conducted in the traditional "lawn-mowing" pattern with the long axis of the overlapping survey lines running down the axis of the sound in a northwest-to-southeast direction. The 1991 survey resulted in the discovery of 10 warships, the largest being the American heavy cruiser USS *Northampton*, lost during the Battle of Tassafaronga on 30 November 1942. Unfortunately, none of the principal Allied heavy cruisers that sank during the Battle of Savo Island or either of the two Japanese battleships that sank north of Savo Island were located in 1991.

Based on the success of the 1991 campaign, however, a second and far more comprehensive effort was conducted the following year. The primary surface ship for the operation was the *Laney Chouest*, a ship chartered by the U.S. Navy to support their submersible Sea Cliff and the unmanned, remotely operated vehicle Scorpio, as well as a deep-towed side-scan sonar that could reach greater depths than the sonar used in 1991.

The sonar search effort in 1992 began where the previous year's effort had ended, extending the search coverage into the deeper depths of Iron Bottom Sound as well as to the north of Savo Island. This effort was highly successful, resulting in the discovery of the Australian heavy cruiser HMS *Canberra* and the American heavy cruiser USS *Quincy*, both lost during the Battle of Savo Island, and the Japanese battleship *Kirishima*, lost during the final naval battle of Guadalcanal.

Once the various warships had been located, the ROV Scorpio and the submersible Sea Cliff were used separately and together to conduct a detailed inspection of each shipwreck. Working with us to conduct this analysis were naval historians Chuck Haberlein and Richard Frank. We had also brought with us three survivors from the various sea battles fought in Iron Bottom Sound. They included American Stewart Murdock from the light cruiser USS *Atlanta*, Bert Warne from the Australian heavy cruiser HMS *Canberra*, and Michiharu Shinya from the Japanese destroyer *Akatsuki*.

The ships lost in Iron Bottom Sound were relatively easy to find owing to the simplicity of the bottom terrain. Some, however, like the heavy cruisers *Vincennes* and *Astoria*, eluded our detection, since they were lost at night and were clearly not where they thought they were when the Japanese attacked.

It was the logistical issues associated with working in such a remote area of the world that proved the most challenging. The initial search effort in 1991 was also hampered by its limited budget, causing us to select a search boat that was unreliable and poorly suited for the mission. The majority of the time we were in Guadalcanal that year, the ship was at anchor while efforts were being made to keep it operating. During the second year's effort, a problem with the ROV Scorpio resulted in a long delay as we waited for critical parts to be flown in from the United States via a commercial carrier that had an unreliable operating schedule. Obtaining permission from the Solomon Islands government was also a lengthy process, requiring separate trips to the island prior to any work being done.

Lusitania

In 1993, the author organized an expedition off the eastern coast of Ireland near the Old Head of Kinsale (Ballard and Dunmore 1995). It was here in May 1915 that the British Cunard Liner *Lusitania* was torpedoed by the German submarine *U-20*, commanded by Lt. Walter Schweiger; resulting in the loss of 1195 civilian lives during the early phase of World War I.

The British and American press portrayed the sinking as a barbaric act of German aggression against the innocent public, including Americans, whose country remained neutral at the time. The German press, on the other hand, argued that the *Lusitania* was carrying war materials from America to Great Britain, and that it had, in fact, been built by the British Government and was armed. Also, it had been warned not to enter the war zone surrounding Great Britain and that if did it would be treated as a ship of war. Following the war, the tale of *Luistania*'s sinking grew more complex, with various conspiracy theories being advanced, including one that suggested Winston Churchill, then First

Sea Lord, had allowed the sinking to take place in hopes of drawing America into the war on the British side.

At the center of the controversy was the secondary explosion that sank the ship. Since *U-20* fired only one torpedo and eyewitnesses reported two explosions, the German's argued that the secondary explosion occurred when the first torpedo struck the war material in the forward cargo hold of the ship, causing a much more powerful and fatal explosion that the sank the ship. They went on to suggest that were it not for that explosion the ship would have survived the attack. Given this controversy, the primary objective of the 1993 expedition was to see if modern deep submergence technology could be used to determine the cause of the secondary explosion. Unlike that of the *Titanic* and the *Bismarck*, the location of the *Lusitania* was well known. She sank within sight of land and the local fishermen who attempt to catch fish living in and around the wreck site have known her location for many years. Numerous dives had been conducted on the ship's submerged wreckage over the years, including those associated with salvage operations.

Prior to conducting the expedition, the trial records of the 1st District Court of New York were studied. It was here that the Vanderbilt family brought suit against the Cunard Line, charging that the Line was partially responsible for causing the death of Alfred Vanderbilt, the great-grandson of Commodore Vanderbilt, the tycoon who had built a vast railroad and shipping empire.

At that trial, the family proved that the *Lusitania* was, in fact, carrying war material aboard the ship, storing it in the ship's forward magazine. The evidence presented at the trial documented what was placed aboard the ship, from which company it had come, and specifically where on the ship it was placed. The Cunard Line admitted that war material was stored in the ship's forward magazine, but said that it was nonexplosive and could not have been ignited by the torpedo that struck the ship.

The objective of the expedition was to create a detailed map of the ship's present hull form and then superimpose over that form the ship's original construction drawing in an effort to determine exactly where the magazine was located. Once located, a remotely operated vehicle would be used to inspect the magazine's outer hull to see if, in fact, a secondary explosion had destroyed it. To carry out this plan, a number of obstacles needed to be overcome. To begin with, the ship was severely damaged when it sank, with the hull bending into a banana-shaped form. Since sinking, the ship has been used by the Irish Navy for depth charge practice. As a result, the present ship's hull form was further flattened, no longer resembling its original hull configuration.

Over the intervening years, the ship's hull has become an artificial reef, attracting numerous fish and local fishermen, who have lost numerous nets on the ship's jagged metal surfaces. The nets are draped over the wreck like spider webs, making it hazardous to work at the site. In fact, during our mapping effort, both our submersible Delta and our remotely operated vehicle Jason became temporarily trapped in the nets that encase the shipwreck. In the case of the Delta, the pilot had to jettison the stern prop to free the sub from the nets, whereas divers were sent down to cut the ROV Jason loose.

Because of reduced visibility, it was easy to become disorientated, making it difficult to know precisely where our vehicles were at any one time. For that reason, a series of high-frequency transponders were placed around the shipwreck, which could be used to precisely navigate the Jason ROV. With a tracking accuracy measured in centimeters and a high update rate, Jason could be placed in "closed-loop" control. This made it possible to run a series of closely spaced lines back and forth over the ship's hull, which lies on its starboard side in 93 m of water, while the surface ship used its dynamic positioning system to maintain station overhead. A down-looking sonar on Jason coupled to a precision depth sensor, was used to determine the hull position in three-dimensional space. Once completed, the ship's original hull design was fitted to its present shape. This made it possible to pinpoint the location of the magazine.

Since the *Lusitania* rests on its starboard side, the forward hull section where the magazine is located actually rises above the bottom. As a result, we were able to use our smaller ROV Homer to inspect not only the port side of the magazine but the starboard side as well. This inspection revealed that the magazine was intact, did not explode, and therefore did not contribute to the sinking of the ship.

Since *Lusitania* was a passenger ship and not a cargo ship and since the first torpedo struck the starboard side of the ship just aft of the bridge, there was no other area within the ship near this area that could carry munitions. Given the intact nature of the magazine, it appeared that there must be another reason for the secondary explosion other than the detonation of contraband war materials aboard the ship. A number of theories have been advanced to explain the secondary explosion but all lack any convincing evidence.

The torpedo clearly struck one of the ship's starboard coal bunkers. Since the *Lusitania* was nearly at the completion of its trip, this space would have been hot, dry, and all but empty of coal. Samples of the coal were recovered from the floor of the ocean around the ship and proved to be bituminous coal, which is highly explosive and responsible for numerous coal dust explosions in mines on land where it is being extracted. For that reason, it is possible that when the torpedo struck the starboard hull, the violence of the explosion could have thrown the remaining coal dust into the air, possibly igniting it to cause the second explosion.

Midway

The Battle of Midway in June of 1942 is believed by most historians to be the turning point in the American war against Japan during the Pacific campaign of World War II. America's Pacific fleet had been badly damaged during the Japanese sneak attack on Pearl Harbor, 7 December 1941, but, fortunately, all of its carriers were at sea at the time. Shortly after the Pearl Harbor attack, the two warring adversaries met head-on in May 1942 in what proved to be the first-ever naval engagement fought exclusively by fleets of fighters and bombers. That engagement was called the Battle of the Coral Sea, as America sought to check the Japanese advance toward Australia and their planned invasion of Port Moresby, New Guinea. A key player in that historic battle was the American aircraft carrier USS *Yorktown*. Struck by a 550-lb bomb that killed 66 of its crew and left a large hollow inside the hull, *Yorktown* survived the attack, although the Japanese were convinced they had sunk her. She limped back to Pearl Harbor for what her captain thought would be a long repair period of many weeks or months. But Admiral Chester Nimitz had other thoughts. Recently gathered intelligence information suggested that the Japanese were planning a major attack on the island of Midway in hopes of drawing America's remaining aircraft carriers into a battle. When *Yorktown* arrived, Admiral Nimitz gave its captain and crew only three days to make repairs before joining the carriers *Enterprise* and *Hornet* in defense of Midway Island. Despite being badly outnumbered, the American forces carried the day, sinking all four of the Japanese aircraft carriers that had attacked Pearl Harbor, *Kaga*, *Akagi*, *Soryu*, and *Hiryu*, before losing *Yorktown*.

The goal of the 1998 expedition off the island of Midway was to find and document the USS *Yorktown* (Ballard and Archbold 1999). From the very beginning, research into the sinking of the *Yorktown* presented us with a difficult decision. There were three separate estimates of where *Yorktown* sank. One location was based on an analysis by famed naval historian Admiral Samuel Eliot Morison. Another estimated location was determined by a group of naval experts at the Naval War College.

These two estimated locations were close to one another, but a third, based on an analysis by historian Chuck Haberlein of the Naval Historical Center, placed the *Yorktown* many miles to the south of the other two estimates. Haberlein's estimated location was based on an analysis of the destroyers that were escorting *Yorktown* back to port after the carrier was attacked by aircraft from the sole remaining carrier, *Hiryu*, before it too was sunk by American planes. During the attack, three bombs from Japanese Val diver-bombers stuck *Yorktown*. While fighting the resulting fires caused by the bomb damage, two torpedoes from torpedo bombers off the *Hiryu* struck *Yorktown* again.

Now seriously wounded, *Yorktown*'s captain gave the order to abandon ship but at first light the next morning of 5 June, she was still floating. The decision was then made to attempt to salvage *Yorktown* and a group of destroyers were dispatched to come to her aid. The destroyers *Balch, Benham, Hughes, Gwin*, and *Monaghan* formed a defensive screen around the *Yorktown* with a radius of 2000 yards. The destroyer *Hammann* was ordered to stand just off the *Yorktown*'s starboard bow in an effort to put out the fires aboard the carrier and to assist in the salvage operation. Unbeknownst to the Americans, a Japanese submarine, *I-168*, had slipped through the destroyer escort screen and at a range of 1200 yards fired four torpedoes at the stationary *Yorktown* that sent her to the bottom.

It was the logbooks of these destroyers, which Haberlein had studied, that suggested a completely different position than those of the Naval War College and Admiral Morison. When all three estimates were taken into consideration, the total search area was greater than 500 square nautical miles, more than the *Titanic* and *Bismarck* search areas combined.

A preliminary survey of the area where the *Yorktown* had sunk revealed that majority of the search area was a flat featureless abyssal plain at water depths of around 5000 m. It was in this area where Haberlein's analysis of the destroyer logs suggested *Yorktown* lay. But the seafloor in the northern portion of the search area where the Naval War College and Admiral Morison thought the *Yorktown* had sunk, was occupied by a large undersea volcano.

In the past, we have always chosen between (1) using a side-scan sonar like the ones used to find the lost ships off Guadalcanal, which "mow the lawn" in flat regions, and (2) using a towed visual imaging sled, like Argo, that was used to find the *Titanic*, which look for a sunken ship's debris field in rugged terrain. A combination of the two may be used when both flat and rugged bottom terrains are present, as was the case with *Bismarck*, where we used both Argo's low-light-level cameras to search within the mountainous areas and its side-scan sonar to search the flat areas.

Another factor to take into consideration when planning a search effort is the amount of funds available. Working in 5000 m of water in the middle of nowhere is very expensive. Fortunately, we were able to convince the U.S. Navy to assist us in this effort. However, we still had to fund a significant portion of the costs associated with using their support ship *Laney Chouset*, the same ship we used on the Guadalcanal project, and their new ATV remotely operated vehicle system. And we also needed a sonar system to find *Yorktown* before deploying the ATV to image it.

Towing a sonar system in 5000 m of water is a very slow process with an average tow speed of less than 2 knots. At first, we considered using the Deep-Tow system at the Scripps Institution of Oceanography but we were concerned we would not have enough time to search the entire 500 square nautical mile search area. Memories of the three failed attempts to search the 100 square mile search area for the *Titanic* were fresh in our minds and we did not have the resources to spend that much time searching with a sonar that could reach out only 600 m to a side of its track line.

We needed a system that could reach out much farther. We then looked into using other systems like the Sea Marc-1 sonar system Bill Ryan used on his attempt to locate *Titanic*, which had a swath width of one and half to three miles; two and a half to five times the width of the Deep-Tow system's swath. But this system or one like it also needed to be towed at great depth and the drag on its tow cable would result in a similar slow speed of less than 2 knots through the large search area. Again, we did not have sufficient resources to guarantee that we would have enough time to search the entire 500 square nautical mile search area.

These calculations forced us to take a chance on a sonar system that was not designed for the search we had in mind. It was called the MR-1 and it had a total swath width of up to 14 km (figure 6.8). That was the good news. The bad news was that its designer, Bruce Applegate from the University of Hawaii, when asked about using his system to find the *Yorktown*, he was not convinced it could even see the *Yorktown* sitting on the bottom. As Bruce put it, "You are looking for a needle in a haystack and my system looks for haystacks."

Unlike Scripps' Deep-Tow system, which had an operating frequency of 100 kHz, the MR-1 system operated at 12 kHz. But one of the fundamental laws of physics that dominates the use of sonars in underwater search efforts is range verses resolution. A low-frequency sonar like the MR-1 has considerable range or swath width but it lacks the resolution of a higher frequency system.

A second consideration was the fact that the MR-1 sonar was towed just 200 m beneath the surface. That made it possible to tow the sonar at a much faster speed of 7 knots. But it also meant that the sonar was being towed approximately 5000 m above the bottom, and for the sonar to detect the *Yorktown*, it had to be towed off to the side, creating even greater distances. In the final analysis, Applegate was afraid that the MR-1 would not see the *Yorktown*.

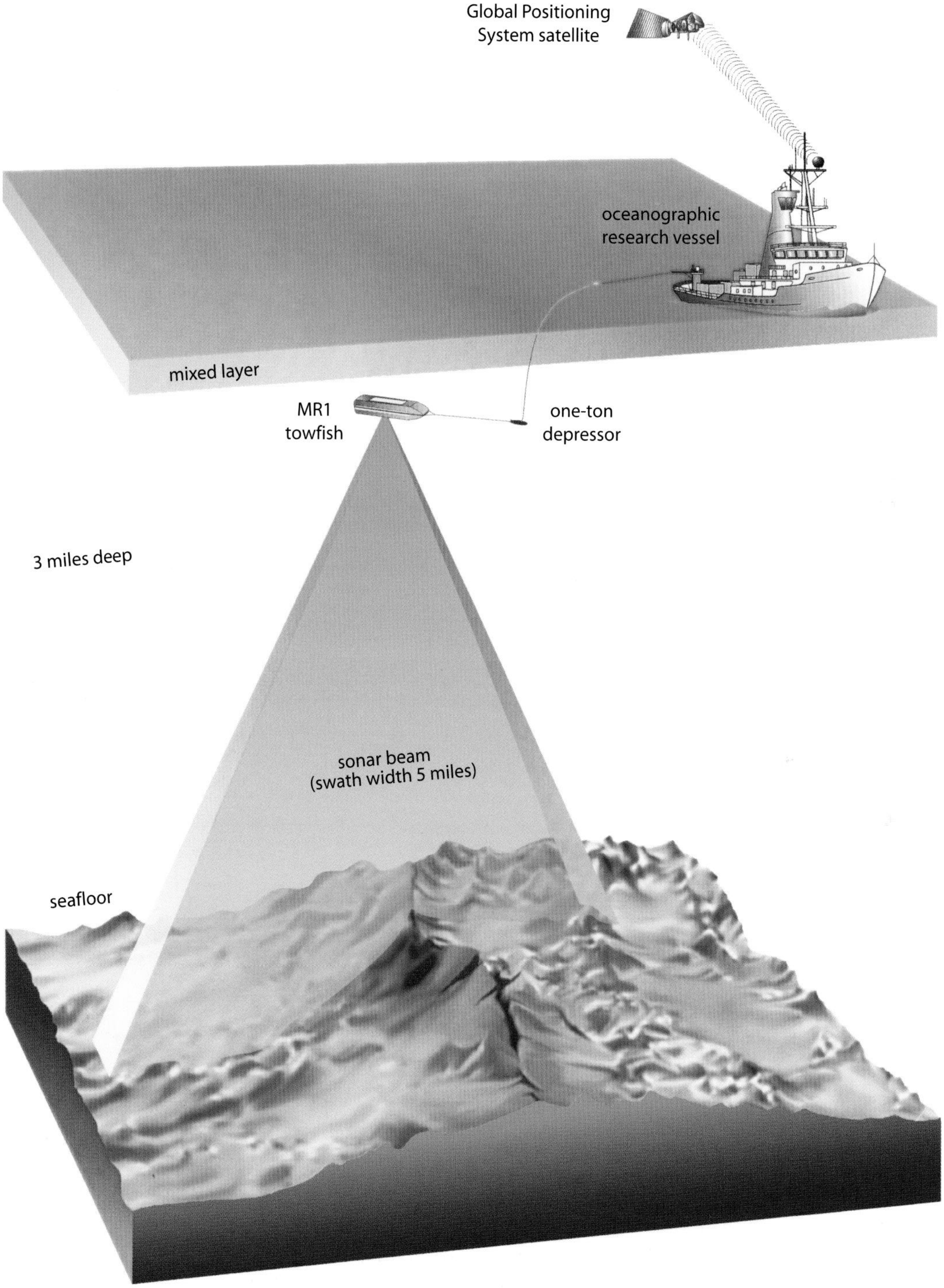

Figure 6.8. Graphic illustrating the University of Hawaii MR-1 sonar system that was used to find the *Yorktown* in more than 5000 m of water. (Graphic created by Paul Oberlander)

Towing speed was also a problem. In 3 miles of water, it takes 8 seconds on average for each ping of the sonar to travel out to the bottom and to return. To receive enough pings in order to detect the *Yorktown*, we had to reduce the tow speed to 3.5 knots; any less than that and the vehicle would become unstable and begin to sway. This negated one of the inherent advantages of using the system in the first place. Also, at this slower speed, the 250-m length of the *Yorktown* would reflect 18 sonar pings, assuming that the sonar would pass parallel to the long axis of the ship and that the ship was intact. That was just above the threshold of detection. On the shipboard sonar print out, such a target would be the size of a rice grain. We had thought of the *Yorktown* as a mighty ship but after running these calculations it had been reduced to a rice grain and that was the best we could hope for. What if we crossed its path at an angle, which was statistically more likely? Equally troubling was the possibility that the *Yorktown* had sunk where the Naval War College and Admiral Morison put it, on an undersea volcano. If that were the case, the sonar would detect numerous rice grains that would have to be resolved using the ATV system, a prospect we did not want to think about throughout the search effort.

After running over these calculations again and again, including those associated with funding the expedition, we came back to using the MR-1, knowing the considerable risk we were taking. We had no other choice. Another factor in those calculations was the fact that the MR-1 was based in Hawaii, where the expedition began, further reducing the cost of using it and insuring that backup support was relatively near to the operating area should a problem arise.

Using the MR-1, it took only 4 days to search the entire 500 square nautical mile search area, which included running two sets of overlapping lines roughly perpendicular to each other (figure 6.9). The overlap was needed since MR-1 had a dead spot or nadir directly beneath the sonar with a width of one nautical mile.

Analyzing the MR-1 records presented a new challenge to using a side-scan sonar in a search effort that we had not encountered before. We had come to understand and accept that using sonars to find something beneath the sea was somewhat of a "black art." Unlike looking at a picture that is relatively easy to understand, what you see on a sonar record is a function of a number of variables, including, most importantly, the sonar's aspect angle. Looking at the *Yorktown* image end on looks entirely different than if the sonar had run parallel to her length. We were used to that from previous sonar system efforts like those conducted in Iron Bottom Sound. What was new was the considerable range between the sonar and a target on the bottom. On our first pass through the search area, we detected an object the size of the *Yorktown* but when we passed by it again on another pass from the other side, the object shifted its position. What we were seeing was two different sides of a small hill. It had the right length but not the right width.

Once the survey was completed there were a number of rice grain-sized targets but only one that was consistently in the same position when viewed from different angles. Ironically, the sonar had passed over this target on the first sonar run through the area days before. Although it had not been seen in that first pass because it was in the sonar's nadir, it was detected by the sonar's depth sounder, which pings straight down into the nadir. Looking at the depth

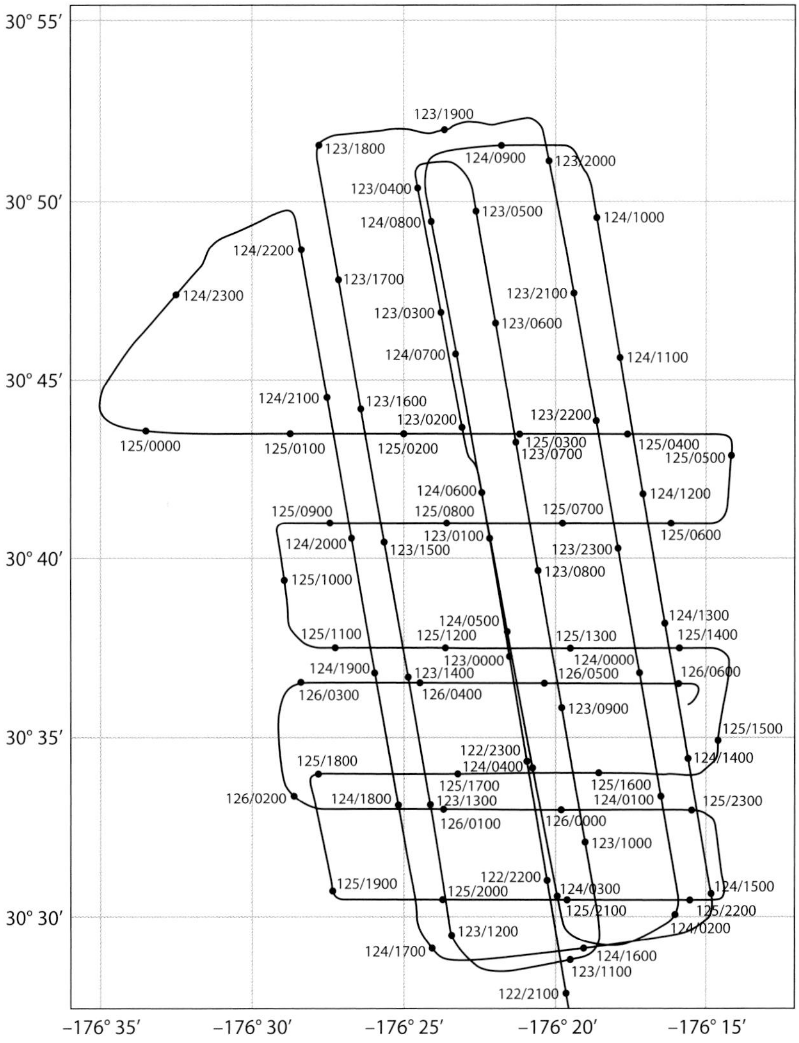

Figure 6.9. Trackline coverage of the MR-1 sonar system during the search effort for *Yorktown*. (Graphic created by Paul Oberlander)

data one could see that the sonar passed over a discreet object 13 m tall. Since the *Yorktown* was 27 m high, it was logical that it would sink approximately halfway into the soft bottom muds of the abyssal plain.

Our gamble appeared to be paying off. We had located the *Yorktown* in a record time of four days. Now all we had to do was use the *Laney Chouest's* dynamic positioning system to place us directly over the target's GPS coordinates and lower the ATV vehicle for a look.

That proved to be much more difficult that we had ever imagined; in fact, we almost failed reaching *Yorktown* with the ROV. The ATV was not a hardened ROV system like Jason. It was an engineering test bed, a prototype developed by engineers more interested in advancing the engineering sciences of remotely operated vehicles than in developing a system that would have a long history of field programs. In fact, it never did. Unlike the Jason system that was operated by the same team of engineers and technicians that had developed it, the ATV had been turned over to the operational Navy and was being run by a team that was dedicated but constantly changing its personnel. This has never been a formula for success. When the operational team is looking in the manuals, you are in trouble. Those troubles began on 5 May on the first lowering of the vehicle when it experienced a major electrical ground fault at a depth of 1500 m, a problem that took a day to fix. Then on its second lowering, it failed again. Finally, after a second fix with more time lost, a third lowering began on 7 May which ended in near disaster. As the vehicle approached the bottom, one of the large glass pressure housings that contained one of our powerful lighting ballasts imploded, setting off a series of chain reactions that severely damaged the vehicle. Replacement parts needed to be flown into Midway from the United States, and that cost additional days lost at sea. Not wanting to waste time while we waited for the parts to arrive, a search was made with the MR-1 for the Japanese carriers to the west but little research had gone into their possible locations and nothing was found in the short amount of time we had.

On 13 May, we arrived back in Midway to pick up the parts we needed and by 16 May the ATV was back on target heading down to what we thought was *Yorktown*. Unfortunately, the dive had to be aborted and it wasn't until 19 May that we had the vehicle back in the water.

After a long lowering, the ATV finally approached the bottom and a soft featureless sediment surface came into view. The first visual indication that our sonar target was indeed the *Yorktown* was when we began to see clumps of mud just as we had seen as we approached the *Titanic*. When a large object the size of the *Titanic* or *Yorktown* slams into the bottom, it sends up a tremendous amount of bottom sediments that fall back to the bottom in the form of large clumps or splatter.

We had passed over the *Yorktown* on our first sonar run on 2 May, but it took 17 days before we had our first visual confirmation that the target was indeed the *Yorktown* (figure 6.10). Only four days were left to document the ship before we had to end the cruise and head back to Midway, we had just made it under the wire. The problems we had expected to encounter in finding her did not materialize, while what we thought would be the easy task of filming her turned out to be the most difficult. More lessons learned.

PT-109

The search for *PT-109* was not driven by the importance of the naval engagement that led to its sinking but by the person who was captaining the ship when it went down, future U.S. president John F. Kennedy (Ballard and Morgan 2002). Kennedy was a young officer facing his first combat experience when he was sent off to the Solomon Islands in 1945 to take command of *PT-109*. Many historians have stated that he went to the Solomon Islands a brash immature son of a wealthy and powerful political figure and returned a mature man ready for the challenges he would late face against the Soviets during the Cold War and the Cuban "Missile Crisis" in the early 1960s.

Following the Japanese failed attempt to retake the airstrip on Guadalcanal, the American forces began pushing their way up the "Slot" toward the Japanese bases at Rabaul on New Britain Island to the northwest of the Solomon Islands. By taking Henderson Field, named after marine pilot Major Lofton Henderson,

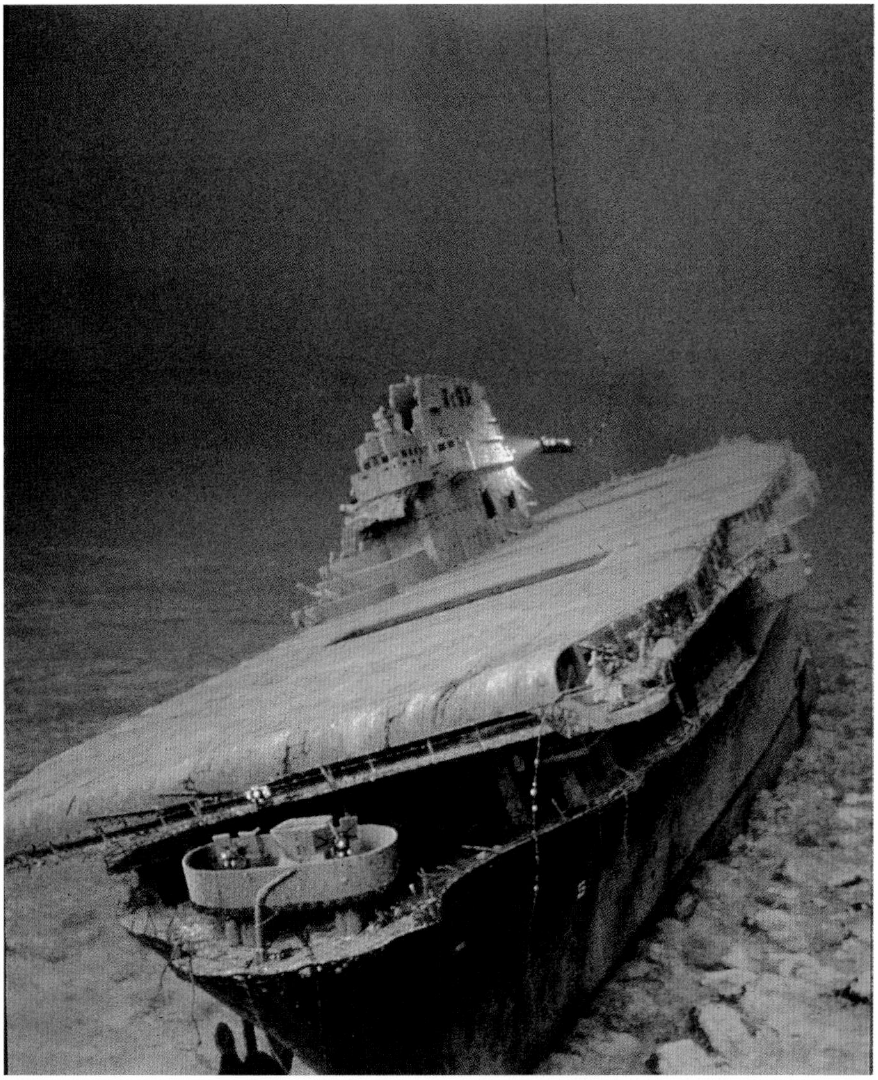

Figure 6.10. Artist rendering by Ken Marschall of *Yorktown* being inspected by the ATV. (Copyright Ken Marschall)

a squadron commander who had been killed at the Battle of Midway, the Americans controlled the airspace over the Solomon Islands. This forced the Japanese to work at night as they sought to resupply their forces attempting to resist the American advance. This nightly resupply effort took on the name "Tokyo Express."

In an attempt to block the "Express," American forces relied upon the "Coastwatchers," Australian and New Zealand allies who had found themselves caught behind enemy lines when the Japanese invaded the British held islands of the Solomons and had committed to helping the Allied Forces retake the islands. Working with local natives, the Coastwatchers were the eyes and ears for the Allies. Distributed along the length of the Slot reaching back to Rabaul, they were able to monitor Japanese activity on land, in the air, and down the Slot by sea and then radio that information to the Allies. It was a Coastwatcher by the name of Reginald Evans, situated on the island of Kolombangara (figure 6.11), who alerted the PT boat squadron on the island of Rendova, where

Figure 6.11. Blackett Strait with island of Kolombangara in the background. Small island on extreme right is Naru while small island to the upper left is Kennedy. (Copyright Ira Block)

PT-109 was based, that the Tokyo Express would be running the night of 1 August 1943. Not knowing which side of Kolombangara the express would be traveling along that night, the Americans dispatched destroyers to the east of the island and 15 PT boats to the small and more restricted western side known as Blackett Strait. It was here that *PT-109*, while on patrol, collided with the Japanese destroyer Amagiri and sank.

On first glance, the search for *PT-109* appeared simple and straightforward, if you discounted its remote location in, at times, violent and unstable country. After all, Evans had witnessed the collision caused by the Japanese destroyer as it ran over *PT-109* that night, cutting it in half and setting the night aglow in the resulting explosion that had sunk her. Blackett Strait was narrow, less than 6 nautical miles across, and relatively shallow, at depths of less than 800 m. The estimated search area around the collision site was less than 5 square nautical miles. Our experience years before in the gently sloping mud bottom of Iron Bottom Sound to the southeast off Guadalcanal suggested that the remains of the boat would be easily detected by side-scan sonar.

Granted, eyewitnesses reported the boat to have been cut in half with the bow floating up onto a reef and lost to natural forces over the years. But the stern section contained three large Chrysler diesel engineers, torpedoes and their metal launchers, as well as gas tanks, machine guns, and a host of other metal objects that should still be there. Plus, there was some thought that wood-boring organisms like those that had eaten the pine deck of the *Titanic* would not eat the mahogany laminated wooden hull of *PT-109*.

As our research continued, these early thoughts about the ease of the search we planned began to evaporate. First were reports from the Naval Historical Center that many researchers believed that when the Japanese destroyer collided with *PT-109* it was not cut in half, that the boat was struck by a glancing blow, which would have rotated *PT-109* away from the cut edge of the destroyer's bow, and that only a small portion of the boat would have fallen to the bottom at the impact site. If this were true, finding such a small hull fragment, which some thought at most would be one of the Chrysler diesel engines and perhaps the aft torpedo launcher and torpedo on the starboard side, would be difficult. Further lessening our chances of locating it was the terrain in the area of the collision.

We chartered a small Australian vessel, the Grayscout, and installed our sonar system, Echo, and ROV system, Argus/Little Hercules, onboard (figure 6.12). Once our Echo side-scan sonar was deployed, a systematic search was carried out down the long axis of Blackett Strait. As soon as Echo reached the bottom it became clear that the bottom of the Strait was much more complex than that seen in Iron Bottom Sound. On both sides of the Blackett Strait were steep limestone scarps and at their base, where the remains of *PT-109*'s stern were thought to lie, were piles of blocky reef talus that had fallen off the scarps. Kolombangara is also a historically active volcano and previous activity over time has included pyroclastics eruptions that have fallen into the Strait, scattering large rock debris. The sonar records also suggested that possible subterranean lava flows had occurred, further complicating the search in this area by masking potential targets. The result was numerous sonar targets that could easily be the small hull section we were looking for. Although a number of these targets were inspected using our towed imaging sled Argus and our small remotely operated vehicle Little Hercules, all proved to be naturally occurring geologic features.

At this phase in the search effort, it appeared that finding any of *PT-109*'s remains was hopeless, sending us back to our original research database. That is when we began to question Coastwatcher Reginald Evans' account of the sinking. Evans had seen an explosion during the early morning hours of 2 August 1943. Initially, he thought it was a Japanese barge that had been hit by American

Figure 6.12. Australian search ship *Grayscout* on location in Blackett Strait. (Copyright Ira Block)

gunfire. But when he reported the explosion to Rendova the following morning he was informed of the loss of *PT-109* that same morning in the same area.

He then reported sighting something floating in Blackett Strait just to the southeast of the collision site. He was too far away to see what it was, only that it was a small object. He had no way of knowing that at that very moment, Kennedy along with 10 of his 12 crew members were clinging to the bow section of *PT-109*. The bow was still full of air and drifting in the current down Blackett Strait toward its southern entrance at Ferguson Passage, the channel *PT-109* had passed through the night before.

Later that afternoon, Evans reported a second sighting of what must have been *PT-109*'s bow section still floating in Blackett Strait, but the current had carried farther southeast. By that time, Kennedy and his remaining crew, fearing detection by the Japanese, who controlled the surrounding islands, had left the bow and swam to Plum Pudding Island, which was later renamed Kennedy Island by the local population. From these three sightings it was possible to obtain a current drift vector and speed.

But Evans reported not seeing the bow section for the next two days. On 5 August he reported seeing it again, this time washed up on the fringing reef of Naru Island after drifting through Ferguson Passage. There it was destroyed by subsequent wave action with nothing remaining when we visited the area.

If all of this was true, there was little to nothing of *PT-109* to be found. But had it disappeared for two days and then reappeared? That did not make sense. Thanks to the hard work of the National Geographic Society, which was making a film about our efforts to locate *PT-109*, their research team had tracked down the two Solomon Island natives, Eroni Kumana and Biuku Gasa, who worked with Reginald Evans and who actually had found Kennedy hiding in the underbrush on Naru Island.

I asked Eroni about the boat Evans reported seeing on the fringing reef of Naru Island after Evans had moved to the small island opposite Naru on the other side of Ferguson Passage. Eroni said he had seen it and that it had contained Japanese rifles. This sounded more like a Japanese supply barge from the Tokyo Express than *PT-109*. What if the reason Evans had not seen the bow section of *PT-109* floating in Blackett Straits two days after its collision with the Japanese destroyer was because it had filled with water and sunk the night after the collision and had not drifted through Ferguson Passage. If that were the case, we could predict its sinking location based on the current vector and speed observed the day before. Based on that extrapolation, we estimated that *PT-109* should have sunk in the southeastern region of Blackett Strait just north and west of Ferguson Passage.

With that information in hand, we conducted our second Echo side-scan sonar search. Unlike the original search area off Kolombangra, the bottom terrain in this area was flat, characterized by large sand waves and no rock outcrops or talus accumulations except along the base of the western scarp bounding Blackett Strait near Naru and Kennedy Islands (figure 6.13).

At the estimated site of its sinking we found only one sonar target that was rectangular in shape, resting in a field of large sand waves. Our initial thought was that it was a sunken barge but, in the absence of other targets, we decided to investigate it with our ROV Little Hercules (figure 6.14).

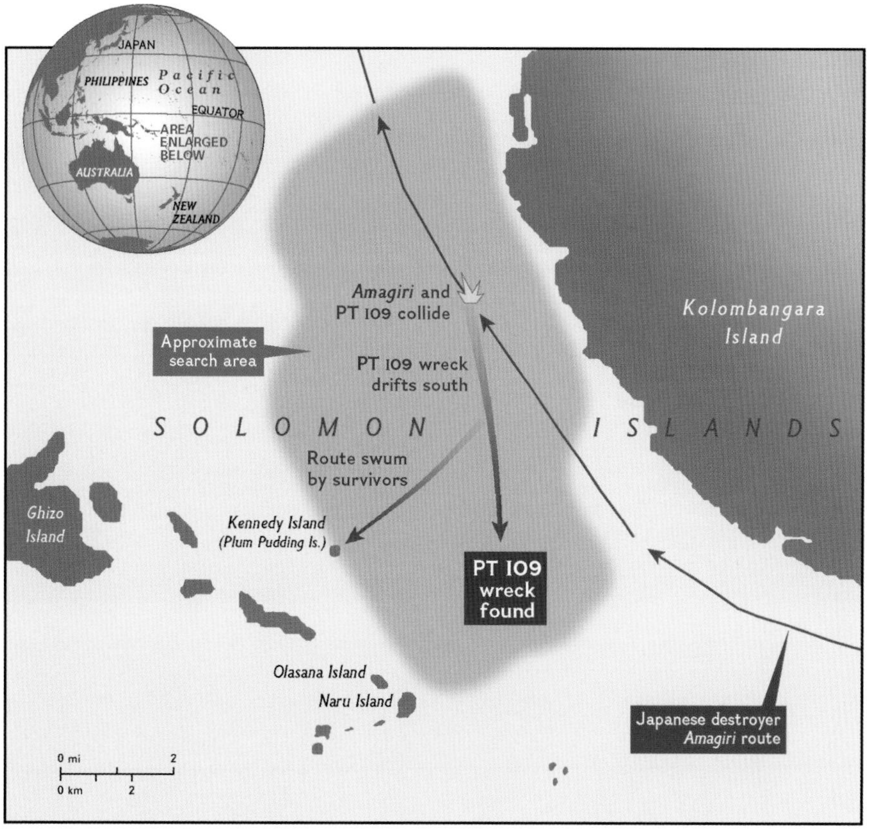

Figure 6.13. Dark blue area denotes area searched by Echo side-scan sonar vehicle. (Copyright National Geographic Society)

The bottom consisted of large, coarse sand swells measuring more than 3 m from trough to crest at a depth of 700 m. It was difficult to maneuver the ship and ROV in the strong currents passing through Ferguson Passage. At the base of one of the sand swells, a long, tubular shape came into view that appeared to be a torpedo launcher with a torpedo protruding from its stern. The torpedo had double counter-rotating propellers characteristic of those carried by *PT-109*. The sonar image recorded by Echo was much larger than the objects seen on the surface of the sand swell, suggesting that a much larger object lay buried beneath the migrating sand swell. The sonar image was equal to the width of *PT-109* and the length and shape of the large diesel fuel tanks and three Chrysler engines.

Next to the torpedo launcher was a small, circular fixture similar to the training mechanism used to rotate the tube into the proper launch position, situated in the right position according to the boat's deck plan. Little Hercules pushed against the fixture and was unable to move it, suggesting that the fixture was rigidly attached to the deck below. Out of respect for the two members of the crew who died on *PT-109*, the U.S. Navy, and through consultation with the Kennedy family, we chose not to excavate around the launcher in an effort to learn more about was buried beneath the sand. The site was to be treated as a war grave and not be disturbed.

Figure 6.14. ROV Little Hercules inspecting *PT-109*'s exposed torpedo launcher (*bottom*) (Copyright Institute for Exploration) while author and ROV pilot Dave Lovalvo analyze the sonar data (*top*). (Copyright Ira Block)

Summary

In summary, we have developed a strategy, through years of research and mission planning, to conduct successful expeditions to find contemporary shipwrecks in the deep sea. This strategy involves not only an understanding of the historical details surrounding each ship's sinking, but a detailed analysis of the oceanographic environment in which each ship sits. The survey to locate the ships requires an understanding of ocean currents and seafloor morphology and geology. This understanding is key to the way a site is approached, and the mapping phase results in a greater understanding of the marine environment. Often we publish papers that deal with the detailed oceanographic contexts surrounding the archaeological sites. This understanding is also key to the establishment of the environmental baseline that represents the archaeological context for each site. Archaeological oceanography involves the investigation of these details.

References

Ballard, R. D. (1987). *The Discovery of the Titanic.* New York: Warner/Madison Press.

Ballard, R. D. (1990). *The Discovery of the Bismarck.* New York: Scholastic/Madison Press.

Ballard, R. D. (1993). *The Lost Ships of Guadalcanal.* New York: Madison Press.

Ballard, R. D. (1995). *Explorations.* New York: Hyperion.

Ballard, R. D., and R. Archbold (1999). *Return to Midway.* Washington, DC: National Geographic Society/MadisonPress.

Ballard, R. D., and S. Dunmore (1995). *Exploring the Lusitania.* New York: Warner/ Madison Press.

Ballard, R. D., and M. H. Morgan (2002). *Collision with History.* Washington, DC: National Geographic Society.

Lord, W. (1955). *A Night to Remember.* New York: R&W Holt.

Uchupi, E., M. Muck, and R. D. Ballard (1988). Geology of the *Titanic* site and vicinity. *Deep-Sea Research*, 35:1093–110.

PART THREE

Deep-water Shipwrecks in the Mediterranean and Black Seas

Searching for Ancient Shipwrecks in the Deep Sea

7

Robert D. Ballard

The search for ancient shipwrecks in the deep sea is fundamentally different from that for contemporary shipwrecks. In the former case, the search is generally focused on a specific ship of historical significance such as the *Titanic*, *Bismarck*, or *Yorktown*. As a result, a great deal of information exists pertaining to its sinking, including historical archives and eyewitness accounts. Commonly, other researchers or historians have written extensively about the ship and in some cases others have already attempted to locate it.

Most importantly, the historic shipwrecks mentioned above were large, weighing in excess of 40,000 tons, and were made of steel, making them much easier to detect. Large debris fields containing additional metal objects or inorganic material that resists deterioration frequently surround these ships and can be used to find them. The impact of the ship with the bottom of the ocean commonly creates a large crater or generates a landslide that increases the acoustic or optical detection potential.

Searching for shipwrecks lost in ancient battles such as the Battles of Salamis, Ecnomus, or Actium might suggest that a similar approach could be taken, although finding a specific ship lost in one of those battles would be highly improbable. Unfortunately, two factors make finding ancient warships difficult. The traditional warships used during these battles were wooden galleys with single, double, or the more common three-tiered triremes. Although they carried a bronze-plated ram, they were kept light for speed as compared to the heavier and more rounded merchant galleys. The warships were also prized in battle. Once rammed, they commonly swamped and many did not sink at all. Should the ship actually sink, wood-boring organisms would quickly attack its wooden members and unless it was carrying a looted cargo, very little would remain to detect.

Equally important was the fact that these battles were fought close to shore. As a result, any shipwrecks resulting from these engagements more than likely landed in water depths measured in the tens to a few hundreds of meters. This would place the shipwreck in the sunlit layer of the sea where biological activity is heightened, long-shore currents that can redistribute sediment and bury the ship are strong, and where fishing activity can be intense. As is shown later in this chapter, the shallow depths of the marginal seas like the Mediterranean and Black Sea have experienced a great deal of long-term bottom trawling activity by commercial fishermen as well as illegal trawling activity by fishermen specifically in search of ancient artifacts. Along the shorelines of Sicily, for example, a dense pattern of long linear ruts can be seen in the soft bottom muds down to depths of 250 to 300 m cut by the heavy doors of otter trawls.

For these reasons, the focus of our work to date has been on finding ancient commercial shipwrecks that were carrying cargo and were lost in deep waters where bottom conditions lead to their long-term preservation. Early in our efforts, we made the assumptions that (1) the ancient mariners were well aware of the ports they wanted to reach, (2) the incentive of making a profit caused them to select trade routes that were safe and as short as possible, and (3) for these reasons, they traveled directly between the major seaports of the ancient world, courses that could take them far from shore and into deep water. In addition, piracy in the ancient world was a common activity of coastal states, particularly during periods of regional conflicts. The pirating typically was carried out near shore, giving ancient mariners further incentive for traveling beyond sight of land. Lastly, storms at sea, which account for many lost ships, are more violent in shallow coastal waters than on the high seas, possibly influencing the ancient mariners' routes.

Carthage to Rome

Our first systematic search for ancient deep-water trade routes began in the summer of 1988 (Ballard et al. 2000a; Ballard 2005) in the central Mediterranean Sea. The primary focus of that initial search effort was the Straits of Sicily, which lies between the ancient seaport of Carthage and the western tip of Sicily near the present port of Traponi, connecting Africa with Europe. In an effort to avoid the concerns of coastal states, all work was carried out in international waters. On average, we worked at a distance of more than 24 nautical miles out to sea.

The first search lines, conducted in 1988 using the Argo imaging sled, took place at the southwestern entrance to the Gulf of Cagliari on the southern tip of Sardinia, since this was the staging area for the expedition. The port of Cagliari was also used in ancient times as a maritime trading center. We surveyed along transect lines parallel to shore in hopes of intersecting the ancient trade routes that connected southern Sardinia with ancient Carthage and the modern seaport of Trapini on the western tip of Sicily, which was also a center for ancient maritime trade. A long linear ridge running northeast to southwest off the Gulf of Cagliari was selected for investigation in hopes that this perched feature would be less affected by bottom sediment transport coming out the Gulf toward deeper waters. Unfortunately, no ancient artifacts or shipwrecks were detected.

The next search lines were surveyed in the Straits of Sicily, but just prior to arriving on site a major storm in the area resuspended the bottom sediments, greatly reducing bottom visibility. Since the search strategy was based on the use of the towed imaging sled Argo, which had been used successfully to visually locate the *Titanic* and *Bismarck*, poor visibility in the Straits made a visual search impossible. For this reason, during the next several days, a number of search lines were searched north and west of ancient Carthage in modern day Tunisia, but those lines did not detect any ancient artifacts or shipwrecks along those lines.

A decision was then made to make a series of long search lines along the northern coast of Sicily, beginning in a region known as Skerki Bank at the northern entrance to the Straits of Sicily, followed by two long search lines along the south coast of Italy, including the entrance to the Gulf of Naples. During the course of this search effort, which involved more than 200 nautical miles of continuous visual inspection of the bottom and acoustic search coverage out 100 m to either side of those lines, only one area was found to contain ancient artifacts and contemporary and ancient shipwrecks. That site was at the northeastern tip of Skerki Bank, at the intersection of the trade routes connecting the ancient seaport of Cagliari in southern Sardinia and the west tip of Sicily and the ancient trade route between Carthage and Ostia, the seaport for Rome.

Over subsequent years, we conducted a number of expeditions to this site, which resulted in the location of hundreds of ancient artifacts, typically isolated amphora (figure 7.1), two long lines of amphora, and eight shipwrecks; five of which were from the Roman period dating from the 1st century BCE to last quarter of the 4th century CE. The range in dates for the isolated amphora dated as far back as the 3rd century BCE, representing continuous use of this deep-water trade route for more than six centuries, covering the span of Roman domination in the Mediterranean Sea.

But why were so many ancient artifacts and shipwrecks found here and not elsewhere during the original 1988 search effort? There are a number of possibilities, all of which may be contributing factors. First and foremost, the site is located at the intersection of two lines connecting ancient Carthage and Rome and southern Sardinia and western Sicily. Second, the site is located on an uplifted block of undersea terrain where the processes of deposition and erosion appear to be limited. This became even more apparent during a survey of the area conducted in 2003. That expedition was carried out aboard the R/V *Knorr* on which was mounted a multibeam mapping sonar. Using this system, we were able to produce a detailed bathymetric map of the Skerki Bank region (figure 7.2).

An inspection of that map clearly indicates that the regional topography is characteristic of a submerged "karst terrain" more than likely formed when the Skerki Bank region was subaerially exposed six million years ago during the Miocene. Such terrain contains numerous depressions or "sinkholes" that greatly limit the horizontal transport of bottom sediments. In addition, the remote location of the site from surrounding landmasses results in low pelagic sedimentation rates. Therefore the site represents an ideal place to find ancient artifacts and shipwrecks that are exposed on the surface of the bottom sediments.

Another reason for their presence at this site was the apparent lack of significant bottom-trawling activities. The average water depth at the site is 800 m, a depth characterized by the absence of light and limited biological activity. For that reason, one would suspect that bottom trawls used by commercial fisherman to catch benthic fish would not work at these depths. Although no bottom trawl marks were seen in the area, extensive trawl marks were seen farther to the east in comparable water depths near the western tip of Sicily, where commercial fishermen base many of their trawlers.

Further indications of extensive deep-water trawling activity around Sicily were observed on the R/V *Knorr* cruise in 2003 (Ballard 2005). Following our mapping efforts at Skerki Bank, a search effort was conducted along the ancient

Figure 7.1. Images of Skerki D taken by Hercules during the 2003 expedition on the R/V *Knorr*. (Copyright Institute for Exploration)

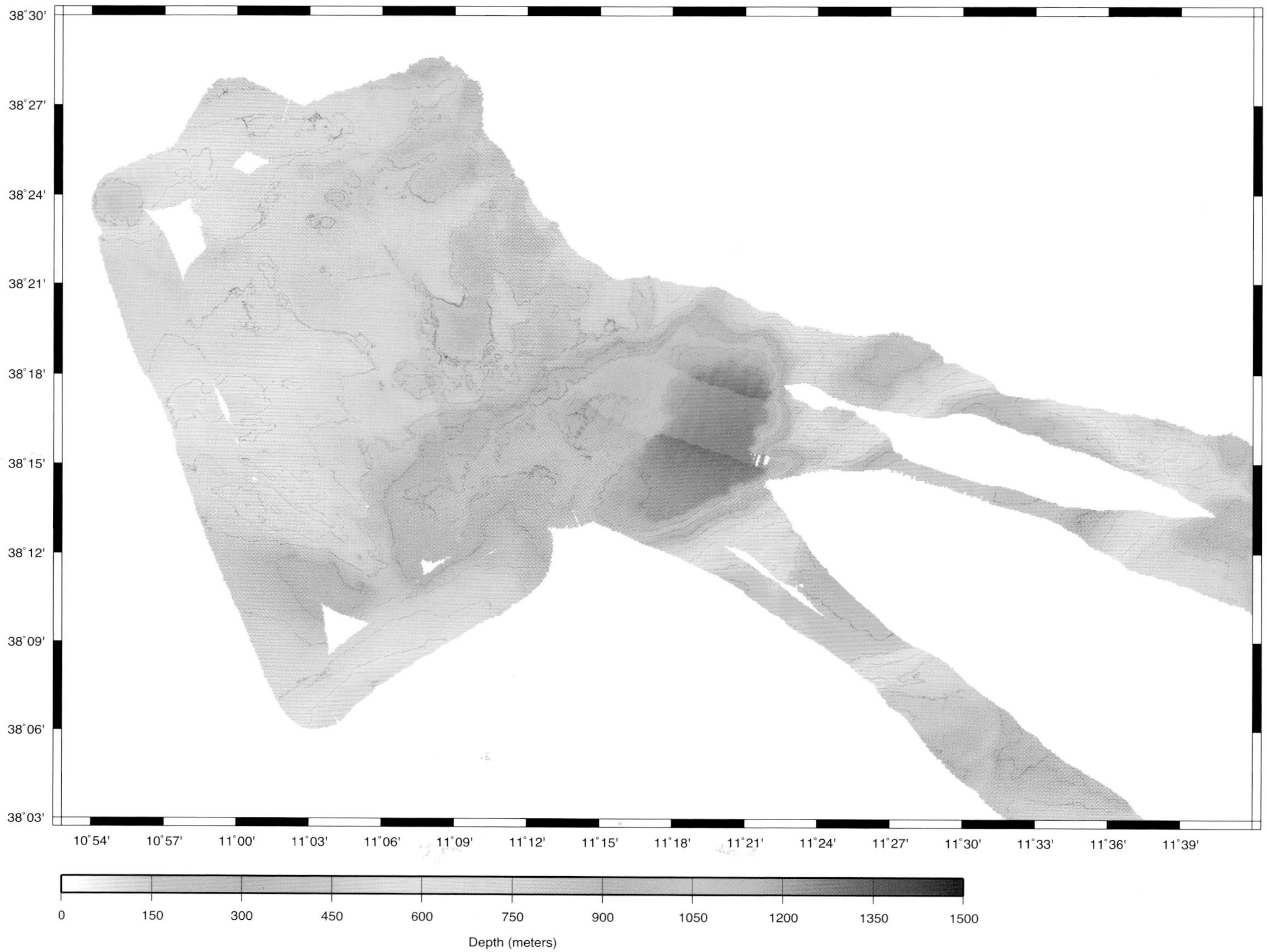

Figure 7.2. Bathymetric map made using the Seabeam sonar system on the R/V *Knorr* during the 2003 expedition to Skerki Bank. (Copyright Institute for Exploration)

trade routes connecting Pantelleria Island and the western approaches to Malta. On this occasion, we used our visual search vehicle Argus. Like Argo, Argus was outfitted with powerful lights and various underwater cameras. The primary system was a high-definition video camera that could be panned up and down. Argus also has a series of thrusters that make it possible to rotate the vehicle to the left and right as it was towed by *Knorr* at the end of a fiber-optic cable. Argus also had a forward-scanning sonar to alert the operator to possible targets in front of or to either side of the vehicle's track line.

During this survey, only one possible ancient shipwreck of unknown age was imaged, but as the island of Malta was approached numerous fishing vessels were observed carrying out bottom-trawling activities below the depth of 1000 m. This observation was further supported by the presence of extensive bottom-trawl marks seen in the region west and north of Malta by Argus' cameras. This extensive bottom-trawling activity was observed not only around Malta and Sicily but also in other areas where we have conducted visual search programs, including off the Gulf of Naples, the coast of Egypt, and in the Black Sea. This activity does not bode well for the future discovery of shipwrecks in the shallow-water regions of the world.

Numerous ancient naval battles were fought near shore that resulted in the loss of countless ships. Based on our knowledge to date, however, the probability of locating those shipwrecks is not high. Another factor that needs to be taken into consideration is the nature of the bottom on the continental shelf compared with the deep sea. During the last Ice Age, sea level dropped more than 120 m (400 ft), exposing a major portion of the continental shelf (refer to chapter 9). As a result, the process of erosion created a hard resistent surface compared to the soft surface of the deep-sea, which is covered by mud. When shipwrecks come to rest on this harder eroded shallow seafloor, they do not penetrate deep into the bottom sediments. Instead, the majority of the shipwreck is exposed to benthic bottom processes in waters having a high concentration of dissolved oxygen. Even if the shipwreck sinks intact and comes to rest lying upright on the bottom, the weakening of its hull fasteners through time will lead to its collapse, resulting in a positive topographic feature that is subject to disturbance by bottom-trawl nets. Continuous trawling activity over time will result in the shipwreck being torn apart and its contents either being recovered in the trawl nets or scattered across the ocean floor only to be encountered by future bottom-trawling activity. When the author visited the Sicilian seaport of Trapani, he was shown a warehouse of artifacts recovered by local authorities from fisherman either by accident or intentionally for sale on the black market.

The intensity and destructive nature of bottom-trawling activity has been further documented during the investigation of contemporary shipwrecks, including those mentioned in the previous chapter. They included our investigations of the *Lusitania, Andrea Doria, Wilhem Gustloff, General Steuben, Gola,* and *Britannic,* all shipwrecks lost in waters less than 200 m deep. The exposed superstructure of these steel shipwrecks are commonly wrapped in fishing nets, making their investigation extremely dangerous and possibly expensive due to the lost of equipment, as is documented in the previous chapter.

As a result, future efforts to locate ancient shipwrecks in shallow water should concentrate in rugged bottom terrains where benthic fishing activity does not take place. Unfortunately, such locations make it difficult to locate shipwrecks using traditional acoustic search systems like side-scanning sonars, since the shipwreck signature is commonly lost in the acoustic background noise so characteristic of strongly reflecting rock outcrops.

An important observation made during the investigation of the Skerki Bank shipwrecks as well as at other sites in deep water was the depth to which these shipwrecks had penetrated into the bottom. Unlike, the subaerially eroded surface of the shallower continental shelf, soft sediments, which have been laid down one particle at a time, cover the vast majority of the deep sea. Although there are major areas of the deep sea where the bottom surface is either coarse-grained sediments, such as areas covered by turbidity flows, or hard bedrock surfaces, no major search efforts for ancient shipwrecks has yet to be conducted in these areas.

All of the ancient shipwrecks discovered to date in water depths greater than 200 m have penetrated into the bottom mud up to their main deck. This is expected given the specific gravity of soft deep-sea sediments. The vast majority of the shipwrecks we have found to date, including contemporary steel shipwrecks, are resting upright on the bottom, including those that were known to have rolled over on the surface, like the *Bismarck* and *Yorktown*. When a ship sinks, air pockets within the ship may affect its initial sinking geometry. But as the ship passes into deeper and deeper water, the compartments either implode and become fully flooded or the trapped air is reduced by pressure to tiny bubbles. As this is taking place, the ship's righting moment causes it to rotate back into a vertical position unless a significant shift in the cargo prevents this from happening, a situation we have yet to observe. Once the ship rights itself, its keel can easily cut into the soft bottom sediments, penetrating into the bottom until it reaches a new state of equilibrium, usually when the main deck is flush with the bottom surface.

As a result, the vast majority of the ship is immediately placed into what are commonly anoxic conditions. With ancient shipwrecks, this not only leads to its long-term preservation but also serves to hold the ship intact, since its frame and planking cannot splay outward as seen in shallow-water shipwrecks, which do not penetrate into the bottom to any appreciable depths. As the exposed wooden superstructures are eaten by wood-boring organisms, the remaining shipwreck beneath the sediments is capped by a small piles of artifacts, limiting its risk to bottom-trawling activity.

A final reason why so many shipwrecks were found in the Skerki Bank area was simply because so few other areas have been investigated by dense surveys like we accomplished.

It is interesting to note that the vast majority of the isolated amphora detected in the Skerki Bank area were intact, jettisoned overboard for reasons other than being broken during their handling. They commonly carried wine, olive oil, or garum. (Garum is a fish sauce. When properly prepared it had a pleasant aroma, looked like aged honey wine, and was often mixed with wine to drink.) The contents of these amphora would have been consumed during a typical ocean passage and the now empty vessels would be thrown overboard, which explains their cheap construction and subsequent lack of value. Since the shipwrecks on Skerki Bank had been visited and sampled on several previous expeditions it was not necessary to recover many of the amphora during our 2003 expedition and for that reason they were removed to a near-by storage area and placed in depressions to protect them for any subsequent bottom-trawling activity that might be carried out by commercial fishermen (figure 7.3).

The insight we have obtained from mapping the ancient deep-water trade route of Skerki Bank will guide us in our search for other such routes around the world. Just like a modern highway, these ancient deep-water trade routes should be littered by debris cast off the passing ships either as unwanted trash or in an attempt to save the ship during a storm. The 2003 trip to Skerki Bank (Ballard 2005) also provided us with an exact opportunity to test Hercules, the next generation of remotely operated vehicle systems designed specifically for the archaeological community (see figures 2.1, 4.1, and 7.4).

Figure 7.3. An amphora being removed to the storage area next to the Skerki D wreck site. (Copyright Institute for Exploration)

Figure 7.4. Image of Hercules at the Skerki D site in 2003, taken from companion vehicle Argus. (Copyright Institute for Exploration)

Ashkelon

Unlike the Skeri Bank program, our efforts off the ancient seaport of Ashkelon, Israel (Ballard et al. 2002), which involved the investigation of two Phoenician shipwrecks from the Iron Age (750 BCE), was the result of a chance discovery by the U.S. Navy's nuclear research submarine *NR-1* and not a concerted search effort carried out by our team. In January 1968, the Israeli Navy lost a diesel submarine named *Dakar*, which disappeared without a trace on its maiden voyage from England, where it was purchased from the British Navy, to its new home port in Israel. After years of analysis various lines of argument were developed, one of which suggested it might have been lost off the coast of the Sinai desert in an area between the Nile delta and Ashkelon, Israel.

As a result of this analysis, the Israeli Navy enlisted the support of the U.S. Navy and its research submarine *NR-1* to assist them in searching for the *Dakar* off the Sinai Region of Egypt. In 1997, prior to the Navy's search effort for the *Dakar*, *NR-1* had participated in a search effort for our Skerki Bank program that resulted in the location of the Roman era shipwrecks previously mentioned. As a result, the crew of the *NR-1* was well aware of our interests in ancient shipwrecks and knew from those efforts what ancient shipwrecks looked like on their submarine's highly sophisticated search sonars.

During their 1997 search effort off the Sinai, the crew of the *NR-1* located three shipwrecks that they then visually inspected and documented with a video camera. Although the tapes were of a poor quality, it was clear that two of the shipwrecks consisted of large piles of amphoras.

When the submarine returned to its home base in Groton, Connecticut, the author was invited to review the tapes collected at the two sites. After reviewing the tapes, the author invited Lawrence Stager of Harvard to also review the tapes. He concluded that the ships could be important, possibly from the Iron Age. This conclusion led to our subsequent expedition in 1999 (Ballard et al. 2002), which documented both of these shipwrecks, proving them to be of Phoenician origin and, in fact, dating from the Iron Age. Both shipwrecks were laden with wine and were located on a possible deep-water trade route connecting ancient Phoenicia with either Egypt or Carthage.

Unlike the Skerki Bank shipwrecks, the Phoenician shipwrecks off the Sinai came to rest at 400 m, where a strong bottom current exists. Although both shipwrecks penetrated into the soft bottom mud to their main deck like the Roman era shipwrecks had done on Skerki Bank, the bottom current subsequently scoured away the surrounding sediments, creating separate pits in which the ships now rest. As this scouring process took place, deeper wooden members of the ship's hull were exposed and were immediately removed by wood-boring organisms. The result is two separate elongated piles of amphora in the shape of the ship's hull (figures 7.5 and 7.6). The mapping and recovery operation carried out in 1999 revealed that the vast majority of the artifacts were unbroken, further attesting to the relatively gentle nature of the ship's sinking and its subsequent impact with the bottom.

After completing our investigation of these two shipwrecks, we attempted to locate additional shipwrecks possibly lost along this apparent deep-water trade route. The side-scan sonar image of our recent discoveries was an elongated pit

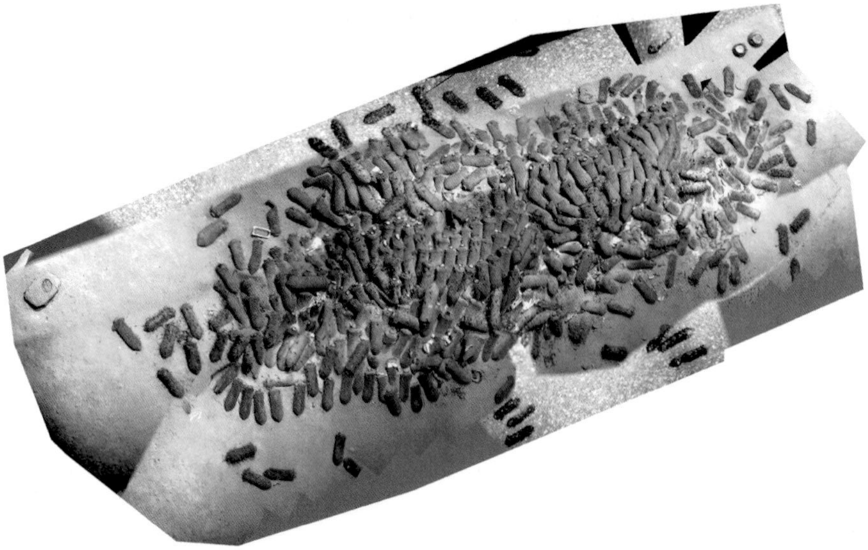

Figure 7.5. *Tanit* (Shipwreck A), circa 750 BCE, was first discovered by the U.S. Navy nuclear research submarine *NR-1* during a search effort for lost Israeli submarine *Dakar*. (Tanit, protector of Phoenician seafarers, was the Iron Age successor of the leading Canaanite goddesses Astarte and "Ashera of the Sea." (Photomosaic created by Hanumat Singh, WHOI)

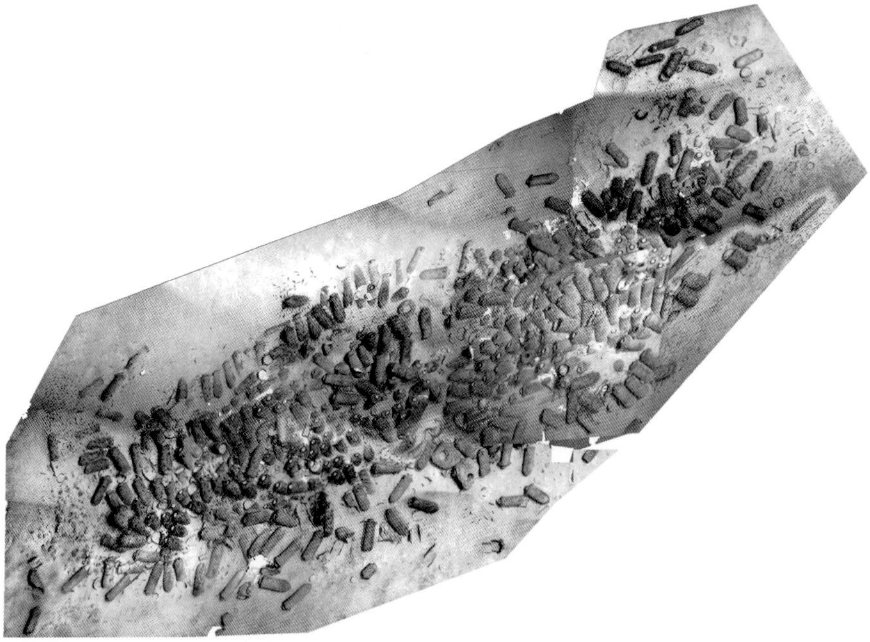

Figure 7.6. *Elissa* (Shipwreck B), circa 750 BCE, was also discovered by the *NR-1* off the coast of the Sinai. (Elissa, Tyrian princess and sister of the Tyrian king Pum'yaton, better known as Pygmalion, fled from mainland Phoenicia to Cyprus and picked up a crew of Phoenicians. Together they set sail for the western Mediterranean. According to legend, she then went on and found Carthage.) (Photomosaic created by Hanumat Singh, WHOI)

with the central pile of amphora having a strong sonar return. Using this sonar signature as a guide, we then traveled east along the east-west trade route and encountered numerous other targets of similar shape, size, and acoustic reflectivity. We then inspected several of these sonar targets with our remotely operated vehicle and discovered a series of gas seeps (Coleman and Ballard 2001) but no new shipwrecks were found. In fact, there were so many gas seeps in this area, we concluded that a search for additional shipwrecks would prove difficult and the search was terminated.

Additional Shipwreck Sites in the Deep Sea

Prior to and since the Navy contacted us about their discovery of the two ancient shipwrecks described in the previous section, we have been contacted by a number of commercial organizations regarding their discoveries of other ancient shipwrecks in the deep sea. In fact, the *NR-1*, on an earlier expedition years before their 1997 search off the Sinai, had come upon an ancient shipwreck in the Straits of Sicily near the coast of Tunisia, but poor visibility hampered our ability to investigate that site in 1988 (Ballard et al. 2000a).

We were also told by a deep-sea search and salvage company that while they were searching for a downed commercial airliner in the Tyrrhenian Sea north of Skerki Bank they had encountered an ancient shipwreck characterized by a large pile of amphora in approximately 3500 m water depth. Unfortunately, they would not reveal its exact location so we were unable to visit that site on subsequent expeditions in that region.

After the U.S. Navy search off the Sinai, the Israeli Navy enlisted the support of another U.S. commercial search and salvage company to continue searching for the *Dakar* in another probable sinking site. That effort resulted in the discovery of additional ancient shipwrecks in 1999 south of Cyprus in more than 3000 m of water. One of them was about 18 m long and consisted of 2000 to 3000 amphora dating to around the end of the 3rd century BCE. Four additional targets were seen using side-scan sonar that appeared to be ancient shipwrecks, one of which was also videotaped by an ROV.

A European company conducting a pipeline survey in the Black Sea has told us of their encounter with ancient shipwrecks but would not give us the precise information that would help guide us to those sites. Odyssey Marine Exploration, while searching for a treasure ship off Spain, located a 5th-century BCE Phoenician shipwreck in more than 750 m of water.

It is clear from all of these reports that numerous shipwrecks were lost in the deep water by ancient mariners who traveled trade routes that carried them far from shore.

Black Sea

The primary reason for conducting a long-term search for ancient shipwrecks in the Black Sea was our hope of discovering ships that had come to rest in the deep anoxic bottom waters of this unique body of water. Since 1976 when marine

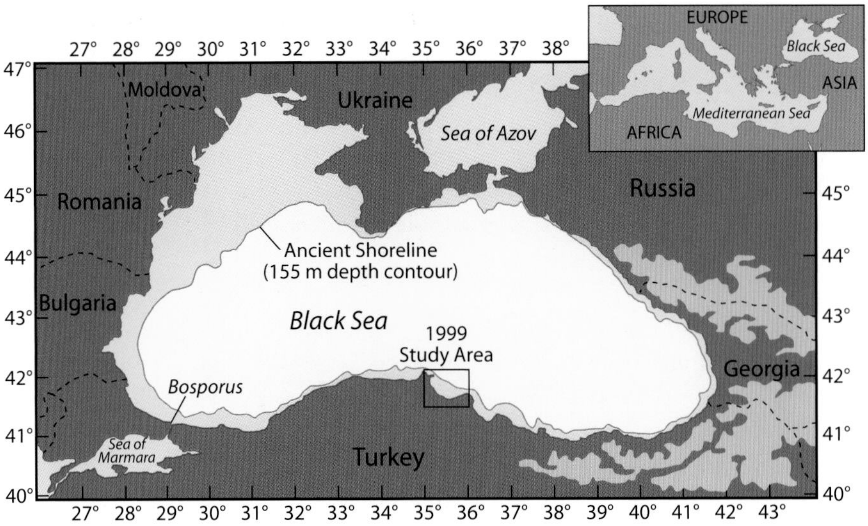

Figure 7.7. Map of Black Sea showing the initial 1999 study area as well as the ancient shoreline depicted by the 155-m depth contour. (Copyright Institute for Exploration)

scientist Willard Bascom published his book *Deep Water Ancient Ships*, the archaeological community has discussed the potential for finding highly preserved wooden shipwrecks in the deep waters of Black Sea. But it wasn't until 1996, after the Cold War had ended, that such a program could get underway.

The Black Sea is a closed deep-water basin with only a shallow outlet at the Bosporus (see figure 7.7). During the Holocene when sea level was lower, the Black Sea was filled with freshwater. But approximately 7500 years ago, as global sea level was rising due to melting polar ice, Mediterranean seawater broke through the Bosporus filling the basin with salt water as the lighter freshwater was quickly displaced. Although salt water continues to flow into the Black Sea as brackish water flows out, this two-way circulation pattern affects only the upper 200 m, isolating the deeper more saline water layers. As a result, the deeper water quickly became stagnant, producing the anoxic condition in which wood-boring mollusks cannot live.

The initial focus of our 1996 program was the north central Turkish port of Sinop. It was hoped that a land-based archaeological survey of this port and the hinterland around it would establish Sinop as an ancient maritime trading center with suspected trade routes branching out to the north toward the Crimea, to the west toward the Bosporus, and to the east. Led by Fred Hiebert of the University of Pennsylvania, this land-based survey showed that agricultural sites appeared around Sinop as early as the sixth and fifth millennium BCE. Evidence of Early Bronze Age coastal settlements of later classical ports and colonies also exist in this area of the Black Sea. Active trading occurred around the entire basin in the third and second millennium BCE and by 800–700 BCE the Black Sea and Sinop had become a major crossroads of the ancient world. With the advent of Greek colonization in the area, trade proliferated across the Black Sea from Sinop to trading partners in the Crimea (Ballard et al. 2001).

Given this well-documented history, undersea surveys began in 1998 and 1999 when a series of side-scan sonar surveys were conducted in the harbor and the shallow, oxygenated waters around the Sinop peninsula to water depths less than 200 m. Unfortunately, the area has been intensely fished by bottom trawlers and no significant sonar targets were discovered. In 2000, a major expedition was conducted off Sinop using a variety of undersea vehicles capable of working in the deepest waters of the Black Sea (Ballard et al. 2001; Ward and Ballard 2004; Coleman and Ballard 2001). These vehicles included the DSL-120 side-scan sonar; the Argus towed imaging sled, and the Little Hercules remotely operated vehicle (ROV).

All three trade routes were explored: the northern trade route from Sinop to the Crimea, the western trade route from Sinop to the Bosporus, and the eastern trade route from Sinop to the ancient city of Trapezus, which is now Trabzon. The initial search effort was carried out on the western route. A series of east–west lines were surveyed in water depths ranging from 95 to 600 m. Four shipwrecks were located in this area, all of Byzantine age. It was surprising to find three of the shipwrecks in relatively shallow water, less than 100 m, resting on a flat sediment surface, since previous experience in the Mediterranean Sea revealed extensive bottom-trawling activity, which would obliderate sites.

The reason for their presence was the complex water chemistry in the Black Sea. The boundary between the upper oxygenated layer and the deeper anoxic layer was thought to occur at approximately 180 m. But when we began discovering shipwrecks with preserved wood above 180 m we hypothesized that the internal dynamics of the basin could influence shipwreck preservation. The three shipwrecks above 180 m (i.e., wrecks A, B, and C) were all in approximately 100 m of water, yet all three appeared to have wooden portions of the hull still present despite their Byzantine age. At an even shallower site of 91 m (site 82), which was thought to have been a subaerial site several thousands years in age, wooden objects were recovered that were 210 to 250 years in age, yet exhibited no wood-boring activity (Ballard et al. 2001). Further examination of the benthic community at these sites suggested that internal waves could periodically introduce deeper anoxic waters into shallower depths, creating an intermediate "co-existence" or "mixed" layer between 85 and 180 m that could prohibit wood-boring animals from living there. This would help to explain the presence of ancient wood at all of these sites. Wreck D, at a depth of 320 m, is clearly within the anoxic layer and for that reason its exposed wooden elements are perfectly preserved (figures 7.8 and 7.9).

Another major different between wrecks A–C and the anoxic wreck D is the nature of the bottom on which the wrecks came to rest. Since wrecks A–C rest on a surface that was previously exposed subaerially and only recently flooded, approximately 7500 years ago, the surface remains hard. Attempts to collect a core at site 82, for example, were unsuccessful due to the hard, compacted nature of the sediment. In addition, rock outcrops surround all of the shallower wreck sites. Wreck D, on the other hand, came to rest on the floor of an ancient lake bottom covered with soft mud.

As a result of the differences in the bottom sediment composition, the shallower wrecks did not penetrate into the bottom while the deeper wreck did. Even though none of the ships appeared to have experienced intense wood-boring activity, the shallower ships have splayed open, perhaps due to the weakening over time of the fastenings that held the ships together, while the deeper wreck penetrated deeper into the bottom and is being held together by the surrounding mud matrix, which keeps it from splaying open.

We returned to the Black Sea in 2003 (Ballard and Ward 2005) with our newly developed remotely operated vehicle Hercules, which was designed specifically to excavate ancient shipwrecks (figure 7.10).

Hercules was employed at the shallower wreck sites (figure 7.11) to map the sites and collect surface samples. Wreck D, which is beyond the limit of Turkish

Figure 7.8. Mast of Wreck D located in the anoxic layer off the northern coast of Turkey near the ancient maritime center of Sinop. Image taken by ROV Little Hercules with companion imaging vehicle Argus seen in the background. (Copyright Institute for Exploration)

Figure 7.9. Wreck D mast, seen upright in the background, surrounded by exposed wooden ship components. Carbon-14 date confirmed that the wreck is from the Byzantine period. (Copyright Institute for Exploration)

territorial waters, was partially excavated by Hercules, and several artifacts were recovered (figure 7.12), as described in detail in chapter 8.

The occurrence of a mixed layer between 85 and 180 m with the absence of a thriving benthic community of organisms should limit bottom-fishing activities and for that reason limit the damage done by such operations. Future expeditions to the Black Sea will hopefully test this hypothesis.

Figure 7.10. Artistic rendering of Argus and Hercules recovering an artifact from Wreck A in the mixed layer of the Black Sea. R/V *Knorr* is using its dynamic positioning system to hold position while an elevator has been dropped next to the wreck site. Hercules can carry the artifact to the elevator which can carry the artifact to the surface. (Copyright National Geographic Society)

Figure 7.11. Author watching in the shipboard control van as the ROV pilot recovers an artifact from Wreck A. (Copyright David McLain)

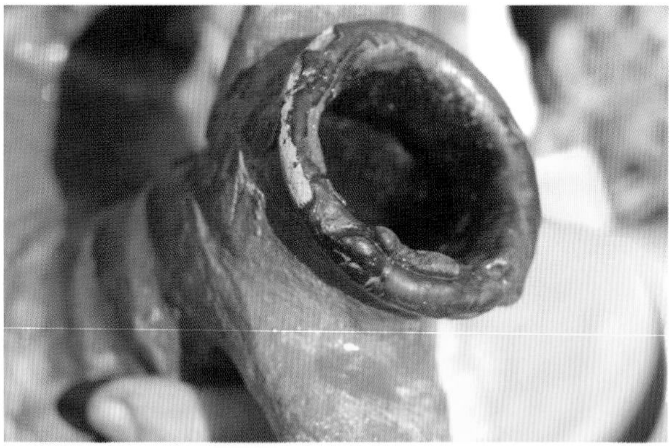

Figure 7.12. Amphora recovered from Wreck D reveals the high state of preservation at the site. Note the pine seal and drip marks.

A final observation based on our exploration of ancient trade routes in the Mediterranean and Black Seas is that the ships we found between Carthage and Rome, between Egypt and Phoenicia, and in the Black Sea date to periods of relative stability. I hypothesize that stable periods in ancient times encouraged long-distance trade.

References

Akal, T., R. D. Ballard, and G. F. Bass (2004).The application of recent advances in underwater detection and survey techniques to underwater archaeology. Conference Proceedings, Bodrum, Turkey, 3–7 May 2004.

Ballard, R. D. (2005). *Deep Water Archaeology, Terra Marique: Studies in Art History and Marine Archaeology in Honor of Anna Marguarite McCann*, 171–6. Oxford, UK: Oxbow Books.

Ballard, R. D. and C. Ward (2005). Searching for deep-water ships in the black Sea. In *Beneath the Seven Seas*, ed. G. F. Bass, 124–6. London: Thames & Hudson.

Ballard, R. D., A. M. McCann, D. Yoerger, L. Whitcomb, D. Mindell, J. Oleson, H. Singh, B. Foley, J. Adams, D. Piechota, and C. Giangrande (2000a). The discovery of ancient history in the deep sea using advanced deep submergence technology. *Deep Sea Research*, Part I, 47:1591–620.

Ballard, R. D., D. Coleman, and G. Rosenberg (2000b). Further evidence of abrupt holocene drowning of Black Sea shelf. *Marine Geology* 170:253–61.

Ballard, R. D., F. T. Hiebert , D. F. Coleman, C. Ward, J. Smith, K. Willis, B. Foley, K. Croff, C. Major, and F. Torre (2001). Deepwater archaeology of the Black Sea: the 2000 season at Sinop, Turkey. *American Journal of Archaeology* 105:607–23.

Ballard, R. D., L. E. Stager, D. Master, D. Yoerger, D. Mindell, L. Whitcomb, H. Singh, and D. Piechota (2002). Iron Age shipwrecks in deep water off Ashkelon, Israel. *American Journal of Archaeology* 106:151–68.

Bascom, W. (1976). *Deep Water, Ancient Ships*. Garden City, NJ: Doubleday.

Coleman, D., and R. D. Ballard (2001). A highly concentrated region of cold hydrocarbon seeps in the southeastern Mediterranean Sea, *Geo-Marine Letters* 21:162–67.

Ward, C., and R. D. Ballard (2004). Deep-water archaeological survey in the Black Sea: 2000 season. *The International Journal of Nautical Archaeology* 33:2–13.

8

The Remote Exploration and Archaeological Survey of Four Byzantine Ships in the Black Sea

Cheryl Ward and Rachel Horlings

A pilot effort to explore shallow- and deep-water environments in the Black Sea as part of a long-term project developed by the Institute for Exploration (IFE) tested survey methodology and equipment for deep-water archaeological applications. The application of traditional and innovative remote-sensing methods supported standard archaeological approaches to site survey in a relatively hostile marine environment and resulted in the discovery of four shipwrecks that date to the 4th to 6th centuries CE, including one of the best preserved seagoing ships from antiquity, a discovery predicted by Willard Bascom (1976). This chapter reviews the maritime survey, describes methodology used to locate four ships in 2000 and data collected from those sites in 2000 and 2003, presents preliminary conclusions about those vessels, and discusses directions and possible implications of future research.

Background

Collaborative efforts under the overall direction of Robert Ballard brought the Institute for Exploration, the University of Pennsylvania, the University of Rhode Island Graduate School of Oceanography, the Massachusetts Institute of Technology, Woods Hole Oceanographic Institution, the Institute of Nautical Archaeology, and Florida State University together for a program of terrestrial and marine survey focused on Sinop, Turkey (see figure 7.7) (Ballard et al. 2001; Ward and Ballard 2004). The Holocene transformation of the glacier-fed Euxine Lake into the Black Sea when it was inundated by salt water from the Mediterranean Sea created a new landscape, including an underwater realm where few organisms could survive. As salt water flowed into the closed basin, it essentially smothered the freshwater below it. Very low rates of internal motion and mixing meant no fresh oxygen reached the deep waters after the influx (Oğuz et al. 1993). At 150 m or deeper, there is insufficient oxygen to support most biological life forms, between 170 and 200 m, a suboxic zone is characterized by low oxygen and low sulfide content, and below 200 m is an anoxic layer with consistently high concentrations of sulfides and low oxygen, which results in conditions that promote preservation of wood and other organic matter (Murray et al. 1989; Codispoti et al. 1991). The speed and intensity of inflow are debated (Aksu et al. 2002; Görür et al. 2001; Ryan et al. 1997; Uchupi and Ross 2000), but evaluation of mollusks shells from a scoop sample collected during the 1999 survey season suggests that the extinction of freshwater mollusks and replacement by saline species took place between 7460 and 6820 BP (uncorrected radiocarbon years), about 7000 years ago (Ballard et al. 2000). In addition to providing convenient harbor facilities today, Sinop played a central role in regional trade from before the time Greek colonies were established in the 8th century BCE (Hiebert 2001:16; Doonan 2004). The project's four brief seasons of maritime survey were based there.

In addition to maritime surveys, a multiyear terrestrial survey led by Fredrik Hiebert, Owen Doonan, and Alex Gantos located hundreds of archaeological sites (Doonan 2004). Terrestrial team members recorded all archaeological sites they encountered but focused particularly on identifying the pattern of settlements in the landscape that existed at the time of the flooding of the lake in order to seek similar landscapes along the now-submerged ancient shoreline. They found a number of small, relatively isolated, Neolithic sites on elevated areas that often overlooked watercourses and, on one of Sinop's highest points, a stratified Bronze Age village with extensive trade connections indicated by ceramic remains (Hiebert et al. 1997). Hiebert (2001) and Doonan (2004) believe that these and other sites from the time of Greek colonization through the medieval period (Kassab Tezgör and Tatlican 1998) indicate that the archaeological remains of people who lived near Sinop in the past show a specialized maritime adaptation to a coastal environment. The maritime survey was designed to seek additional evidence of that adaptation. David Mindell directed underwater surveys of Sinop's anchorage, conducting a side-scan sonar survey in waters less than 60 m deep near Sinop harbor in 1998 and returning in 1999 to examine several dozen anomalies through video images provided by a remotely operated vehicle or ROV (Mindell et al. 1998). Few anomalies proved to be of archaeological origin, but a late 18th-century CE iron anchor, a large storage jar, and the remains of a 19th-century steamship were identified. Work northeast of Sinop at depths up to 150 m focused on a search for the ancient coastline of the Black Sea (Ballard et al. 2000).

In 2000, the team worked 15–30 km west of Sinop, seeking information about the submerged landscape and potential trade routes that might be indicated by the remains of shipwrecks or jettisoned cargo. Historical and archaeological studies on land indicated long-distance exchange dated to at least the mid-5th millennium BCE and that the most intense period of seaborne exchange was between the 2nd and 7th centuries CE in the period of late antiquity (Hiebert et al. 1997; Hiebert 2001; Doonan 2004). In 2000, remote sensing (side-scan) surveys were conducted and targets or anomalies were investigated using ROVs with video and still photographic capabilities.

Four shipwrecks dated to the 4th to 6th centuries CE (Ward and Ballard 2004) and a site originally interpreted as evidence for human habitation in the Neolithic period (Ballard et al. 2001) were located. In 2003 the team returned to four of five sites with an ROV especially designed for deep-water archaeological investigations that require precision documentation and subsurface testing. Project goals set and achieved by the team included demonstrating the ability to conduct standard archaeological survey in deep water and testing the hypothesis that deep-water shipwrecks exist in the Black Sea and are far better preserved than shipwrecks in the upper marine waters. Limitations on data acquisition and processing associated with this new ROV system do not permit us to generate complete detailed descriptions of the sites but do enable particularistic examination of each ship as well as a discussion of the potential contributions of deep-water archaeology to the study of maritime societies, exchange, ancient ships, and seafaring.

The 2000 Season

In 2000, the M/V *Northern Horizon*, a vessel with dynamic positioning capability, served as the research platform (Coleman et al. 2000). After preliminary bathymetric data were examined to determine where ancient waterways or hills may have been located prior to the marine inundation, survey paths were laid out by the team to search for features such as relic stream beds in the submerged landscape and shipwrecks. Acoustic targets acquired by the DSL-120 phased-array, deep-towed side-scan sonar system (Singh et al. 2000) were investigated using the ROV Little Hercules and optical towsled Argus, both developed by IFE (Coleman et al. 2000; Coleman 2002). Argus carried lights and cameras, including a 3-chip video camera, an electronic still camera, and a 35-mm color still camera, and thrusters to control heading. It is controlled from the ship as it locates acoustic targets originally identified by the DSL-120 with a scanning sonar mounted directly on the towsled. To limit the effects of ship motion and cable drag on the ROV, Little Hercules is tethered to Argus essentially decoupling the ROV from the ship. Little Hercules carries cameras capable of providing extremely high-quality images, obstacle-avoidance sonar, sensors for pressure, depth, and compass heading, and thrusters for lateral and vertical movement. Outstanding visual images permitted preliminary examination of sites, but no measured plans or complete photomosaics were produced and measurements provided here were estimated by comparison to objects of known dimension.

At each site, pilots maneuvered Little Hercules at a sufficient elevation to avoid the site while remaining close enough to investigate artifact and feature details. The archaeologist directing the investigation guided pilots and determined which areas and objects to focus on. Argus hovered above and behind Little Hercules, providing light, recording video and still images, and providing a more comprehensive view of the wreck area. No artifacts were recovered in 2000, but sediment and wood samples were collected for analysis and radiocarbon dating at site 82 and site D (Ballard et al. 2001). Project staff examined more than 200 acoustic signatures identified in DSL-120 side-scan sonar tracklines, and ROVs subsequently inspected 52 anomalies considered to be candidates for ancient settlement or other archaeological sites.

Five targets met survey objectives (Ward and Ballard 2004). Site 82, originally interpreted as a preflood habitation site (Ballard et al. 2001), and four ancient shipwreck sites were explored and visually recorded. Sites A, B, and C are at depths of 85 and 95 m, and site D is located in the anoxic layer at 324 m. Sites A, B, and C exhibit the classic mounded deposit of an undisturbed ancient ship carrying a cargo of transport amphora; site D consists of a wooden sailing ship sitting upright on the seabed, buried in sediment to deck level (Ballard et al. 2001; Ward and Ballard 2004).

The 2003 Season

In 2003, the National Science Foundation and other funding permitted an expanded team of oceanographers, archaeologists, engineers, and a conservator to return to the Black Sea on R/V *Knorr* with Argus, and a new ROV named Hercules, built to IFE specifications with tools designed for subsurface testing and survey. Hercules and Argus worked in the same configuration tested by Little Hercules and Argus in 2000, but Hercules' capabilities increased the volume of data acquired and provided enhanced capabilities at depth. After a visit to site 82, now considered to represent a geological rather than an archaeological feature, Hercules and Argus investigated sites B and C, and conducted mapping and sampling work before continuing on to site D. At each site, Hercules conducted sonar and subbottom profiler acoustic surveys and acquired electronic still images, but no precisely measured dimensions or site plans were constructed. The descriptions and analysis that follow draw on video and still images from both seasons and on data from artifacts recovered in 2003.

Sites A, B, and C

Sites A, B, and C, shipwrecks at depths of 85 and 95 m, appear as mounds of exposed transport amphoras. These shipping jars are best known to archaeologists today as containers for bulk shipping and storage of liquids such as wine, garum (fish sauce), and olive oil, although a variety of products were carried (Haldane 1991). When merchants loaded ancient ships with shipping jars (figure 8.1), they stacked jars vertically in interlocking tiers to ensure minimal movement in transit and to allow for maximum stowage in the cargo hold

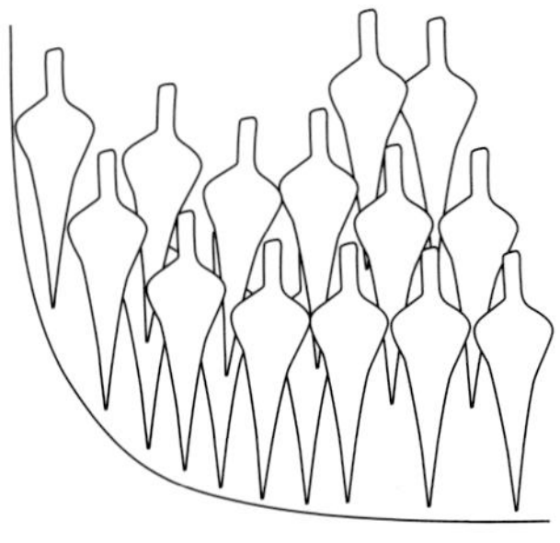

Figure 8.1. A reconstructed lading plan for the late antique shipwrecks based on patterns of dispersal on the seabed as well as excavations of other ships with transport amphora cargos.

Figure 8.2. Anchovies, algae, sediment and shipping jars were common sights at sites A, B, and C. (Copyright Institute for Exploration)

(Casson 1994; Gianfrotta et al. 1997; Grace 1949). Most Mediterranean shipwreck sites with shipping jars exhibit similar lading patterns, but many reflect disruption caused by the collapse of the hull, a process often accelerated by the presence of wood-eating organisms such as *Teredo navalis*, commonly called the shipworm. As the hull disintegrates, jars fall away from the central cargo area and create a sloped mound that traps sediment and often promotes growth of, for example, Poseidon grass. Wood preservation on the surface of most sites is minimal, although portions of the hull buried in near-anaerobic conditions remain. In the Black Sea, similar processes were evident at sites A, B, and C (figure 8.2).

Each site includes timbers and objects identifiable as modern or recent debris deposited since the wrecking of the ships. Such debris is representative of the long cycle of site formation under similar processes operating during the last 15 centuries. Sites A, B, and C resemble Mediterranean shipwreck sites in terms of site formation processes, but sites A and B have considerably more wood protruding from the mound of jars than is usual on Mediterranean shipwreck sites of similar age. Fluctuations in the anoxic water layer usually found below the depth of these sites likely accounts for the presence of the wood today, but the origin of the wood is unknown. We suspect some of the timbers are original and others are more recent deposits, but at present, we lack evidence in the form of scientific dating or technological features to identify any timber visible on any site's surface as belonging to the original ship.

Site descriptions and observations rely on video and still imagery from the 2000 and the 2003 field seasons and on artifacts and samples raised in 2003 (Horlings 2005). Partial photomosaics were generated from electronic still images acquired during site visits, but as yet, no complete mosaic was produced for any site and no detailed plans have been created. Instead, Horlings created preliminary site plans for sites A, B, and C using video footage, still images,

TABLE 8.1.
ESTIMATED DIMENSIONS OF SHIPS A–D AND ESTIMATED SITE AREAS

Shipwreck site	*Length (m)*	*Width (m)*	*Elevation (m)*	*Section area (m^2)*	*Total exposed site area (m^2)*
Site A	18.0	10.0	1.0–1.5		180
Main	13.0–14.5	5.0	1.0–1.5	65.0–73.0	
Small	3.0	3.0	0.5	9.0	
Site B	14.0–16.0	12.0–13.0	2.0		210
Site C (2000)	8.0–8.5	7.5	0.5		64
Section 1	5.0	3.0–3.5	0.5	15.0–18.0	
Section 2	3.0	3.0	0.5	9.0	
Section 3	4.5–5.0	3.0	0.5	14.0–15.0	
Site C (2003)	8.5	7.0	0.5		60
Section 1	4.5	2.0	0.5	9.0	
Section 2	4.0	2.5	0.5	10.0	
Section 3	5.0	3.0–3.5	0.5	15.0–18.0	
Site D	12.0–14.0	3.5–4.0	–		42–56

and partial photomosaics of the wreck sites. We relied on known dimensions of carrot-shaped jars, about 0.88 m long, to estimate approximate dimensions of each site (table 8.1). Relative placement of objects was determined by examining different image sources and angles but does not reflect direct measurements. All dimensions reflect estimates based on video and still images.

Site A

Site A covers approximately 180 m^2 in two areas (figure 8.3). Area 1 is a large, low, oval-shaped mound made up of orange, carrot-shaped shipping jars. The ends of area 1 are arbitrarily labeled end 1 and end 2. Area 1 measures approximately 13 to 14.5 m in length, 5 m width, and 1 to 1.5 m in elevation. Area 2, about 3 × 3 m, also consists of carrot-shaped jars and is approximately 4 m from end 2 of area 1. All visible shipping jars on site A are carrot-shaped, and approximately 0.88 m long. Many jars are nearly upright, occur in clusters (figure 8.4), and seem to reflect the original lading pattern. Shipping jars are not distributed uniformly; several areas on site A have few or no jars visible, and other areas have jars concentrated, stacked in several layers, and densely packed. There are no objects on the surface between areas 1 and 2. Clusters of twigs and branches on the site's surface are almost certainly examples of modern debris. Other objects, such as a plastic sack and what may be a wine bottle, also are modern. Accumulation of modern debris is common on shipwrecks of any age. When only visual data are available, it can be difficult to differentiate between objects of modern and ancient origin.

Site A also incorporates timbers and other objects with a grayish-white substance. Some timbers feature what appears to be notching, shaping, or other purposeful modification, but it cannot be determined from available data whether these timbers were associated with the original ship. One light gray

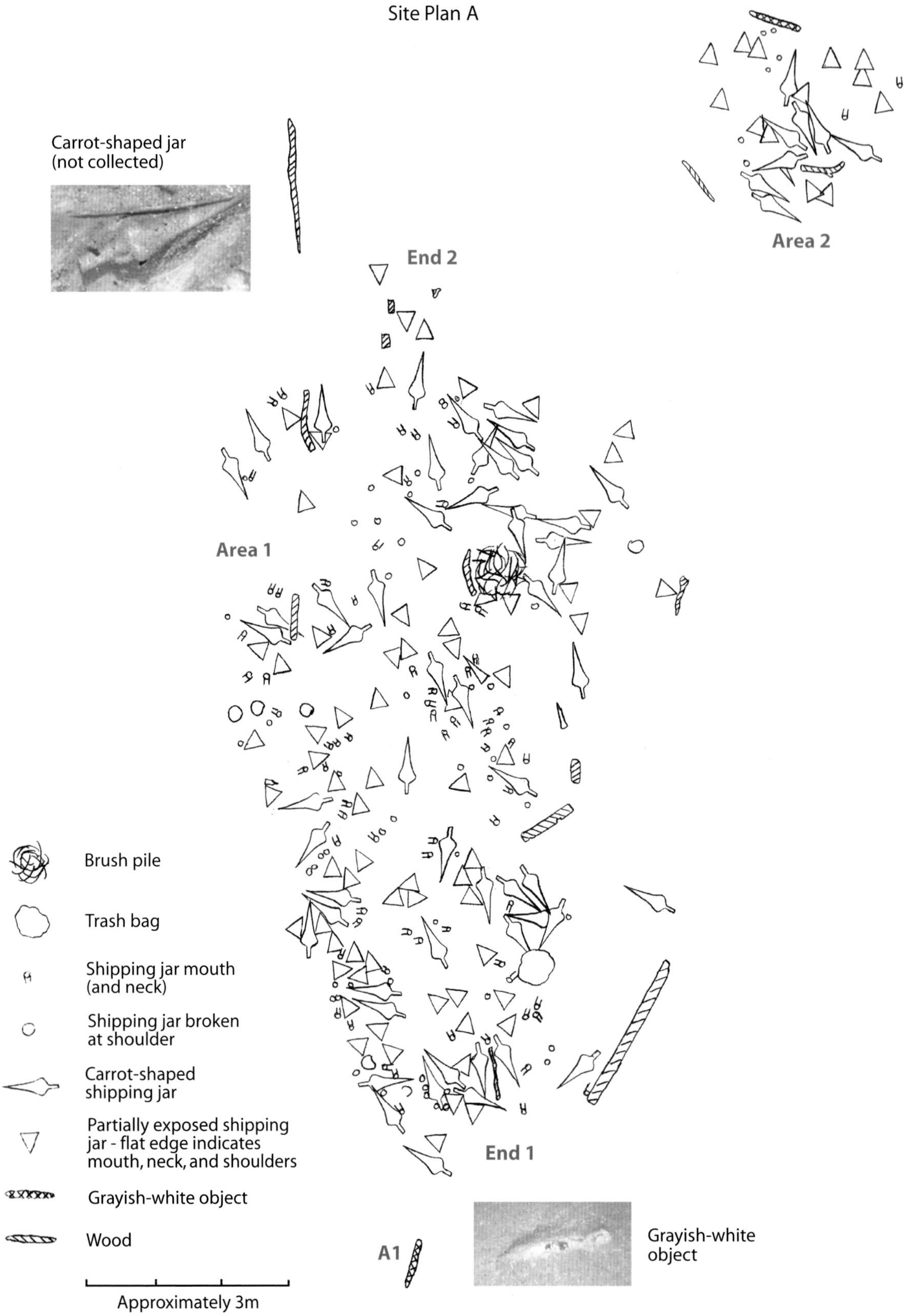

Figure 8.3. Shipwreck site A plan, refer to text for descriptions.

Figure 8.4. Lading patterns remain visible on the surface of site A. (Copyright Institute for Exploration)

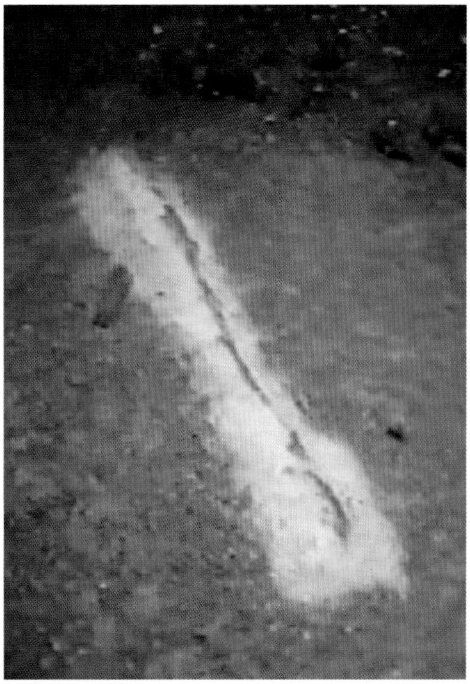

Figure 8.5. Light gray objects at sites A (*left*) and C (*right*) are probably timbers undergoing bacterial decay. (Copyright Institute for Exploration)

object (figure 8.5a) approximately 0.75 m long, is located at end 1, area 1, about 1 m from the shipping jars. The coloration stands out because it is brighter than the sediment around it, but the cause of the color difference is unknown as no samples of this material were obtained. Similar coloration on site C objects photographed in 2000 that are identifiable as timbers in photographs from 2003 suggests the substance may be related to wood decay processes and that these light gray objects are timbers rather than, for example, lead, a gray metal with white corrosion products.

Site B

Site B (figure 8.6) is a 210-m^2, oval mound with a waist measuring about 14–16 m in length with a maximum width of 12 m, rising nearly 2 m above the

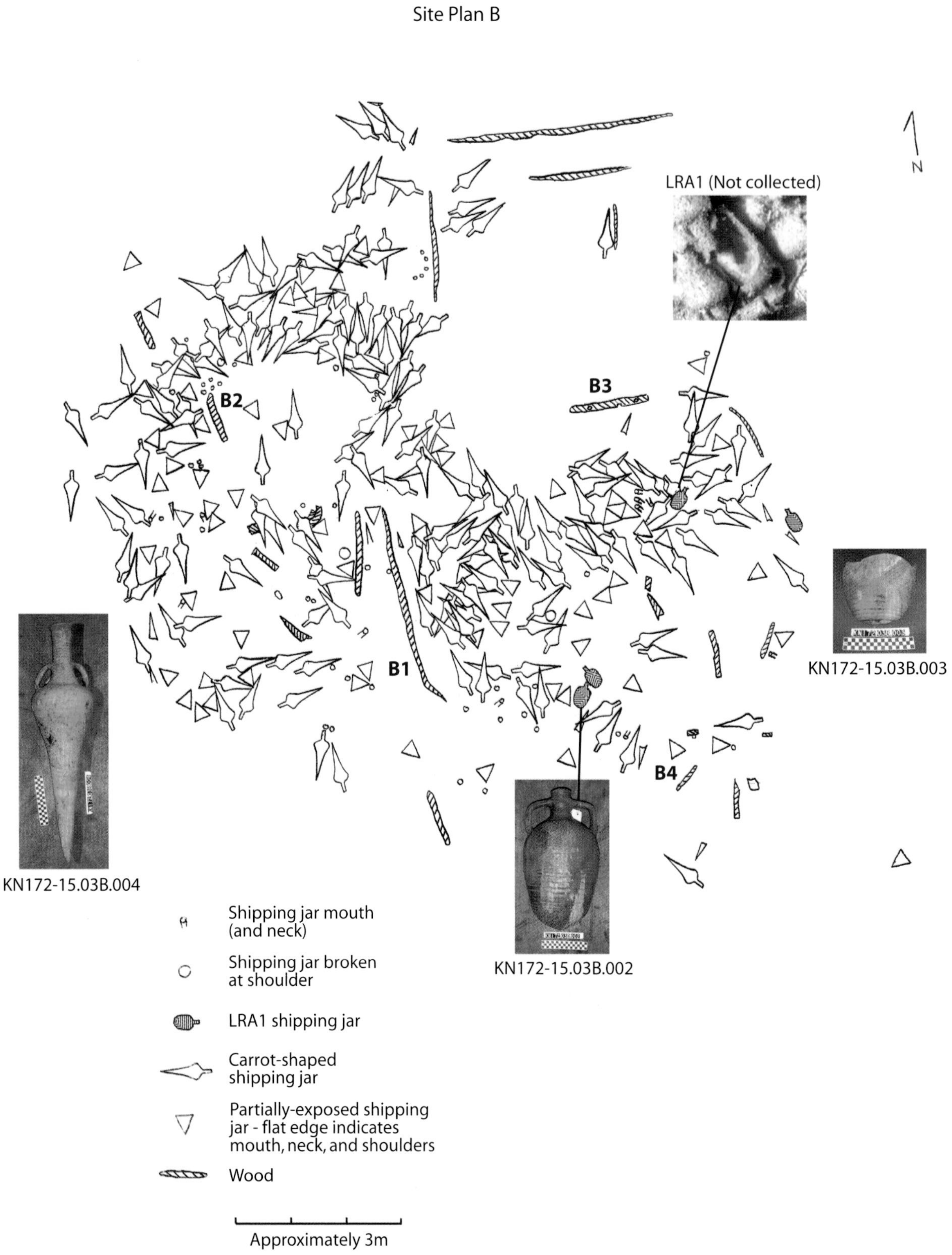

Figure 8.6. Shipwreck site B plan, refer to text for description.

seafloor. Stacked, carrot-shaped jars like those at site A create internal elevation differences of up to 0.5 m. The surface of site B includes many more broken shipping jars than are visible at sites A and C. The elevation of the mound at B is such that it has trapped twigs, branches, modern debris and trash, including what appears to be an oil filter. Almost all shipping jars on the surface are carrot-shaped, with slight variations in dimension and shape. The carrot-shaped jars are distributed unevenly across the site, some in concentrations and some scattered individually between stacks of higher and lower elevations, and few concentrations of partially buried jars are visible. None of the shipping jars on the surface are sealed, and many are broken. Several jars contain an unidentified, compact, white substance that has drawn away from the walls (figure 8.7, upper left). Because all the jars containing this substance are located in the center of the site and consequently out of the reach of Hercules' manipulators, none were recovered for analysis. LRA1 (Late Roman Amphora 1) shipping jars are present on the eastern half of the site (figure 8.7, upper right). Of the five unambiguous examples, four are essentially complete, while the fifth is badly broken. Several mostly buried jars and other large shards in the eastern half of the wreck site may be from LRA1 jars, but are too deeply buried or fragmentary to allow positive identification.

Site B also contains substantial amounts of wood. Shaped timbers with both rectangular (width somewhat greater than thickness) and plank-like (width at least two times thickness) cross sections are present. At least three timbers, labeled B2, B3, and B4 on the plan in figure 8.6, display intentional modification (figure 8.7, lower right). Other timbers and wood fragments on the surface and protruding from mound sediments may have been modified, but it was difficult to distinguish between intentional modification and erosion. In 2000, we recorded timber B1, approximately 3.7 m long with a plank-like cross section, near the center of site B (figure 8.6). Both ends of the timber were narrower than the body, and the southern end rose at a slight angle. We hoped that we could determine whether this was an intentional modification when we returned, but timber B1 had disappeared by then, serving instead as a reminder of the dynamic forces present on the seafloor and of the difficulty in assigning origins to objects based solely on the interpretation of images.

Site C

Site C was visited briefly in 2000 (Ballard et al. 2001; Ward and Ballard 2004) and in 2003. This site is significant because it contributes to understanding the stability of sites in deeper water and demonstrates the capabilities of side-scan sonar in locating archaeological sites. Three clusters of mostly buried, upright shipping jars with a few scattered jars on the surface raise the site only slightly above the seafloor (less than 0.5 m). The only type of shipping jar visible on the site is the carrot-shaped jar, and, as on sites A and B, slight variations between individual examples can be identified. Very few broken jars were visible on the site in either 2000 or 2003. Between September 2000 and August 2003, the site's appearance changed (figure 8.8). For example, in 2000, 108 jars were visible, but only 89 were exposed in 2003. In 2003, only 60 m^2 of artifacts were visible, while approximately 64 m^2 were exposed in 2000.

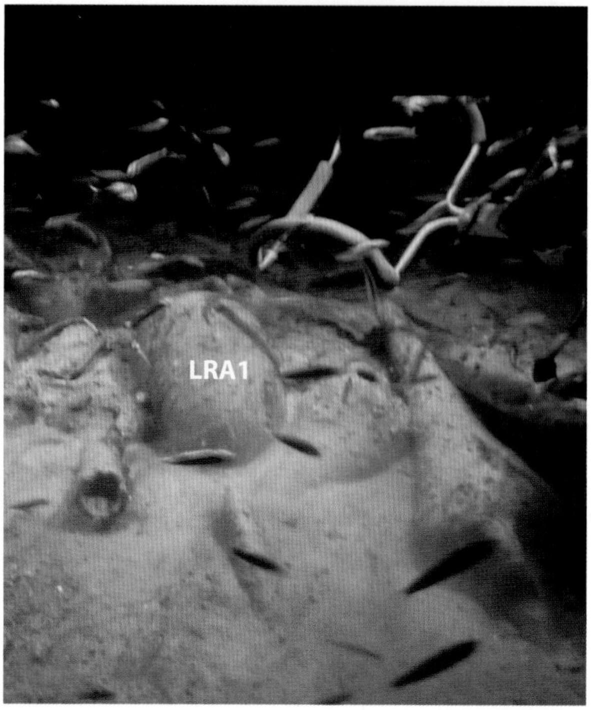

Figure 8.7. (*Upper left*) Some broken jars at site B contained a compact, white substance that pulled away from the edges of individual jars as it solidified. (*Upper right*) Site B included both carrot-shaped and LRA1 shipping jars and (*lower right*) timbers with notches and other modifications. (Copyright Institute for Exploration)

As at sites A and B, the slight elevation of the jars above the surface trapped branches, twigs, timbers, and other modern debris. Video images from 2000 indicate that at least one mostly buried timber that may be intentionally modified was associated with shipping jars at site C, but the timber was not exposed in 2003 so no better images are available for studying it further. Six light gray objects visible in the 2000 footage and still images (figure 8.5) range from 0.3 to 1.9 m. The light gray coloring appears to be the result of disintegration of the exterior. No gray objects are visible in 2003 images, though two timbers on the wreck site (C1 and C3) are the same sizes and in relatively the same positions as were two of the gray objects in 2000 (C4 and C2). The correlation between the locations of gray objects (2000) and timbers (2003) supports the hypothesis that the gray objects are in fact wooden objects undergoing a chemical or biological reaction to the ambient environment. Too few details are visible on most of the gray objects to allow for identification of original cross sections and shapes, but those that are identifiable appear to be unmodified logs with severely degraded exteriors, suggesting that the same identification is likely for similar objects on other sites, especially site A.

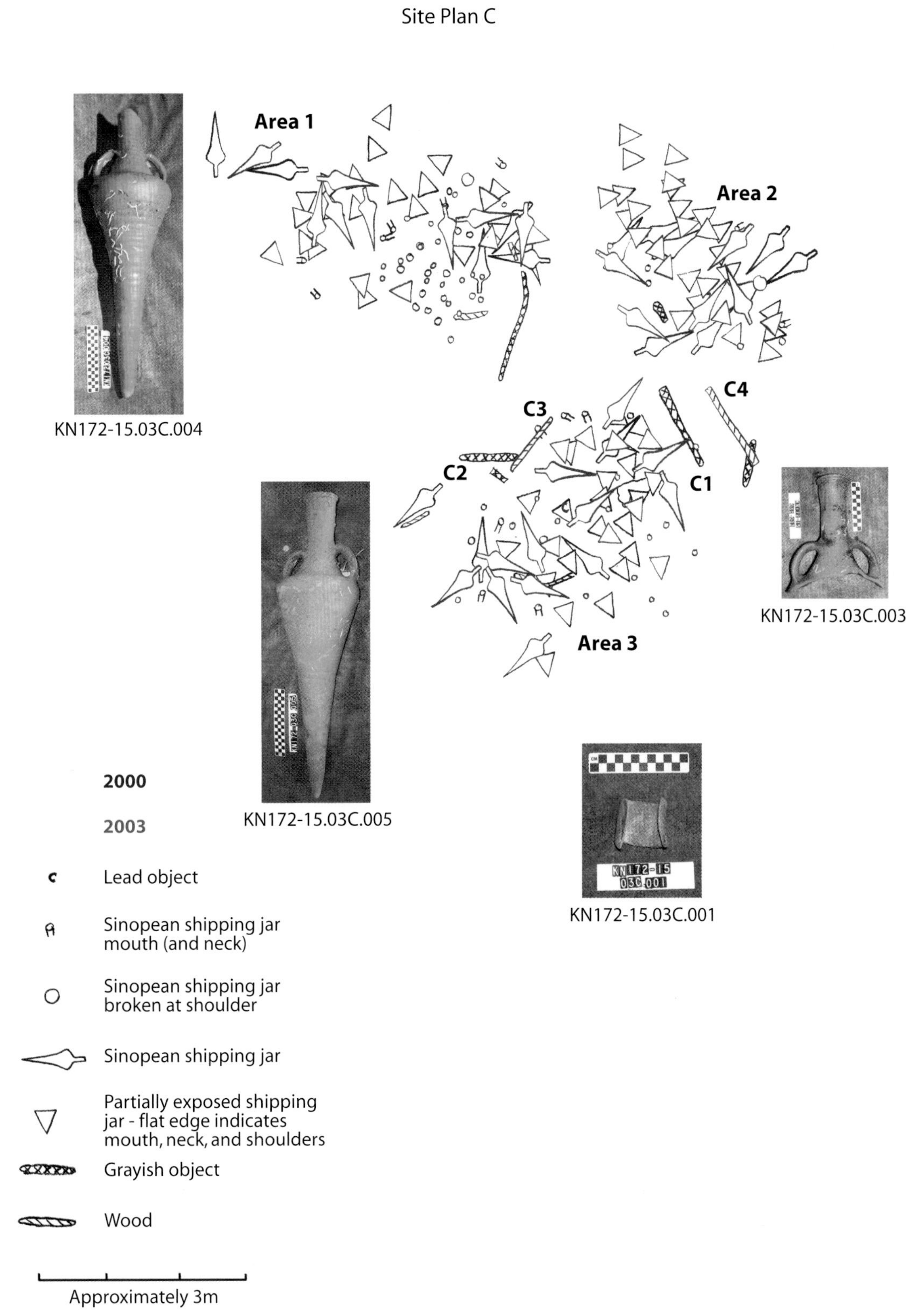

Figure 8.8. Shipwreck site C plan, refer to text for details.

Site D

Site D appeared as a long, slender, and upright feature in acoustic survey data acquired by the DSL-120 during the 2000 season at a depth of 324 m. Under the lights of Argus, Little Hercules approached the site as scientists and pilots in the control van saw, for the first time in about 1500 years, a ship's wooden mast standing about 11 m above the seabed (figure 8.9). The ship is buried to its deck in sediment (figure 8.10) and the anoxic environment preserved elements rarely found on shallower shipwreck sites. A fir (*Abies* sp.) wood sample from the rudder support or bollard produced a radiocarbon date of 1610 ± 40 BP (Beta-147532) calibrated to 410–520 CE (Ballard et al. 2001), and the recovery in 2003 of three shipping jars of a type produced locally in Sinop during the 5th and 6th centuries confirms the ship's antiquity. Ship D may be the best preserved ancient shipwreck yet discovered. Visual survey in 2000 provided a record of the site's appearance used to generate preliminary descriptions of its features and preservation (Ward and Ballard 2004; Ballard et al. 2001). Preliminary analysis of data from 2000 determined the orientation of the hull

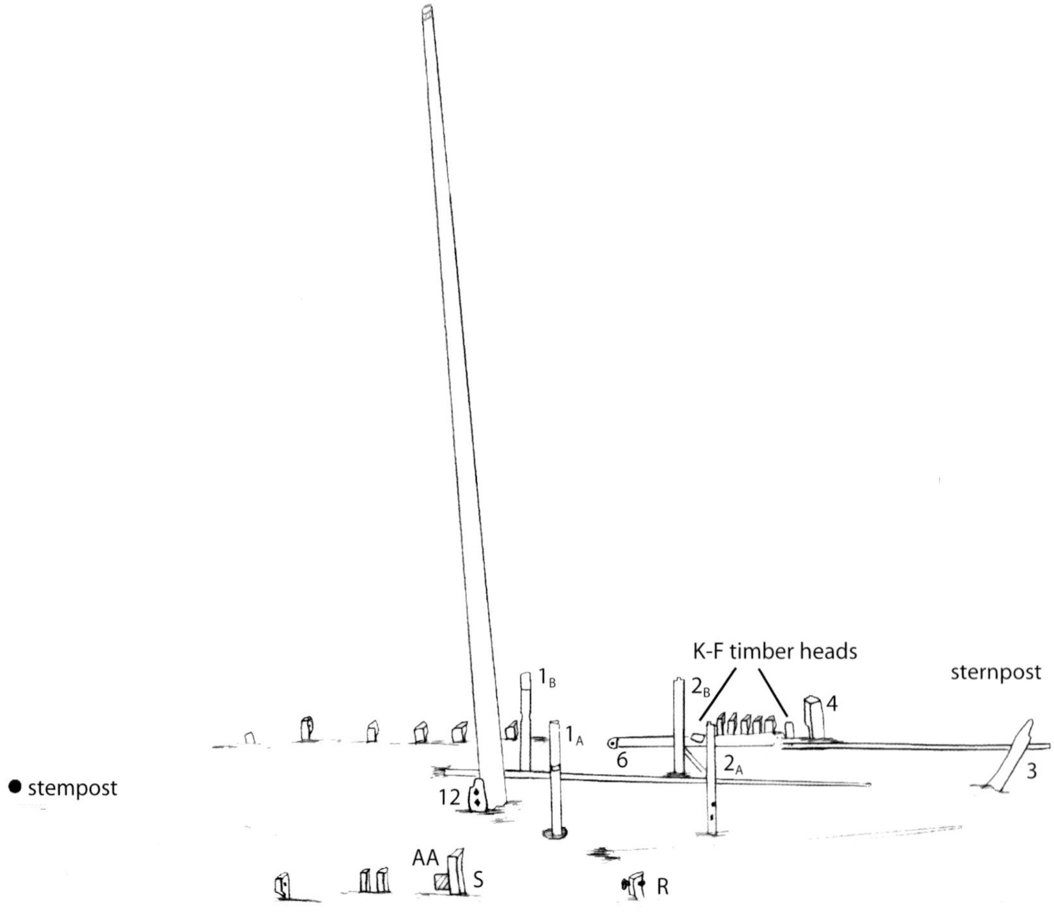

Figure 8.9. Sketch plan of ship at site D.

and identified the vessel's sternpost, a starboard rudder support or bollard, 18 top timbers with holes for pins and one pin, spars, a beam at midships, the mast and its partner, two pairs of stanchions, and a handful of treenails in an area 12–14 m long and about 4 m wide (Ward and Ballard 2004).

In 2003, survey goals included using Hercules and Argus to acquire sufficient data to generate an accurate site plan, test subbottom profiling and acoustic survey methodology, conduct subsurface testing to locate and examine the ends of the vessel, recover diagnostic artifacts and samples to assist in identifying cargo and hull components, and investigate ship structure. Because time on site was limited, not all goals were achieved, but we did gain substantial new information about the ship at site D. Two fixed laser beams 10 cm apart were part of the Hercules tool set and provided rough estimates of dimensions at the site.

We arrived at site D mid-afternoon on 3 August and departed at sunrise 6 August 2003. The site looked much as it did in 2000, with the addition of a 12-oz, white and red beverage can. Navigation points were established and provided rough estimates of distances. Before approaching the ship closely, Hercules pilots settled the ROV on the seabed and deployed a suction dredge powered by a hydraulic pump. Two separate suction/jetting nozzles permitted a work routine to be established. The more dextrous Kraft Predator arm grasped one nozzle and began to pull it away from the ROV body while suction generated by the other nozzle moved sediment in the water column out of the work area, through and behind Hercules, permitting a constant view of the site while the ROV was stationary. Positioning the vehicle so sea currents suctioned sediment away from the site was the primary consideration in

Figure 8.10. A group of six top timbers (F–K) includes one that seems to be out of alignment, perhaps because the heavy spar fell on it. (Copyright Institute for Exploration)

approaching the ship's starboard side at midships, since each time the vehicle moved, clouds of sediment rose into the water column and required about 30 min to clear before work could again begin.

After exposing some hull planking between two top timbers there, we moved the ROV to the stern where it cleaned the sternpost and removed another 0.80 m of sediment around its base. Although we exposed a total length of about 1.5 m in an area about 0.50 m in diameter, no other timbers were visible. The starboard edge of the sternpost's inner face is broken off, and a large crack is visible lower on the post (figure 8.11, top). The scarf and tenon in the upper scarf table of the exposed sternpost and holes for metal fasteners (figure 8.11, bottom) suggest other timbers originally were attached to it. Time constraints prevented the examination of other components (rudder support, top timbers, spars, stanchions) in the stern, and we instead moved to the port side slightly forward of midships.

In 2000, the only artifact other than the ship itself was photographed here, a ceramic jar with one handle visible just forward of a large beam that is immediately before the mast. The lack of hull planking on outboard surfaces of the beam and a nearby top timber was puzzling in 2000 and remained so in 2003 as no planking was encountered even after excavating to a depth of about 1 m below the beam's lower surface. The beam, once swept clean, proved to be about 0.25 × 0.15 m and was covered with adze marks, emphasized by the swelling of waterlogged tissues, and more visible today than when the ship sailed. The top timber immediately aft of the beam was coated with pitch. Below the beam, Hercules excavated the first 0.5 m of sediment with its alternating dark and light gray layers of soft sediments, and below that, more compacted uniformly lighter gray sediment that incorporated leaves and twigs, exposing pale shipping jars with dark rims of shiny pine pitch.[1] We exposed seven small transport amphoras in an area approximately 2 m long, 1.5 m wide, and 1 m deep. They lay as the ship's crew had arranged them, on their sides and aligned parallel with the hull's long axis. None of the jars were sealed when found, and no trace of stoppers remained in the three jars we raised to the surface. No hull components were exposed in the subsurface test of the cargo hold, and no deck components were identified.

Exploration of the bow and long sediment ridges that were interpreted as possible spars from the 2000 images followed. We set Hercules down in alignment with the mast and sternpost at a point just beyond the proposed limits of the bow and began excavating one of the more prominent ridges. It proved to be sediment rather than a spar, and no spars or deck components were encountered in this area. Nearby, at a depth of about 1 m, we reached the stempost, about 0.12 × 0.30 m, and visually recorded its scarf and details such as a tenon in the table and a finely bevelled edge on its inner face (figure 8.12). Like the sternpost, the stempost now lacks the timber that once extended its curvature above deck level. No other timbers or fastenings were visible.

In the few hours that remained, we returned to the starboard midships area to direct Hercules in the exposure of the only hull planking identified (figure 8.13). We excavated inside and outside the planking and identified a mortise-and-tenon fastening in its upper edge. As in other test pits, it was clear that significant parts of the hull were missing. Whatever had been attached above

Figure 8.11. The sternpost (*top*) was excavated to a depth of about 1.2 m; (*bottom*) black stains surround holes that probably held iron nails. (Copyright Institute for Exploration)

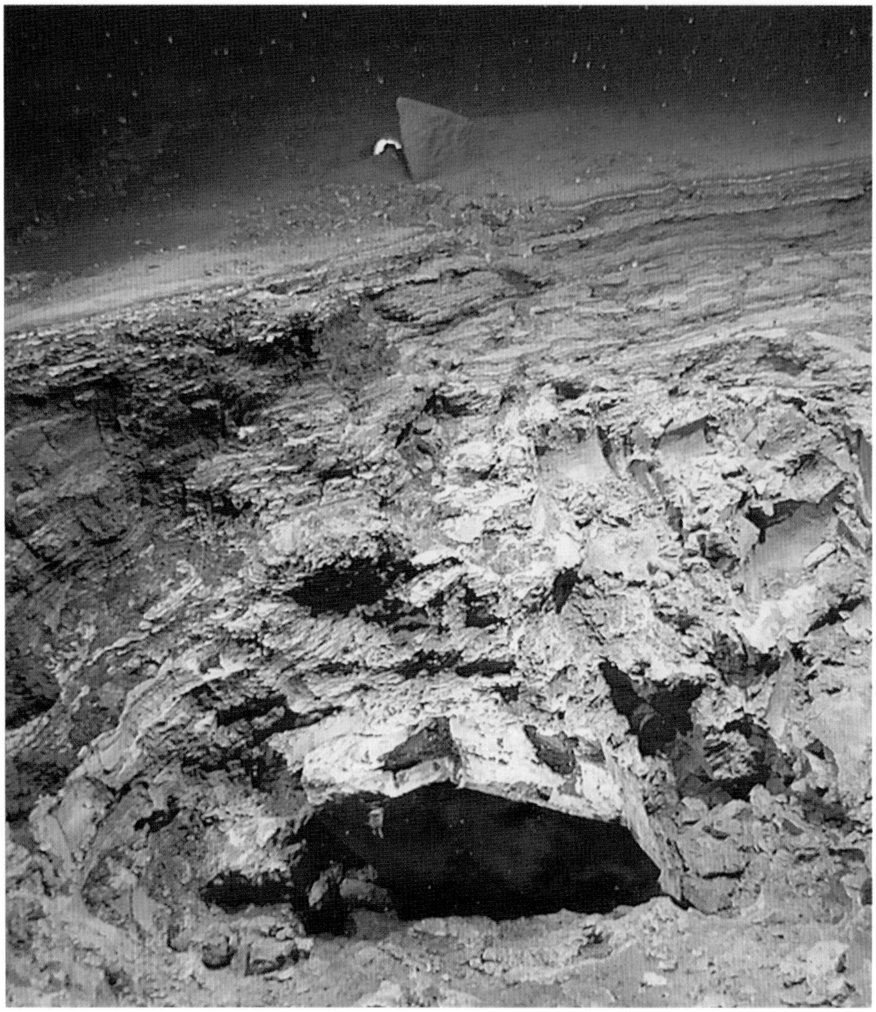

Figure 8.12. The stempost was buried deeply in the sediments. (Copyrigth Institute for Exploration)

Figure 8.13. Our final test area on the ship confirmed that it was at least partially built using mortise-and-tenon joinery similar to Mediterranean ships. Two top timbers, splayed in opposite directions, are about 1.5 m apart. (Copyright Institute for Exploration)

this line of planking, and even to this particular plank's after end, was absent. An oak plank[2] fragment recovered by Hercules is coated with pine pitch more degraded than that remaining in the shipping jars. It incorporates part of a mortise-and-tenon fastening.

Artifact Assemblage

All recovered artifacts are curated by the Sinop Museum. No artifacts were collected from site A in 2000, and the team did not return to site A in the summer of 2003. Sampling at site B included artifact recovery and a scoop sample of sediments from the southern edge of the site. Collected artifacts (figure 8.6) include part of a broken, pitch-lined LRA1 shipping jar (KN172-15.03B.003), a complete carrot-shaped shipping jar (KN172-15.03B.004), and a complete LRA1 jar (KN172-15.03B.002). The jar interiors were lined with pitch; preliminary palynological analysis identified high concentrations of pine pollen in pitch samples. None of these shipping jars bear any sort of stamp or maker's mark. Sediment sample KN172-15.03B.005 included *Abies* sp. (fir) and *Pinus* sp. (pine) wood fragments and hardwood twigs.

Artifacts recovered from site C (figure 8.8) included three carrot-shaped shipping jars, one complete (KN172-15.03C.005), one with a broken mouth and toe (KN172-15.03C.004), and the upper portion of a third jar (KN172-15.03C.003). All were coated with pine pitch on the interior. Sediment sample 03C.002 contained *Populus* sp. (poplar) and *Salix* sp. (willow) wood fragments. An unidentified lead object (KN172-15.03C.001) is approximately 0.5 cm thick and 5.5 cm wide. Formed into an incomplete oval 6 cm long, the object appears to have been made of two layers of lead hammered together. Some scratches are visible on its surface, and impressions in the lead appear to have been made by metal tools, though no diagnostic features are visible.

All of the shipping jars at sites A and C, and most of those at site B, are of a type commonly known as "carrot-shaped" (Kassab Tezgör 1999; Kassab Tezgör and Dereli 2001). In antiquity, a pottery near Sinop manufactured shipping jars of several types, including a carrot-shaped form in reddish-orange fabric with quartz and pyroxene inclusions (Garlan 1998; Kassab Tezgör et al. 1998; Kassab Tezgör and Tatlican 1998). Similar jars may have been made elsewhere, but it is likely that the jars at A, B, and C were made near Sinop as each of the recovered examples has diagnostic pyroxene inclusions. According to recent studies of Black Sea transport amphoras, carrot-shaped Sinopean shipping jars were most popular in the 4th and early 5th centuries CE (Kassab Tezgör 1996; Kassab Tezgör and Tatlican 1998). Although the shipping jars recovered in 2003 are similar to those from Demirci, their proportions are not identical. Kassab Tezgör and Dereli (2001) illustrate seven Demirci types (A–G), of which the closest parallels are types C and D, both with diameters of 0.28 m and respective lengths of 0.80 and 0.94 m. The jars in our sample are approximately 0.88 m long and have a maximum diameter of 0.25 m. Capacity, measured by filling a wet jar to the base of the neck, ranged from 5.7 to 6.3 liters, and the jars weigh between 5.4 and 6.24 kg.

Site B includes a few Late Roman Amphora 1 shipping jars, similar to those found on the Yassı ada 7th-century ship in Turkey (Bass and van Doorninck 1982; van Alfen 1996) and elsewhere in the eastern Mediterranean and beyond. There are many variations of the LRA1 (van Alfen 1996) and its origins, though uncertain, likely include southwestern Asia Minor and the Antioch region of Northern Syria (Peacock and Williams 1986). Production of LRA1 jars dates from the 4th to the 7th century CE (van Alfen 1996). Although only five LRA1 jars are visible on the surface at site B, at least two forms are present. KN172-15.03B.002, recovered from midships, is slightly wider at the shoulder than at the base, and of a type dated by Peacock and Williams (1986) to the later 5th and early 6th centuries. Zemer (1978) dates a similar jar to the 6th to early 7th centuries, a date also assigned to the type by van Doorninck (2002) in his study of the Yassı ada assemblage (c. 625 CE). A second LRA1 variation photographed but not collected at site B is slightly smaller with straighter sides and may represent van Alfen's type VI (1996).

At site D, we collected three shipping jars of the *patê claire,* Demirci kiln type identified by Kassab Tezgör and Touma (2001). Dated to the end of the 4th through 6th centuries CE, they likely were made in or near Sinop. Demirci-type shipping jars from site D have the characteristic yellowish-green clay color with pyroxene inclusions linked to the Dermirci kiln mentioned earlier (figure 8.14, top). Their overall length was inconsistent (0.515, 0.55, and 0.57 m), as was maximum diameter (0.188, 0.218, and 0.205 m). The neck of each recovered jar also happened to be decorated by different means (smooth, horizontal finger ridges, and spiraling ridges). It is typical for this type for the mouth to be poorly finished, as each jar was lined with gleaming pine pitch that extended about a centimeter beyond the rim, forming a smooth edge (figure 8.14, bottom). Kassab Tezgör and Touma (2001:109) note a Demirci-type shipping jar at Dibsi Faraj in north Syria with spiraling ridges on the neck, and several other examples are known from that site and from Ras Ibn Hani, near Ugarit. In 1997, the remains of a small boat or dispersed shipwreck in shallow water at Karakum on Böz Tepe yielded 12 similar jars (Kassab Tezgör et al. 1998) but little else.[3]

In addition to transport amphoras and the lead object, we collected a number of organic samples. A scoop sample from midships at site B produced large numbers of insects and insect frass, fig seeds, and a few weed seeds. In addition to the sample of fir from the rudder support, a second wood fragment, identified as *Quercus* sp. (white oak group), was acquired in 2000 but its original location on the ship is not known. Identification of a 2003 plank sample as oak, pine, fir, and poplar or willow twigs and wood fragments caught in the suction device provides a range of woods local to the Black Sea and harvested there from ancient times. Oak leaves encountered in the sediment around shipping jars also were trapped in the suction device. The leaves may or may not be part of the ship's original contents; a plastic bottle and aluminum can also are present on the surface of site D, reminding us of the continual processes of deposition and movement. Preliminary palynological analysis of sediment samples from the jars reflects the forested environment and, for those jars lined with pitch, indicates a pine origin but does not provide clues to jar contents. Macrobotanical analysis of jar contents likewise is inconclusive, but scheduled evaluation of sediment samples for tannins and lipids may be more informative.

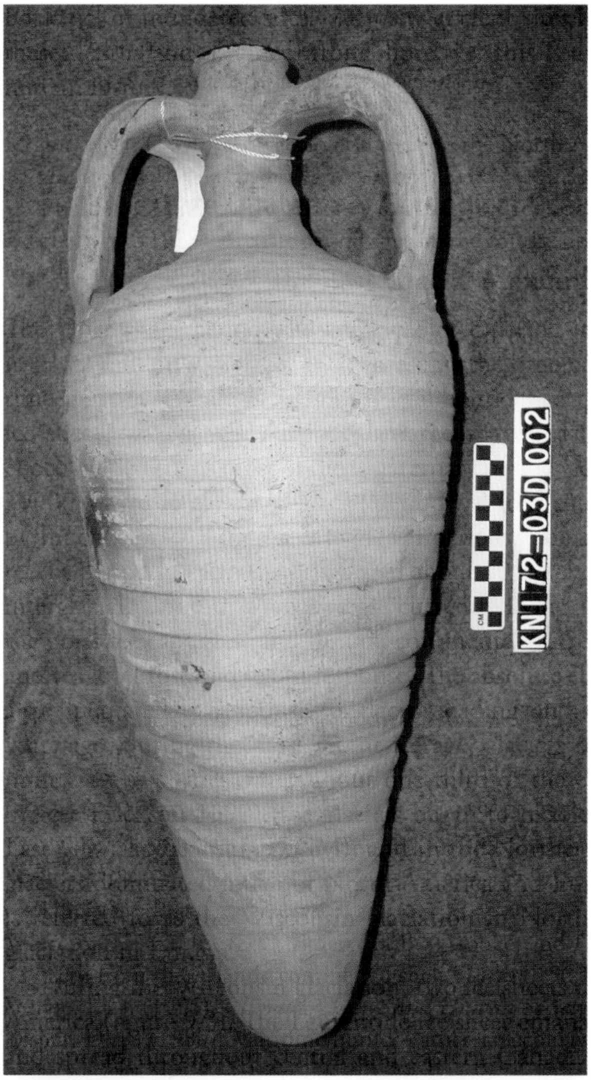

Figure 8.14. (*Top*) Demirci-type jars on ship D gleamed with pine pitch applied some 1500 years ago, both over the lip and inside (*bottom*). (Photographs courtesy of Dennis Piechota)

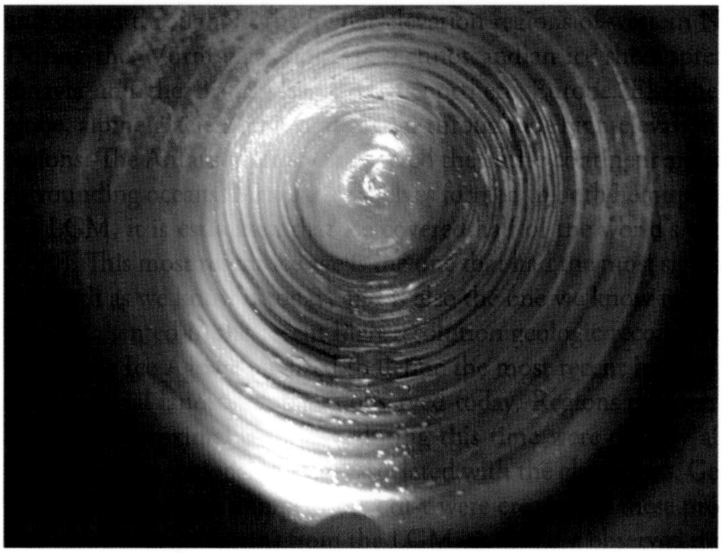

Discussion

The identification of three sites at depths of 85–95 m permitted us to examine shipwrecks that resemble Mediterranean sites in many ways but have some distinct differences. Each of the sites is better preserved than most sites of comparable age in the Mediterranean and includes more substantial timbers on and near the surface than is usual in sites with comparable exposure in the Mediterranean Sea. The shipwrecks have not been disturbed by divers, mooring activities, or fishing, and provided opportunities to test remote sensing equipment and survey procedures. Visiting the sites twice also permits comparison of site features. Large and small timbers are in different locations or missing entirely on sites B and C, and the area and number of exposed artifacts at site C was 8% smaller on the second visit. Site C is subject to the most dynamic surface environment, although it is the most deeply buried site. It is likely to be the best preserved of the shallower shipwrecks. Uniformity of carrot-shaped transport amphora styles suggests that the ships sank at about the same time. Exactly when that was is more problematic as kiln studies suggest that this shape was the predominant type during the 4th and 5th centuries, but LRA1 jars that probably date to the late 6th century are present at site B.

Ceramic styles and a radiocarbon date suggest ship D sank in the late 5th or early 6th century. The 2003 visit provided new data about the ship as well as artifacts from its final lading. The ship seems to be more deeply buried at the bow, which makes it impossible to determine whether the mast is canted forward, as seems possible from images acquired in 2000. In 2000, the orientation of the ship was determined on the basis of two pairs of stanchions whose spacing and height reflected images of a 3rd-century mosaic representation from Tunis (Basch 1987) and a 2nd-century ship on Trajan's column and in mosaics at Ostia (Ward and Ballard 2004). In addition to stanchions aft of the mast in these images, a stanchion, possibly one of a pair, is shown aft of the mast on a ship in an unusual 5th- or 6th-century mosaic at Kelenderis, Turkey[4] (Zoroğlu 1994; Friedman 2003).

The Kelenderis ship is depicted arriving at a harbor, towing behind it two ship's boats on lines attached to sturdy posts beside each quarter rudder. The sternpost protrudes only a little above the highest line of planking, and the stem is higher. A stay runs forward from the top of the mast to the bow, and a second line may represent the halyard tackle, a backstay, or the port shroud. Although the mosaic is damaged at this point, it is possible that the tackle is attached at a point on the vessel's centerline, as is the case for lateen-rigged modern dhows (Facey 1979, for example). The mast is lashed to a mast partner and, at a slight distance aft, to a pair of stanchions. It does not appear to be canted.

The quadrilateral sail is spread on a yard portrayed as longer than the ship's hull, and a line of reef points at an angle to the sail's foot dangles in the wind. A low structure centered on the mast supports what seems to be the furled foot of the sail or a spare sail or shade furled on a spar. A problem in defining the feature is that the bundle is portrayed as being on the port side of the mast, like the standing rigging, and between the backstay and the mast, an impossible position for the foot of the sail. The sail lacks brailing lines but the artist has carefully shown a line of reef points that is at an angle to the yard, a common characteristic of lateen sails and significantly different from the checkerboard-patterned sails with lines of brails prominently shown in many slightly earlier representations (Basch 1987, for example). Although the sail on the small sailing boat is more obviously quadrilateral as Friedman (2003) has observed, in the authors' opinion, the angle of the line of reef points on the large ship's sail and the similar portrayal of the line of reef points on the small boat may indicate an effort to portray a settee or Arabic lateen sail, particularly if the line in the stern is understood as a halyard tackle, points emphasized by Pomey (2006) and Roberts (2006).

Points of similarity between the Kelenderis mosaic ship and ship D include the presence of a mast partner lashed to the mast and to a stanchion immediately behind the mast, an arrangement that would be facilitated by the stepped face of ship D's mast partner and the position and shape of notches in the forward pair of stanchions (Ward and Ballard 2004). The mast on ship D has a squared cavity at its top with a remnant of line or of a feature analogous to a mast band intended to support a top. This band is on ship D in the same position as indicated by two white lines on the Kelenderis ship that intersect with the lines and yard and are just below a curved element at the top of the mast. Friedman (2003) has suggested the Kelenderis feature is a parrel; the feature on ship D does not display the most important characteristic of a parrel because it is in a fixed rather than sliding position. The structure around the mosaic ship's mast may be the open deck structure or yard cradle proposed in Ward and Ballard (2004). A major difference is that ship D at 12–14 m in length was likely much smaller than the Kelenderis vessel, whose size is implied by its two ship's boats.

A mosaic from a Roman house on the Capitoline (Basch 1987) illustrates another prominent feature of ship D. The ship's deck and bulwark are shown with unusual clarity. Top timbers alternate with darker panels and are linked by a cap-rail; they also support a line of planking just at or slightly above deck level. On ship D, some of the top timbers that once outlined the deck remain. About 0.25 m below the heads, treenails about 0.03 m in diameter protrude about 0.025 m outboard of the top timber, suggesting they once fastened the top timbers to planking. There are no traces of panels or cap-rails on ship D, but it is easy to imagine a similar deck enclosure.

Other fastenings on ship D include slightly larger treenail heads visible on each side of the mast partner. These treenails seem to pass through the partner transversely; unfortunately, we were not able to acquire detailed diagnostic images here in 2003 so responsible speculation about their purpose is limited. Examining this timber from below deck would be illuminating, as its configuration is difficult to explain using the images available. Although the only mast to be published from near this time period has no features that would account for treenails in a mast partner (Riccardi 2002), the mere presence of the mast after some 1500 years may indicate an unexpected method of securing it within the hull. Certainly the ~0.25-m beam before the mast is a partner beam that was locked into the side of the hull, utilizing a notch cut into its end and a second notch cut into the outer face of the top timber just aft of the partner beam.

Unfortunately, the planking, and many other timbers, that could tell us much about precisely how this ship was assembled are missing. A few clues were noted. In addition to mortise-and-tenon joinery used in the stempost and sternpost, the uppermost preserved strake included mortise-and-tenon fastenings, probably unpegged like those in the 7th-century Yassı ada ship (van Doorninck 1982). In the sternpost (figure 8.13, bottom), holes surrounded by black stains testify to metal nails, probably iron, but no other timbers are present. No nails were visible in the stempost, but far less of it was exposed as it is more deeply buried than the stern.

It is difficult to imagine what forces would rip planking from the treenails in the top timbers without leaving a trace behind. The force required to remove the sample we acquired was significant, and it had no treenails. It is easier to imagine nails dissolving in the corrosive environment of the deep Black Sea, allowing timbers to loosen and over time drift away, but the absence of so many hull components suggest that the ship suffered greatly in wrecking, upon striking the bottom, or over time. Perhaps the missing planks took the missing top timbers with them when they fell away from the ship. The neatly stacked shipping jars testify to a long period of hull stability even though we found no planking in the area. Wherever we did encounter wood, it was resilient and firmly attached to the hull, so we remain perplexed about the current condition of the ship.

Conclusions

The four shipwrecks identified by the Black Sea project constitute a compact sample when viewed from geographic, economic, and chronological perspectives. The vessels all date to the early Byzantine period (4th–7th centuries CE), and all are slightly north and west of Sinop. Ship D is perhaps one of the best preserved shipwrecks from antiquity, and, if excavated, would provide vital and unique information about the operation of ships and the lives of those who sailed upon them because of the extraordinary preservation of organic materials in the deep Black Sea.

The proximity of ships A, B, and C suggests a pattern of loss, perhaps related to microenvironmental weather conditions associated with the Sinop peninsula, but variations in the shipping jars indicate that we are looking at multiple events, not a single sinking in a storm. Although we cannot predict the intended destination of any of these ships, they each took on cargo carried in transport amphora manufactured in Sinop, and it likely was their last port of call. Did these ships sink just hours out from the town on the route west to Byzantium? Or were they caught in bad weather between picking up a cargo at a vineyard and returning to Sinop or heading out to sea for the Crimea? Our survey cannot provide these answers, but it has demonstrated the utility of side-scan sonar in searching for dispersed targets on a smooth bottom, including one almost entirely buried and one marked only by its mast.

Notes

1. Although the pitch is mixed with other materials, retains its smooth and shiny surface, and seems to be of a wax-like consistency, palynological examination by Dawn Marshall at Texas A&M University indicates that it is a pine product and not a wax.

2. We thank Robert Blanchette, University of Wisconsin at Madison, for his identification of wood from the Black Sea shipwrecks.

3. Three olive stones in the jars, and a lack of pitch lining, prompted the suggestion that the jars carried olive oil or olives (Kassab Tezgör et al. 1998: 441), but without examining the jar walls for lipids, it is not possible to determine if that was the case. Small numbers of olive stones are frequently found dispersed throughout shipwrecks as well as in sediments near shore.

4. I thank Zaraza Friedman for calling this image to my attention.

References

Aksu, A.E., R. N. Hiscott, P. J. Mudie, A. Rochon, M. A. Kaminski, T. Abrajano, and D. Yaşar (2002). Persistent Holocene outflow from the Black Sea to the eastern Mediterranean contradicts Noah's flood hypothesis. *GSA Today* 12:4–10.

Ballard, R. D., D. F. Coleman, and G. Rosenberg (2000). Further evidence of abrupt Holocene drowning of the Black Sea shelf. *Marine Geology* 170:254–61.

Ballard, R. D., F. T. Hiebert, D. F. Coleman, C. Ward, J. S. Smith, K. Willis, B. Foley, K. Croff, C. Major, and F. Torres (2001). Deepwater archaeology of the Black Sea: the 2000 season at Sinop, Turkey. *American Journal of Archaeology* 105:607–23.

Basch, L. (1987). *L'musée imaginaire de la marine antique*. Athens: Institute Hellénique pour la préservation de la tradition nautique.

Bascom, W. (1976). *Deep Water, Ancient Ships*. Garden City, NJ: Doubleday.

Bass, G., and F. H. van Doorninck, Jr. (1982). *Yassi Ada: A Seventh-Century Byzantine Merchantman* I. College Station, TX: Texas A&M University Press.

Casson, L. (1994). *Ships and Seafaring in Ancient Times*. London: University of Texas Press.

Codispoti, L. A., G. E. Friederich, J. W. Murray, and C. E. Sakamoto (1991). Chemical variability in the Black Sea: implications of continuous vertical profiles that penetrated the oxic/anoxic interface. *Deep Sea Research,* 38:S691–S710.

Coleman, D. F. (2002). Underwater archaeology in Thunder Bay National Marine Sanctuary, Lake Huron—preliminary results from a shipwreck mapping survey. *Marine Technology Society Journal* 36:33–44.

Coleman, D. F., J. B. Newman, and R. D. Ballard (2000). Design and implementation of advanced underwater imaging systems for deep sea marine archaeological surveys. In *Oceans 2000 MTS/IEEE Conference and Exhibition*, I, 661–5. Piscataway, NJ: IEEE Oceanic Engineering Society.

Doonan, O. (2004). *Sinop Landscapes: Exploring Connection in the Black Sea Hinterland*. Philadelphia: University of Pennsylvania Museum of Archaeology and Anthropology.

Facey, W. (1979). *Oman, a Seafaring Nation*. Muscat: Ministry of Information and Culture.

Friedman, Z. (2003). *Ship Iconography in Mosaics: An Aid to Understanding Ancient Ships and their Construction*. Thesis, Department of Maritime Civilizations, University of Haifa, Haifa.

Garlan, Y. (1998). Fouilles d'ateliers amphoriques à Nisiköy et a Zeytinlik (Sinop) en 1996 et 1997. *Anatolia Antiqua* 6:407–22.

Görür, N., N. Çağatay, Ö. Emre, B. Alpar, M. Sakınç, Y. İslamoğlu, O. Algan, T. Erkal, M. Keçer, R. Akkök, and G. Karlık (2001). Is the abrupt drowning of the Black Sea Shelf at 7150 yr BP a myth? *Marine Geology* 176:65–73.

Gianfrotta, P. A., X. Nieto, P. Pomey, and A. Tchernia (1997). *La navigation dans l'Antiquité*. Aix-en-Provence: Editions Edisud.

Grace, V. (1949). Standard pottery containers of the ancient Greek world. *Hesperia*: Supplement 8:175–89.

Haldane, C. (1991). Recovery and analysis of plant remains from some Mediterranean shipwreck sites. In J. Renfrew (Ed.), *New Light on Early Farming*, ed. J. Renfrew, 214–23. Edinburgh: Edinburgh University Press.

Hiebert, F. (2001). Black Sea coastal cultures: trade and interaction. *Expedition* 43: 11–20.

Hiebert, F., D. Smart, A. Gantos, and O. Doonan (1997). From mountaintop to ocean bottom: a comprehensive approach to archaeological survey along the Turkish Black Sea coast. In *Ocean Pulse: A Critical Diagnosis*, ed. J. Tanacredi and J. Loret, 93–108. New York: Plenum.

Horlings, R. (2005). *Deepwater Survey, Archaeological Investigation and Historical Contexts of Three Late Antique Black Sea Shipwrecks*. Thesis, Department of Anthropology, Florida State University, Tallahassee.

Kassab Tezgör, D. (1996). Fouilles des atelier d'amphores à Demirci près de Sinope en 1994 et 1995. *Anatolia Antique* 4:335–54.

Kassab Tezgör, D. (1999). Types amphoriques fabriqués à Demirci prés de Sinope. *Production et commerce des amphores anciennes en Mer Noire*, ed. Y. Garlan, 117–23. Provence: Publications de l'Université de Provence.

Kassab Tezgör, D., and F. Dereli (2001). Rapport de la fouille de Demirci-Sinop 2000. *Anatolia Antiqua* 9:215–25.

Kassab Tezgör, D., and I. Tatlican (1998). Fouilles des atelier d'amphores à Demirci près de Sinope en 1996 et 1997. *Anatolia Antiqua* 6:423–42.

Kassab Tezgör, D., and M. Touma (2001). Amphores exportées de mer Noire en Syrie du Nord. *Anatolia Antiqua* 9:101–15.

Kassab Tezgör, D., I. Tatlican, and H. Özdaş (1998). Prospection sous-marine près de la côte sinopéenne: transport d'amphores depuis l'atelier et navigation en mer Noire. *Anatolia Antiqua* 6:443–9.

Mindell, D., B. Foley, and S. Webster (1998). Black Sea survey: cruise report. Manuscript on file at the Massachusetts Institute of Technology and the Institute for Exploration, Mystic, CN.

Murray, J. W., H. W. Jannasch, S. Honjo, R. F. Anderson, W. S. Reeburgh, Z. Top, G. E. Friederich, L. A. Codispoti, and E. Izdar (1989). Unexpected changes in the oxic/anoxic interface in the Black Sea. *Nature* 338:411–3.

Oğuz, T., V. S. Latun, M. A., Latif, V. L. Vladimirov, H. I. Sur, A. A. Markov, E. Ozsoy, B. B. Kotovshichkov, V. N. Eremeev, and U. Unluata (1993). Circulation in the surface and intermediate layers of the Black Sea. *Deep Sea Research* I, 40:1597–612.

Peacock, D.P.S., and D. S. Williams (1986). *Amphorae and the Roman Economy*. London: Longman.

Pomey, P. (2006). The Kelenderis ship: a lateen sail. *International Journal of Nautical Archaeology* 35:326–30.

Riccardi, E. (2002). A ship's mast discovered during excavation of the Roman port at Olbia, Sardinia. *International Journal of Nautical Archaeology*, 31:268–9.

Roberts, O.T.P. (2006). The rig of the Kelenderis ship reconsidered. *International Journal of Nautical Archaeology* 35:330–1.

Ryan, W.B.F., W. C. Pitman III, C. O. Major, K. Shimkus, V. Moskalenko, G. A. Jones, P. Dimitrov, N. Gorur, M. Sakinc, and H. Yuce (1997). An abrupt drowning of the Black Sea shelf. *Marine Geology* 138:119–26.

Singh, H., J. Adams, D. Mindell, and B. Foley (2000). Imaging underwater for archaeology. *Journal of Field Archaeology* 27:319–28.

Steffy, J. R. (1994). *Wooden Ship Building and the Interpretation of Shipwrecks.* College Station, TX: Texas A & M University.

Uchupi, E., and D. A. Ross (2000). Early Holocene marine flooding of the Black Sea. *Quaternary Research* 54:68–71.

van Alfen, P. (1996). New light on the 7th-c. Yassi Ada shipwreck: capacities and standard sizes of LRA1 amphoras. *Journal of Roman Archaeology* 9:189–213.

van Doorninck, F. H., Jr. (1982). The hull remains. In *Yassi Ada: A Seventh-Century Byzantine Merchantman,* ed. G. Bass and F. H. van Doorninck. I, College Station, PA: Publ?.

van Doorninck, F. H., Jr. (2002). Byzantine shipwrecks. In A. Laiou (ed.) *The Economic History of Byzantium from the Seventh through the Fifteenth Century,* I, ed. A. Laiou, 899–905. Dumbarton Oaks Studies 39. Washington, DC: Dumbarten Oaks.

Ward, C., and R. Ballard (2004). Black Sea shipwreck survey 2000. *International Journal of National Archaeology* 33:2–13.

Zemer, A. (1978). *Storage Jars in Ancient Sea Trade,* Haifa: National Maritime Museum Publications.

Zoroğlu, L. (1994). *Kelenderis,* I. Ankara: Dönmez..

PART FOUR

Submerged Landscape Archaeology

Archaeological and Geological Oceanography of Inundated Coastal Landscapes: An Introduction

9

Dwight F. Coleman

Unexplored regions on the submerged continental shelves and associated near shore environments could hold the clues to unraveling some of the mysteries about the expansion of human populations following the last Ice Age (Flemming 1985; Renfrew and Bahn 2004). Exploration for and discovery of submerged terrestrial sites has occurred only during the last few decades. Many known prehistoric underwater archaeological sites were accidentally found by local divers or fishermen in relatively shallow water (Flemming 1985). Very few site discoveries resulted from systematic archaeological survey work, and virtually no definitive in situ cultural sites have been found in deeper water farther offshore. Conducting prehistoric archaeology under water is costly and difficult for a variety of reasons and not many funding agencies sponsor this kind of research. By initiating and carrying out research and exploration in underwater prehistoric archaeology, the potential for answering significant anthropological questions exists. How, where, and why did Ice Age humans migrate and populate new regions on the planet? What was the nature of their subsistence? What types of environments did they prefer to occupy? How widespread was their occupation? State-of-the-art oceanographic technologies can and should be used to help answer these and other questions by systematically surveying, exploring, and intensely studying the submerged environments inundated by the latest transgression.

It is well established that prehistoric human populations flourished during and following the last Ice Age. However, details about the nature of post-Pleistocene human evolution and migration remain poorly known and controversial. There is a strong consensus that humans from this period may have followed coastal routes and favored these environments for a variety of reasons, including abundance of food, ease of transportation, presence of sheltered areas, and a wealth of natural resources (Butzer 1971). As sea level rose in response to melting glaciers, these coastal environments were inundated, and the evidence for this ancient human habitation now lies primarily under water, at depths down to 120 m on the continental shelves. By fully characterizing the submerged landscape and reconstructing the paleo-geography of the continents, including identifying paleoshoreline features, underwater caves, ancient river channels, presently submerged hills, wetlands, and rocky shelters, an archaeological baseline can be established to identify regions in the ocean for further exploration for potentially important archaeological sites. Discovery of new sites that now lie under water could be extremely beneficial to anthropological science and human history, especially in the New World, where our knowledge of prehistory is limited.

Nearly 3% of the earth's dry land surface, representing about 4 million square kilometers of habitable land, was flooded between approximately 20,000 and 5000 years ago (refer to figure 9.1 for timescales). During this time, both humankind and the natural environment experienced rapid and extreme change. The earth's climate warmed and sea level rose more than 100 m in response to the melting glaciers (Fairbanks 1989). Complicating this picture, the deglaciation of the continents caused the elevation of the land to adjust isostatically. This adjustment of the crust results in the land moving vertically relative to sea level, but this effect is regional not global. This isostatic adjustment is still occurring today in certain regions. Land subsidence and uplift occurs naturally, but on very long timescales. Other local and regional effects on the land/sea interface include tectonic and volcanic processes that can both catastrophically and gradually change the landscape.

The last 20,000 years in the earth's history includes the end of the last Ice Age and spans the transition from the late Pleistocene epoch into the Holocene epoch (figure 9.1). This was a period of rapid change in the earth's climate. During this time period, a few noteworthy climatic episodes occurred. The Younger Dryas event, a brief episode about 11,000 years ago when climate was very cold and dry, is characterized by a less abrupt rise in sea level between two very rapid rise pulses (Fairbanks 1989). The "Little Ice Age" was another brief, cold episode from about the 16th to 19th centuries CE when the global average temperature was about 1°C colder than today, spawning a short readvance of glaciers (Mann et al. 1999). In human history, the last 20,000 years represent a time of rapid and extreme change in culture and population. The cultural periods represented for the Old World include the late Paleolithic, Mesolithic, and Neolithic of the Stone Age, transitioning to the Chalcolithic, Bronze Age, and Iron Age in more recent times (figure 9.1). For the New World, we see humans enter for the first time around 13,000 years ago (Flemming 1985), followed by the Paleo-Indian, Archaic, Woodland, and Contact periods. The rapid advance of cultures in the Old World and spread of humans into the New World are arguably the most significant processes in human history during this time period following the last Ice Age.

Throughout the late Quaternary, human population on earth grew rapidly, especially in Africa during the Paleolithic (Reich and Goldstein 1998). Around the time of the agricultural revolution, near the Pleistocene–Holocene transition, the population growth rate increased from about 0.0015% per year to 0.1% per year (Renfrew and Bahn 2004). Today the global population growth rate is about 2% per year, with more rapid growth in less-developed countries

(Miller 1988). In the United States during a 50-year span from 1960 projected to 2010, human population in the coastal zone (defined by counties bordering coastal features of the Atlantic Ocean, Gulf of Mexico, Pacific Ocean, and Great Lakes) will have grown by nearly 60% from 80 million to 127 million individuals (Culliton et al. 1990). Today, about 38% of the world's population lives within 100 km of the coast and about 49% of the world's population lives within 200 km of the coast (CIESIN 2002). A separate statistic indicates that

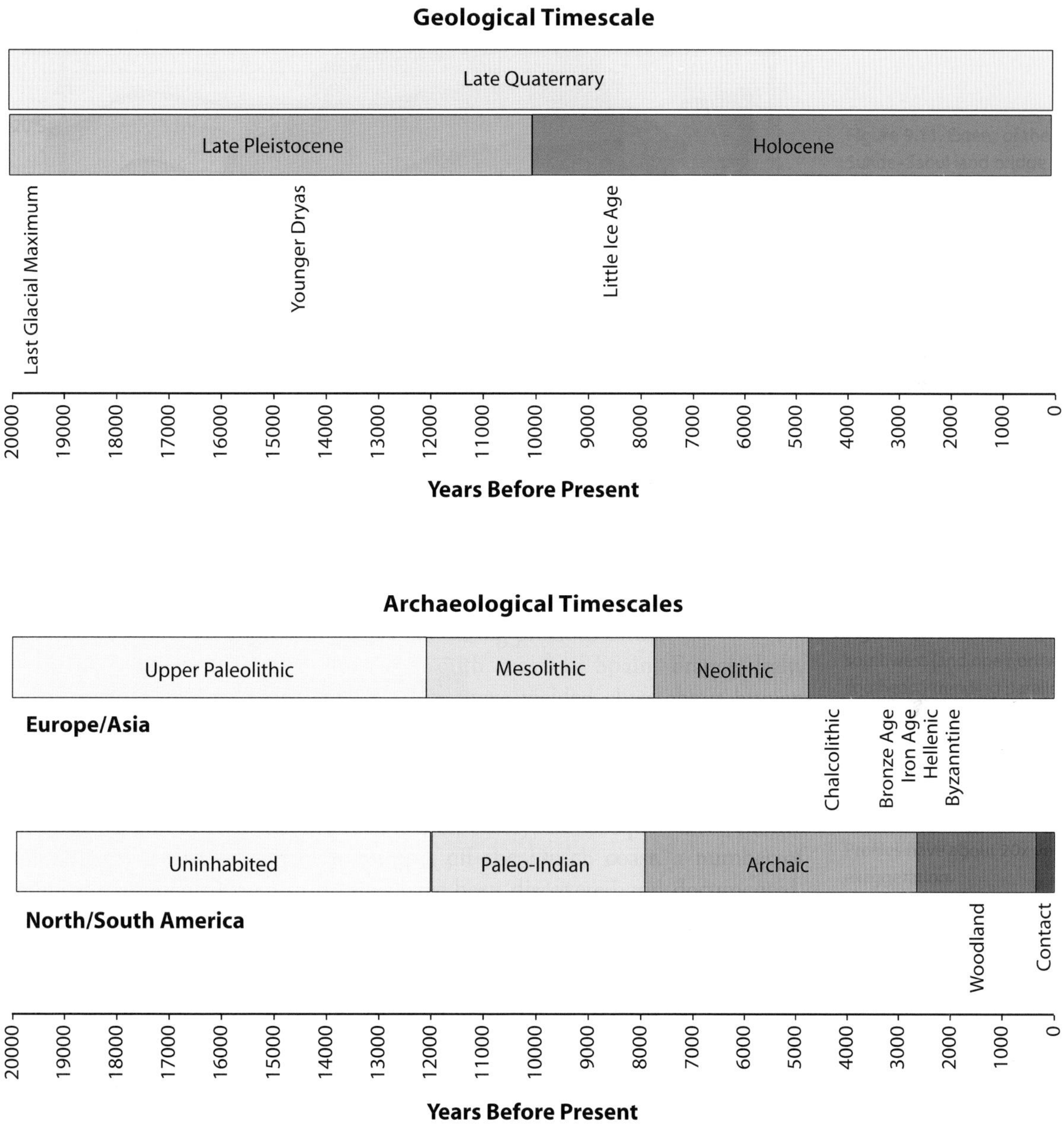

Figure 9.1. Geologic and archaeological timescales for the late Quaternary. Boundaries between adjacent periods are approximate and depend on locality. For example, the early Chalcolithic period is defined by the first appearance of copper metallurgy at a particular locality, which varies significantly from place to place.

about 25% of the world's population lives below the 50-m elevation and about 36% of the world's population lives below the 100-m elevation. So by today's statistics of the world's population distribution (based on calculations from CIESIN 2002), a very large percentage of the people who inhabit the earth either live close to the coast or at low elevations, or both (table 9.1, figure 9.2). These are the people who would be most affected by rising sea level. Although there are no valid statistics available on population distributions for prehistoric times, one can speculate that it would be similar to the distribution of today, in terms of population percentages, except at a much smaller absolute scale. If this were true, according to modern statistics, nearly half the people who lived

TABLE 9.1.
PERCENTAGE OF POPULATION WITHIN DEFINED GEOGRAPHIC SETTINGS

	Elevation (meters above sea level)					*Distance from coast*	
Continent	*0–5*	*5–10*	*10–25*	*25–50*	*50–100*	*100 km*	*200 km*
Africa	3	2	4	5	7	33	48
Asia	7	4	9	10	12	39	49
Europe	4	2	5	8	17	39	55
North America	5	3	7	8	10	44	53
Oceania	13	5	7	10	13	85	95
South America	5	2	6	6	8	49	66
World	6	3	7	8	11	38	49
Cumulative	6	9	16	24	36		

NOTE: For each continent, the percentage of the population that lives within the defined geospatial zone is listed. The global cumulative population by elevation zone is also computed for display in Figure 9.2. Data were computed from raw population density grids, based on 1995 census information. SOURCE OF DATA: CIESIN, Columbia University, 2002. National Aggregates of Geospatial Data: Population, Landscape and Climate Estimates (PLACE).

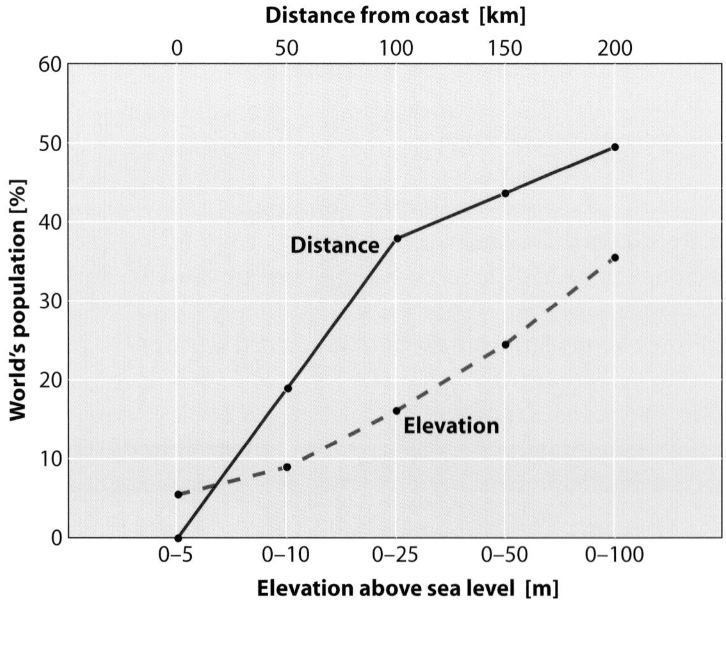

Figure 9.2. Percentage of world's current population within particular geographic settings. Dashed line represents population percentages within defined elevation ranges. Solid line represents population percentages within defined distances from the coast. Data source as in table 9.1.

during the last Ice Age would have settled near the coastal zone, just like modern humans.

At the time of glacial melting, modern human populations flourished and had spread along coastlines throughout the Old World and into the New World. As humans adapted to the warming climate and rising sea level, they organized socially, settled in new coastal environments, and became more technologically advanced (Yesner 1980; Renfrew and Bahn 2004). Coastal sites of human occupation from this time now lie beneath the oceans and marginal seas. Many coastal settlements that were inundated by rising seas, either gradually or catastrophically, must exist today as underwater archaeological sites. By studying these sites, archaeologists and marine scientists will learn a great deal about both human and natural history and their environmental association. The preservation potential is greater for submerged archaeological sites compared to terrestrial sites, but little is known about their abundance and distribution. Due in part to rapidly advancing marine technologies, archaeologists and oceanographers can now locate and precisely survey and sample these submerged cultural sites, opening up a new frontier in marine archaeology. However, a comprehensive methodological approach for this new interdisciplinary field that incorporates both archaeological and oceanographic techniques does not exist. By merging oceanographic mapping, geophysical prospecting methods, and archaeological survey methods, a new methodology can be created and verified.

Exploration of the submerged continental shelf, up to about 120 m water depth, for inundated archaeological sites relies on geophysical prospecting tools and oceanographic survey techniques. With the support of recent technological advancements, a new scientific discipline has emerged that combines the fields of archaeology, oceanography, geophysics, and deep submergence vehicle engineering. This new field of *archaeological oceanography* is truly multidisciplinary in nature and reaches far across both social and physical sciences as exemplified throughout this book. The collection and analysis of artifacts from documented underwater contexts will advance our understanding of prehistory and anthropology. The analysis of marine geophysical data and geological samples will advance our understanding of recent and diverse earth system processes. In combination, the multidisciplinary nature of this young science will advance our understanding of the cultural evolution of humans in association with their changing natural environment.

The marine environment is an ideal setting for archaeological research for many reasons. Firstly, millions of linear kilometers (depending, of course, on how precisely one measures the length) of the habitable continental coastline were inundated by the latest marine transgression. Undoubtedly, landward of these submerged coastlines, there is an abundance of undiscovered archaeological sites, representing a new and untouched scientific landscape. There is high preservation potential for both organic and inorganic ancient material. In addition, and perhaps most importantly, the discovery of submerged terrestrial sites along "land bridges" between continental landmasses or along drowned coastlines could definitively prove the existence of suspected early human migration routes. Lastly, archaeological oceanography will directly benefit from advances in marine technology for surveying, imaging, and sampling. This final point is important for the protection of submerged cultural sites in that sites could be digitally imaged with such high resolution that a minimum amount of sampling would be required for adequate site characterization.

The tools and techniques of marine geological mapping have been continually advancing during recent years, with improved resolution and data processing capabilities. These techniques have been applied to underwater archaeology (Muckelroy 1978), but not extensively. Many archaeological surveys in shallow water, particularly in the Mediterranean (Bass 1975), were on sites previously and randomly discovered by fishermen or local divers. In addition, detailed geophysical investigations of archaeological sites, particularly in the deep sea, can often be cost-prohibitive. For this reason, there has not been a great deal accomplished to date.

In the following chapters, two case studies will be presented. The case studies focus on two diverse regions that were affected differently by the last glaciation—the southern New England continental shelf region and the Great Lakes region. The research presented in these case studies will be synthesized and put into the context of archaeological oceanography. This new scientific discipline is rapidly emerging and scholars in the more traditional fields of archaeology and oceanography are realizing its importance.

Many submerged prehistoric sites have been documented in shallow water in near-shore environments. A majority of these are from Neolithic times (Archaic in the Americas) or later and became inundated during a steady and slow rise in sea level. The case studies that follow present the potential for examining older sites (12,000–8000 BP) in deeper water and currently situated many kilometers offshore. This time span represents the period of the most rapid rise in sea level following the last glaciations, yielding the greatest preservation

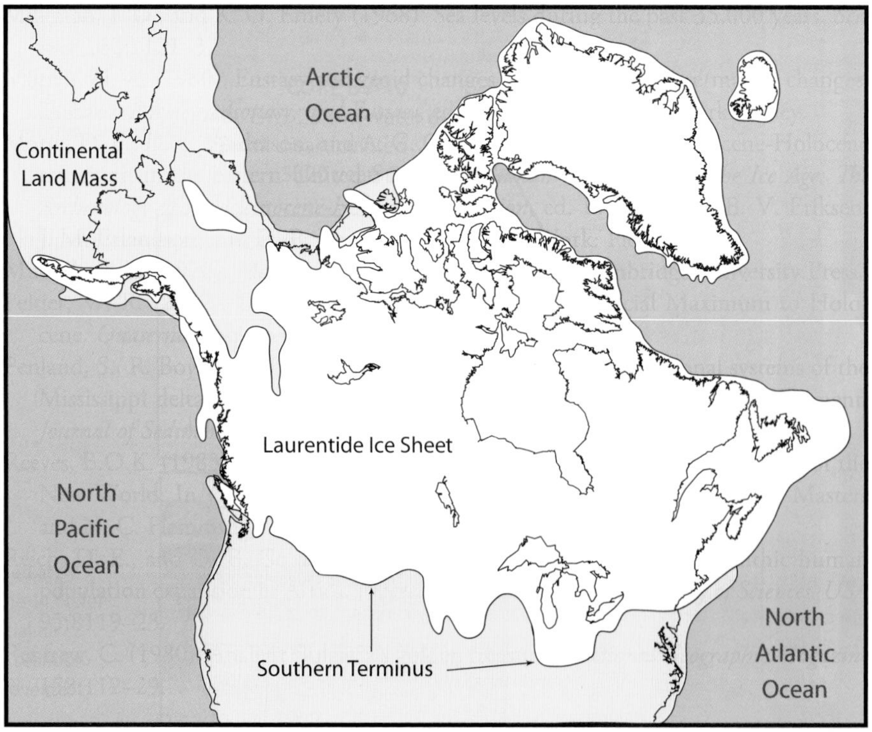

Figure 9.3. Extent of ice cover in North America during the Last Glacial Maximum (modified from Dyke et al. 2002). Yellow regions indicate the extent of exposed land when sea level was lower during this time.

potential of inundated sites. Very few ancient sites have so far been located in these depths and distances from shore, yet this is a very significant period in human history.

Late Quaternary Geological Oceanography

The Last Glacial Maximum

The Quaternary period, the latest geologic period in the Cenozoic era, began nearly 2 million years ago, and represents the time of ice ages. Throughout this time period, several glacial/interglacial episodes transpired, as the continental ice sheets waxed and waned over large spatial scales during the Pleistocene epoch, which spans most of the Quaternary period from its beginning (Erickson 1990; Whyte 1995). Each glacial period was long in duration and was followed by much shorter duration interglacial periods. The Holocene epoch, the most recent period in earth history, began about 10,000 years ago and represents the interglacial period in which we are currently living. During the Pleistocene, as the polar ice caps grew and shrank, continental glaciers advanced and retreated, and sea level rose and fell in response. The absolute elevation of the earth's crust also fluctuated in response to glacial loading and unloading. About 20,000 years ago, during the most recent ice age, glaciers reached their southern terminus. At this time, when about one-third of the earth's surface was covered by ice (Erickson 1990), the glaciers began to recede. This time represents the Last Glacial Maximum (LGM), and in the Northern Hemisphere continental glaciers dominated much of North America and Eurasia. This latest glaciation is referred to as the Wisconsin glaciation in North America and the Würm glaciation in Eurasia.

During the Wisconsin glaciation, two ice sheets dominated northern North America (figure 9.3). The Laurentide ice sheet emanated from the Arctic Circle and spread throughout central and eastern Canada and the northern United States. The Cordilleran ice sheet emanated from the Canadian Rocky Mountains and spread throughout the higher-elevation regions of western North America. During the Würm glaciation, the Fennoscandian ice sheet spread from Scandinavia and the British Islands into northern Europe. Elsewhere around the globe, alpine glaciers grew and spread throughout high-elevation midcontinent regions. The Antarctic ice sheet covered the entire continent and spread into the surrounding oceans. Massive ice shelves formed in both hemispheres and during the LGM, it is estimated that ice covered half of the world's oceans (Erickson 1990). This most recent ice age is the one that had the most significant effect on the earth as we know it today, and is also the one we know most about because it is represented in the recent high-resolution geologic record. The glacial geology of the Ice Age (capitalized to define the most recent ice age) is recorded in the terrestrial landscape that is observed today. Regions that were immediately beneath or proximal to the ice during this time were directly affected by erosional and depositional process associated with the glaciations. Geomorphologic features of the earth's uppermost crust were created by these processes. Glacial and periglacial landforms from the LGM, as they are observed today, have been further modified by eustatic (adjustment of sea level) and isostatic (adjustment of crustal elevation) processes.

Characteristics of Quaternary climate are recorded primarily through ice and sediment cores. These provide high-resolution indicators of past climate, and major climate transitions are well defined in the core record. Ice cores, from the Greenland and Antarctic ice sheets, for example, directly record changes in the earth's global atmosphere, and represent a very precise climate record for the most recent past. Sediment cores from inland lakes record regional climate changes and changes to the terrestrial landscape. These can be used to resolve climate transitions further back in the geologic record. Deep-sea sediment cores directly record events even further back in the geologic record. These marine sediment records indicate both oceanographic and climatic change and their interrelationship. The scientific disciplines of paleoclimatology and paleoceanography decipher the earth's past climate and geographical conditions using evidence primarily found in ice and sediment cores.

Environmental change during and following the LGM was extreme (Adams et al. 1999). Throughout the late Quaternary, abrupt changes in climate brought about abrupt changes in glaciers, which brought about abrupt changes in land elevation and global sea level. All of these changes resulted in dramatic change to the coastal environment, primarily by changing the interface between water and land. Through years of ice age geologic and climatologic research, there is a solid understanding of the causes and effects of late Quaternary environmental change since the LGM. However, how this has influenced human habitation along the changing coastlines remains poorly understood. Catastrophic environmental change would influence human populations quite differently than gradual environmental change. If the changes were not catastrophic, then they would still be abrupt on a geologic timescale, and have a significant impact on people. Despite the rate of change in sea level, or the rapidity of flooding, it remains a fact that sites of human occupation in coastal settings now lie underwater. The rate of submergence, however, would significantly influence the state of preservation, the volume of overburden, and the redistribution of material.

Eustasy

Eustasy refers to variation in the globally averaged absolute elevation of the sea. This is quite different from relative changes in sea level for specific regions. However, to better understand relative sea level change, a firm understanding of absolute sea level change is necessary. Eustatic change in sea level results from change in the total volume of water in the ocean basins. A significant control on this volume is change in global ice volume. When polar ice caps and large continental ice sheets grow, global sea level is lowered. When they melt, global sea level rises. Both relative and absolute sea level change are recorded in the geomorphologic and stratigraphic record. Features of the geomorphologic record important to determining sea level history are the relative positions of ancient shorelines (van Andel 1989). Fossils in the marine stratigraphic record can be used to determine an oceanic oxygen isotope curve, which can be used to determine the volume of water stored in continental ice sheets (Chappell and Shackleton 1986). Absolute dating, typically by the radiocarbon method and

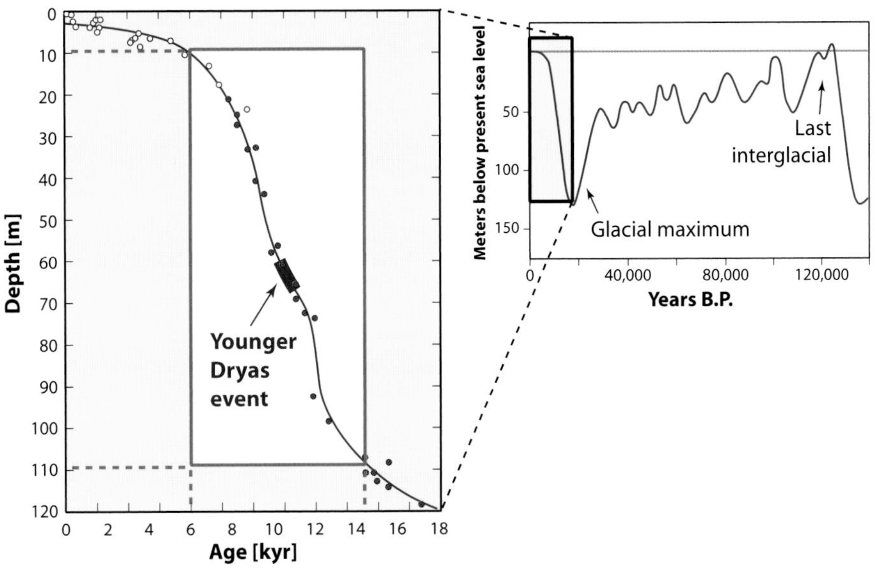

Figure 9.4. Sea level rise curves for the late Quaternary period (right; after van Andel 1989) and since the Last Glacial Maximum (lower left; after Fairbanks 1989). Region in red box represents very rapid rise in sea level of about 100 m between about 6000 and 14,000 BP.

the uranium-series radioisotopic method, of the shorelines and cores, puts the correlation in a temporal reference frame. A eustatic sea level curve, such as for the Quaternary period, can then be created (figure 9.4, right) that is based on this correlation and chronology.

For a finer-scale eustatic sea level curve for the late Quaternary period since the LGM, cores from submerged coral reefs in Barbados were analyzed and dated (Fairbanks 1989). These particular coral formations always grow within a few meters of sea level. Corrections had to be made for the uplift of the Barbados platform and for the conversion from radiocarbon years to calendar years. The curve reveals a relatively slow and gradual rise in global sea level from about 18,000 BP until about 14,000 BP, followed by a steady and very sharp rise from about 14,000 to 6000 BP, then followed by another more gradual rise until the present day (figure 9.4, left) (Fairbanks 1989). During the sharp rise, a relatively brief period of slower rise is represented by the Younger Dryas event. In all, according to these data, about 120 m vertical rise of the global ocean has occurred during the last 18,000 years. This curve is widely cited and considered the standard sea level rise curve for the late Pleistocene and Holocene epochs. Adjustments must be made to this curve to apply it to the shoreline positions and sea level rise history for particular sites, once local and regional isostatic and tectonic effects are determined. Interestingly, as illustrated in figure 9.4, the sea level drop during the LGM is greater than at any other time during the previous 100,000 years. Therefore, around the same time as the expansion of human population and peopling of the New World, sea level was at its lowest. The steepest part of curve in figure 9.4 is between about 14,000 and 6000 BP. During this 8000-year time interval sea level rose about 100 m. That represents a rate of more than one meter per century, which is quite rapid by geological standards. This time period spans from the late Pleistocene into the early Holocene and includes several cultural periods in both the Old and New World. In Europe and Asia, the late Paleolithic, Mesolithic, and early Neolithic time periods are represented and in the Americas, the Paleo-Indian and early Archaic periods are represented.

Hypsometry refers to the measure of the amount of land at particular elevations relative to sea level. Presently, about 71% of the earth's surface is covered by ocean water and 29% is continental land (figure 9.5). The bimodal distribution peaks represent the average elevation of land above sea level (more than two-thirds between 0 and 1000 m) and the average depth of the ocean below sea level (more than one-third between –4000 and –5000 m). This bimodal distribution is due to the fact that there are basically two types of crust on earth, less dense continental crust and more dense oceanic crust. As illustrated in figure 9.5, the cumulative distribution of land area above sea level would

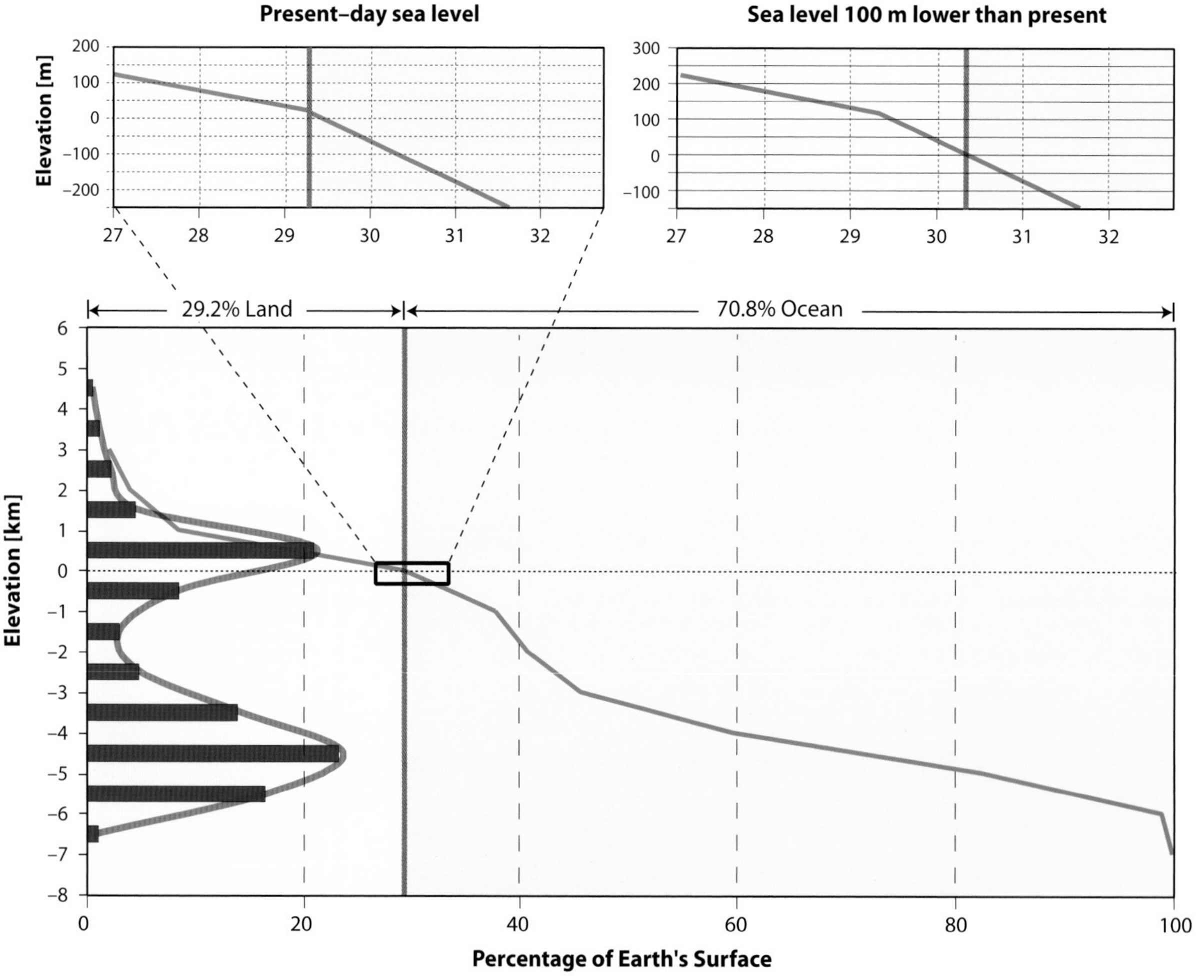

Figure 9.5. Modern hypsometric curve illustrating the effect that lowering sea level by 100 meters increases the total land area by about 1%. Histogram shows the frequency distribution of land elevations (blue bars and red line). The cumulative distribution is indicated by the green line.

substantially increase if sea level were about 100 m lower than at present, such as during the LGM. The result would be an increase in total land surface area by an additional 3% (or about 1% of the total earth's surface area). This represents about 5 million square kilometers of coastal land. Much of this additional land surface would have been covered in ice during the LGM; nevertheless, an enormous amount would have been exposed and habitable. This statistic does not take into account fluctuations in land elevation due to isostasy, however, but it crudely illustrates the point that inundated land masses represent an extremely large area. The fact that about 3% of the continents were flooded by eustatic rise in sea level can stand alone in illustrating the importance eustasy had on humans.

Isostasy

The earth's rigid lithosphere, which floats on the weaker asthenosphere, will deform due to loads contributed by the earth's crust (including continental and oceanic crustal rocks and sediment), water, and ice. The elastic lithosphere flexes under the load, and the plastic asthenosphere flows to accommodate its flexure (figure 9.6). The basic result of the isostatic adjustment is a net vertical increase or decrease in crustal elevation. The amount of vertical displacement

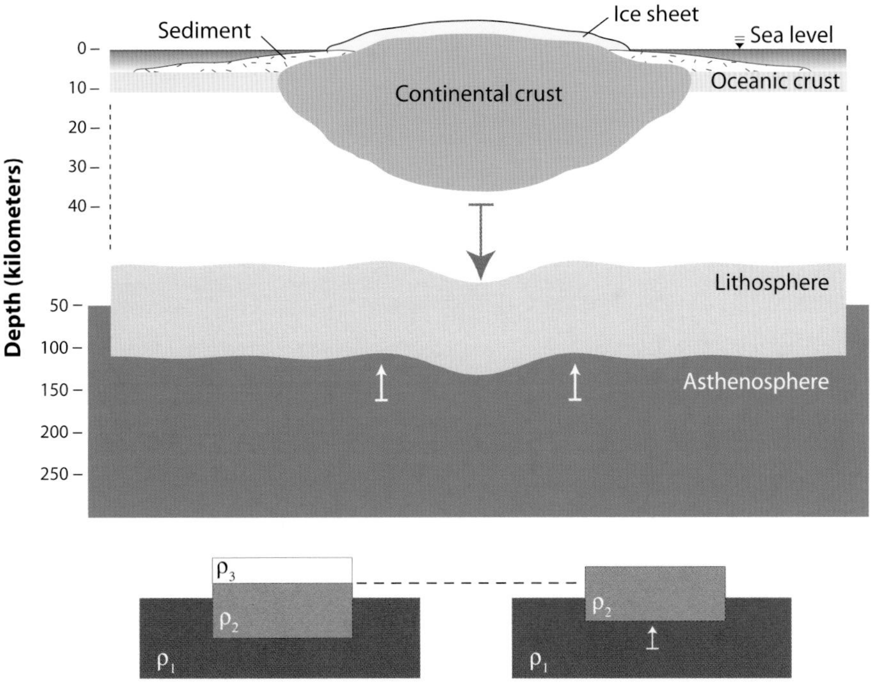

Figure 9.6. Isostasy, or the gravitational equilibrium among the earth's upper layers. The earth's crust, which is part of the rigid lithosphere, but separated here for illustration purposes, causes the asthenosphere to deform to accommodate density fluctuations caused by adding or removing ice or redistributing sediment. The effect is vertical adjustment of the upper layers, based on Archimedes' principle, as illustrated in the lower part of the figure. ρ represents the density of each layer. For $\rho_1 > \rho_2 > \rho_3$, if the top layer is removed, the middle layer would rise and float at a higher elevation.

is a function of the density and thickness of each layer, based on Archimedes' principle. Removal of ice sheets, erosion of the continental crust, and deposition of sediment on the shelf are processes that cause the isostatic equilibrium to adjust. This adjustment is responsible for localized and regional shifts in the elevation of the land relative to the ocean. In some cases, this amounts to significant displacements. Such is the case for Fennoscandia, where a very large amount of ice was removed from a continental landmass, which is surrounded to the north and west primarily by oceanic crust (figure 9.7). At the point of maximum vertical displacement, more than 800 m of uplift occurred as a result of glacial unloading (Mörner 1980). There is a region of zero net vertical displacement, and, as illustrated in figure 9.7, this follows near the present-day northwestern Scandinavian shoreline. Landward of this region, ancient shorelines have been uplifted, and seaward of this region, ancient shorelines remain submerged. Because the earth's asthenosphere must flow beneath the uplifted lithosphere, the effect is to further lower the shorelines offshore, according to a sinusoidal-shaped power law function. This example of isostatic adjustment of Scandinavia is extreme, however.

Isostatic uplift of the continents due to melting glaciers causes changes in the land/sea interface. As the land rises, the shoreline does as well, and a new shoreline forms at a lower level. This is not the case for inland bodies of water, such as the Great Lakes, where the lakes themselves would also adjust isostatically with the land, so there would be little change in shoreline elevation, unless the lake level was fluctuating. On continental margins, if the rate of isostatic uplift was perfectly synchronous with the rate of sea level rise, the shoreline would not change at all. This is almost never the case. The disparity between these rates

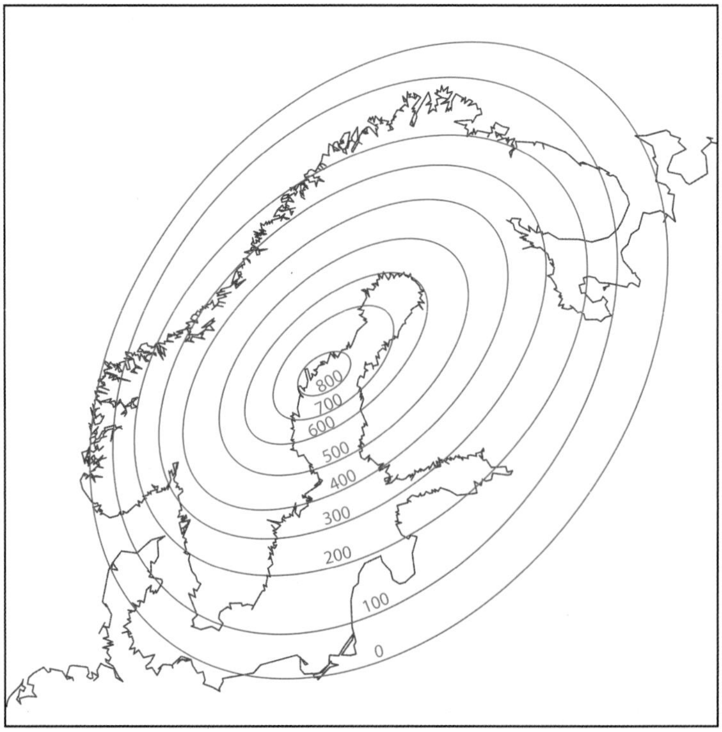

Figure 9.7. Postglacial isostatic rebound of Fennoscandia. Uplift in meters above sea level (Fischer 1995, after Mörner 1980).

results in a net relative sea level rise or lowering. This is commonly the case, especially in higher latitudes where isostatic uplift is significant. It is very difficult to determine absolute sea level change, except in regions where the land elevation has been stable, because it is difficult to distinguish between isostatic and eustatic effects on shoreline elevations. For archaeological purposes, however, the *relative* change is more significant. If the sea rises and falls relative to the land, or the land rises and falls relative to the sea, then the net effect is the same.

Tectonics, Volcanism, and Other Geologic Processes

Tectonic and volcanic processes can significantly affect the elevation of land and, therefore, relative sea level. Mass wasting events that coincide with volcanic eruptions or earthquakes can redistribute rock and sediment, thereby shifting the load on the earth's crust. Similar to the way glacial loading and unloading can cause isostatic adjustment of the crust, sediment loading and unloading due to volcanic and tectonic processes can also have an isostatic effect.

In addition to the effect on landscape and environment, volcanic and tectonic events can significantly influence sites of human activity located near the event sources. Notable volcanic eruptions that have wiped out the population of entire regions have also preserved archaeological sites in the deposits. Pompeii and Herculaneum, ancient cities near the present day Bay of Naples, Italy, were both destroyed and preserved by volcanic ash and pyroclastic flow deposits from the eruption of Mt. Vesuvius in 79 CE (Harris 2000). Another notable eruption is that of the volcano on Thera (Santorini) in the Aegean Sea, during the 17th century BCE. Here, the late Bronze Age Minoan city of Akrotiri lies covered by volcanic deposits from the eruption.

Some other natural processes can influence relative sea level rise and land elevation. Nontectonic isostatic subsidence and uplift can occur in regions of extremely high sedimentation and in regions where mass wasting is prevalent. For example, in the coastal deltaic environments of large river systems, these features can collapse, and occasionally sink under water due to the massive deposition of river sediment over time. Processes such as these have occurred in the Nile delta region of the Mediterranean Sea, on the Egyptian coast. Parts of the ancient city of Alexandria have subsided into the sea and exist there today as underwater archaeological sites (Goddio 1998).

The evolution of shorelines and coastal geological features also depends on the position of the coast on the continental margin relative to significant tectonic features. Passive continental margins, such as those surrounding the Atlantic Ocean behave much differently than active continental margins, such as those surrounding the Pacific Ocean—likewise for the passive margin associated with North Africa in the southern Mediterranean Sea and for the active margin associated with southern Europe in the northern Mediterranean Sea. For coastlines on the leading edge of a continent, like along the California coast, the land is subjected to more vertical dynamics of subsidence and uplift (van Andel 1994). For coastlines on the trailing edge of a continent, the vertical movements of the land relative to the ocean are often less pronounced. These dynamics are functions of the controlling plate tectonic processes and generally occur during long time periods.

Global-scale magmatic activity can also influence relative sea level. These processes are worth mentioning here, but not in great detail. The internal dynamics of the earth's outer core and mantle govern the rate at which magma is supplied to shallower levels in the earth's crust, and this can, in turn, govern global eustatic sea level changes. During the Cretaceous period, for example, a large magma supply caused rapid seafloor spreading throughout the mid-ocean ridge system, increasing its volume, and displacing seawater out of the ocean basins (van Andel 1994). This process did not influence sea level during the Quaternary period, however.

Inundation

Relative sea level rise causes low-lying coastal plains to become inundated. The rate of inundation, which equates to the rate of transgression, strongly influences the adaptation of the coastal environment and the destruction or preservation of coastal geomorphologic features (Belknap and Kraft 1981). Preservation of fragile features of the coastal landscape, including man-made structures, also depends on the rate of inundation. This rate also influences the formation of coastal features such as beaches and lagoons. Gradual inundation would tend to destroy features through the process of erosion, especially by wave activity and shoreface erosion. Rapid inundation could preserve features, depending on the total vertical rise of the sea or lake relative to land. The horizontal distance of the inundated region is a function of the slope of the coastal feature (figure 9.8). A large amount of flat, low-lying land will be inundated with a small rise in sea level, whereas steeper terrain would not be inundated nearly as much with the same sea level rise.

The preservation of coastal features following a marine transgression depends on a number of different factors. Belknap and Kraft (1981) have extensively studied the continental shelf and coast off Delaware to examine the evolution and preservation of coastal geologic features in response to the latest Quaternary transgression. These authors concluded that shoreline elements on the outer shelf have a significantly higher degree of preservation than features on the inner shelf. This is primarily due to the rate of relative sea level rise, which was determined to decrease with time. Other factors that control preservation of geologic features are wave energy, amount of erosion, sediment supply, and tidal range (Belknap and Kraft 1981). For the Delaware coast, the shoreline elements include primarily barrier beaches and lagoons. The barrier and dune sediments are less likely to be preserved than the lagoon sediments. A study of the Louisiana coast, at the Mississippi River delta, revealed sediment sequences of barrier beach systems, such as dunes, spits, and bars, do not survive marine transgressions, but evolve and transform into low-relief sandy offshore shoals (Penland et al. 1988).

If the Belknap and Kraft model is correct, as they predict, then preservation of coastal features should be found far offshore on the distal shelf, in deep water. This location also happens to be where the late Quaternary sedimentary deposits are thinnest (Uchupi et al. 2001). Therefore, this same region could represent the location where sites of human occupation are best preserved and potentially exposed. This is a hypothesis that needs further testing, however.

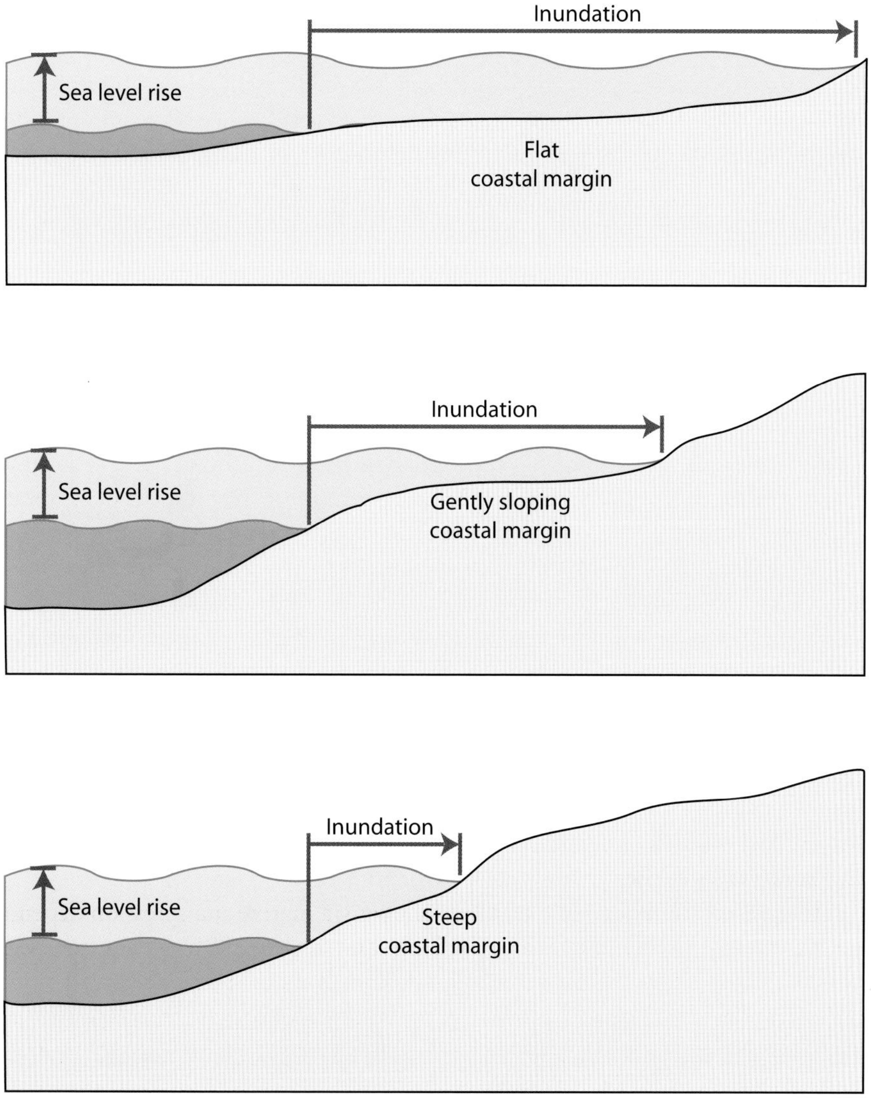

Figure 9.8. Inundation of the coastal margin due to eustatic sea level rise. For relatively flat margins (*top*), a large area is inundated due to a small rise in sea level. For gently sloping margins (*middle*), substantially less area is inundated with the same rise in sea level, and, likewise, for steeply sloping margins (*bottom*), even less area is inundated with the same rise.

Prehistoric Coastal Archaeology

Early Human Populations

The Quaternary period, as discussed in the previous chapter, represents not only the time of ice ages, but also a productive time in human evolution and population expansion. In the early Pleistocene, about 1.7 million years ago, *Homo erectus*, a precursor to modern humans, began to emerge in East Africa (Wells 2002). Material clues to this species' past are rare, however. About 100,000 years ago, *Homo sapiens* began to emerge in Africa with skeletal remains looking more like that of modern humans. Wells's (2002) report on genetic studies (from male Y-chromosome DNA analyses) indicates that all modern humans are descended from a common group of ancestors in Africa around this time. In addition, Neanderthal populations emerged and expanded throughout Europe, and *H. erectus* spread into Southeast Asia (Wells 2002). By late Paleolithic times, around 30,000 years ago, anatomically modern *H. sapiens* were the only surviving hominid species (Wells 2002). *H. sapiens* had different physical features compatible with their technological advancements, showed less evidence of physical injuries, and enjoyed a longer life span compared to other early hominids (Haviland 1997). The Upper Paleolithic is also characterized by widespread migration of humans. By about 50,000 years ago, humans migrated out of Africa and into Indonesia, New Guinea, and Australia, which was a joined land mass resulting from glacial sea-lowering (Wells 2002). However, it was not until about 13,000 years ago that humans migrated into the Americas (Flemming 1985).

The maritime landscape favored coastal adaptations of early humans for a variety of reasons. In Mesolithic times, people who had not yet adapted agriculture were organized socially primarily in mobile hunter–gatherer groups (Renfrew and Bahn 2004). The amount and variety of subsistence resources in coastal environments is far greater than in other environmental settings. All rivers, streams, and tributaries eventually flow to larger bodies of water. Therefore, the coastal environments that surround these larger bodies of water are commonly crossed by rivers, which are the primary freshwater resource. Other natural resources found in coastal environments include shellfish, fish, and marine mammals. There is evidence that shellfish were collected and used by humans living along the Mediterranean coast of southern France between 235,000 and 400,000 years ago (Flemming 1985). Also, many lithic resources are found in coastal settings, where erosion of bedrock and glacial depositional processes produce stones that were used to manufacture tools. In addition, coastal regions afford ample opportunity for travel by personal watercraft (such as the canoe), as these relatively low-lying regions are typically very accessible. This enabled early humans to expand their hunting and gathering regimes by traveling through rivers and coastal lagoons and along the shores of bays (Yesner 1980).

Land Bridges and Migration Routes

The dispersal of human populations and their pattern of settlement depend on their ability to adapt to certain environmental conditions. This ability for adaptation depends on their culture and technological advancement (Butzer 1971). The material aspects of human culture are a significant determining factor in their adaptation to changing environments. These include economy, technology, settlement, and land use (Butzer 1971), aspects that are all interconnected.

Human migration out of Africa, Europe, and Asia and into North/South America and Australia had to include either the use of personal watercraft or the following of routes across presently submerged land bridges (Wells 2002) (figure 9.9). This is a necessary fact as it would be otherwise impossible to get to these places. The only place where humans could have crossed into the Americas is from Siberia across the Beringia land bridge into Alaska, now occupied by the Bering Strait. The only place where humans could have crossed into Australia and New Guinea is across the Sunda–Sahul land bridge from Indonesia. The appearance of the first humans in these remote regions (compared to Africa and Eurasia) definitively corresponds to times of lower sea levels when the shallow

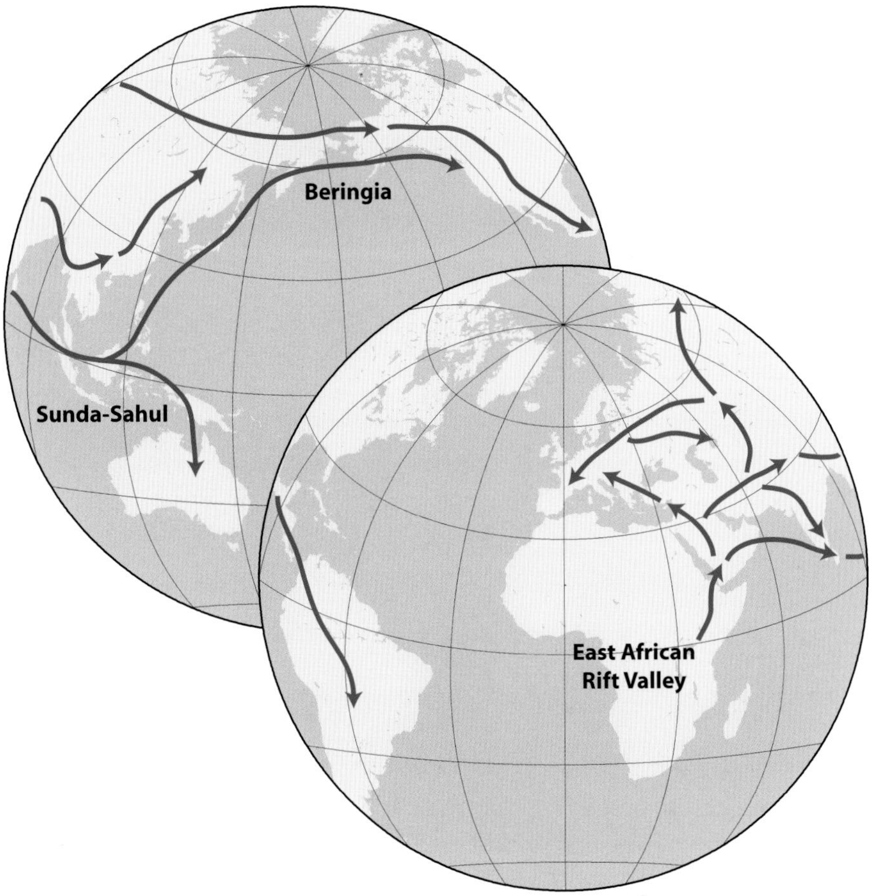

Figure 9.9. Possible migration routes of ancient human populations out of East Africa, throughout Europe and the Middle East, into Asia, and eventually into Australia via the Sunda–Sahul land bridge and into North and South America via the Beringia land bridge (modified after Wells 2002). Data for each path is based on Y-chromosome DNA studies (Wells 2002).

bodies of water that separate these land masses were exposed dry-land surfaces. As illustrated in figure 9.9, the dispersal pattern emanating from Africa to the farthest reaches of the world (barring Antarctica) required land-bridge crossings. Aside from Beringia and Sunda-Sahul, land bridges were most likely used to get people to the British Isles, Japan, and most of the Indonesian island chain. Possible land bridges could have been used by humans crossing the Mediterranean Sea as well (Flemming 1985).

In addition to human populations, the introduction of new animal species into New World environments was made possible by the exposed land bridges (Butzer 1971). Perhaps human populations followed animal migration routes. Like many indirect lines of evidence, human migration across the land bridges is unsubstantiated, and therefore is a controversial subject among anthropologists. This controversy is due to the fact that no underwater in situ sites with material evidence for human occupation have been found along suspected submerged migration routes, such as on the Beringia or Sunda-Sahul shelves (Flemming 1983). One could always argue that humans used boats for crossings, but this seems like a losing debate. Until a prehistoric human site is found on a submerged land bridge, the debate will continue.

The Beringia land bridge (figure 9.10) occupied large portions of the Chukchi and Bering Seas during times following the LGM. As sea level rose, the

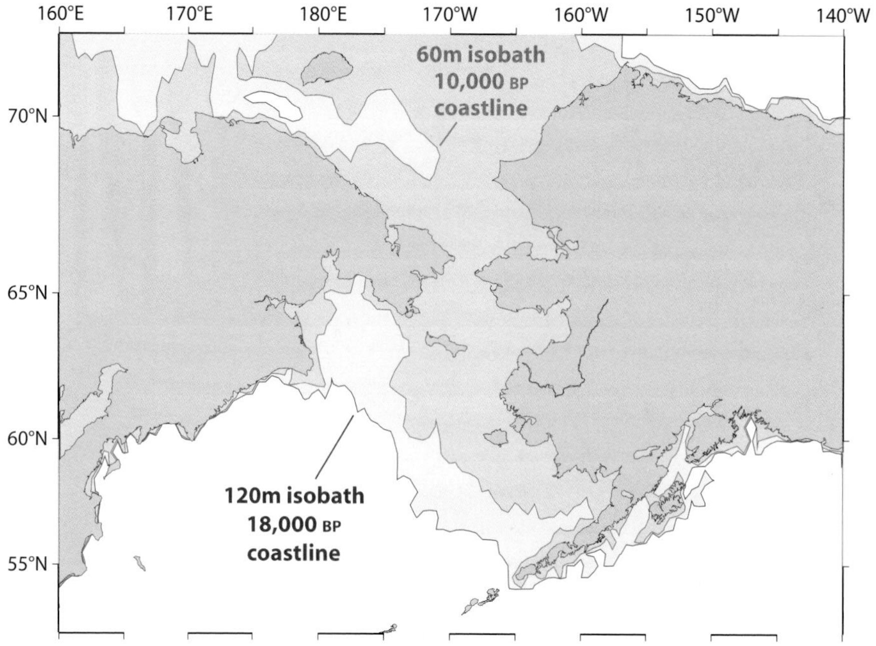

Figure 9.10. Extent of the Beringia land bridge 18,000 years ago (blue, 120-m depth contour) and 10,000 years ago (red, 60-m depth contour) based on the eustatic sea level rise curve (Fairbanks 1989).

total area of the land bridge diminished. As illustrated in figure 9.10, at 18,000 years BP, according to the Barbados sea level rise curve (Fairbanks 1989), sea level was about 120 m lower and at 10,000 BP sea level was about 60 m lower than the present day level. At either time, an enormous expanse of the submerged shelf was exposed, especially to the north beneath the Chukchi Sea. At 10,000 years BP, still more than 1,000,000 additional square kilometers of dry land was exposed between Siberia and Alaska. In fact, due the shallow nature of the Bering Strait, a land bridge existed until only about 5000 years ago, permitting humans to migrate without the use of watercraft.

Due to more extreme water depths, the Sunda–Sahul land bridge never actually connected dry land masses together (figure 9.11). However, large portions of the shallow shelves surrounding the present day islands were exposed to join together large groups of islands and narrow the gap between the Sunda and Sahul shelves. When humans first entered Australia around 50,000 years ago (Wells 2002), the islands were actually farther apart than at later times, yet closer together than today. Therefore, the use of personal watercraft was a necessity to permit crossing between adjacent land masses.

Underwater Coastal Archaeological Sites

Numerous prehistoric sites have been found in coastal environments throughout the world. For purposes here, only sites in Europe and North America are discussed. A number of these sites, especially in the Mediterranean region, are associated with ancient harbors and seaports that have become submerged during approximately the last 5000 years, and exist today in relatively shallow water. Again, for purposes here, only sites that became submerged prior to 5000 BP are discussed. Prior to 5000 BP sea level rise was very rapid, then subsequently tapered off. During the

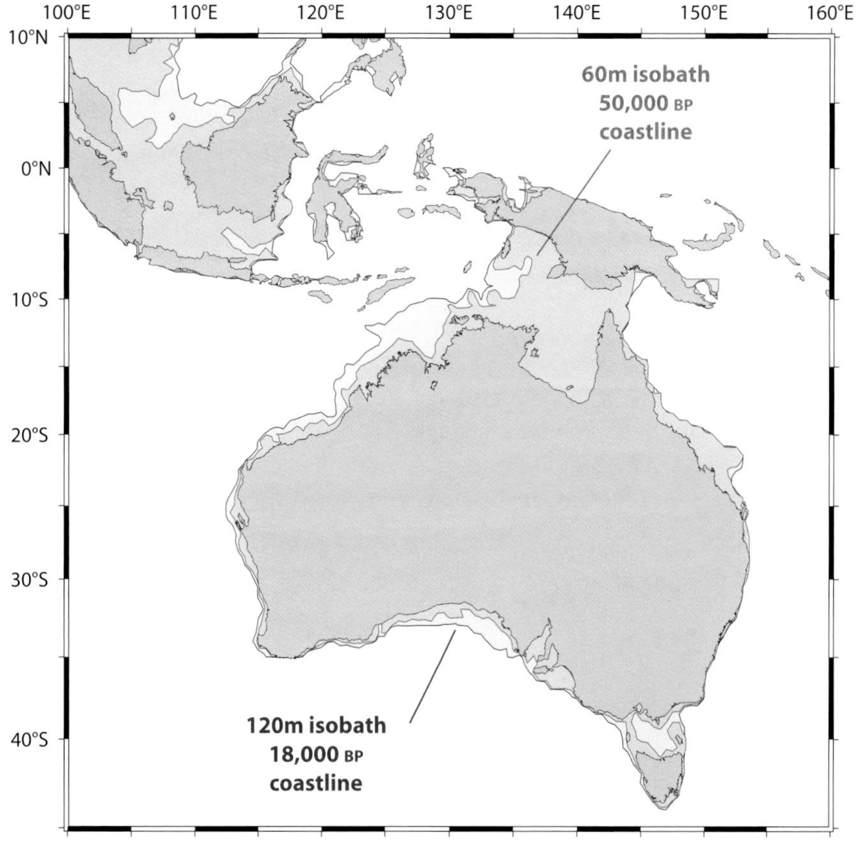

Figure 9.11. Extent of the Sunda–Sahul land bridge 18,000 years ago (blue, 120-m depth contour) based on the eustatic sea level rise curve (Fairbanks 1989) and 50,000 years ago when sea level was higher prior to the Last Glacial Maximum (red, 60-m depth contour) based on the Quaternary sea level rise curve (van Andel 1989). No dry land connection could have existed based on the extreme water depths between islands

last 5000 years sea level rise was much more gradual, and coastal and submerged archaeological sites are more abundant than during previous times.

In the Mediterranean Sea, along the south coasts of Spain, France, Italy, and Greece, a number of submerged caves have yielded clues about human occupancy during times of lower sea level (Flemming 1983). These caves, with entrances that lie under water along limestone sea cliffs, provided shelter for Paleolithic and later human populations. Many of them contain lithic artifacts and hearths, and are adorned by rock art, especially the Cosquer cave (Clottes and Courtin 1996). In northern Europe, off the Dutch coast, a number of significant submerged prehistoric sites have been discovered and documented. Inundated Mesolithic sites in the North Sea contain well-preserved remains of human inhabitants and associated artifacts, including fragile fishing implements manufactured from bone and antler (Verhart 1995).

In British Columbia, Canada, off the Queen Charlotte Islands, researchers mapped a submerged terrestrial landscape of paleo-river channels and paleoshoreline features. Associated with these features were lithic artifacts and faunal remains that indicate humans occupied this coastal environment around 10,000 BP (Josenhans et al. 1995; Fedje and Josenhans 2000). Also in North America, inundated Paleo-Indian and Archaic period sites have been discovered in the Gulf of Mexico, off the Florida coast and near the submerged Sabine River valley, off the Texas and Louisiana coast (Stright 1990). These discoveries include lithic artifacts and shell midden deposits found near submerged paleoshoreline features. Other Pacific and Atlantic inundated sites have been found on the shallow shelf, including mostly isolated finds, but some of the most significant sites were discovered in Florida. Two inundated limestone sinkholes in Florida were investigated and a diverse assemblage of submerged archaeological and faunal material was discovered and dated to between 10,000 and 13,000 BP (Stright 1990). In Warm Mineral Springs, human burial remains, faunal remains, and lithic tools were found in a high state of preservation due to the anaerobic conditions at a depth of about 13 m in the sinkhole (Clausen 1975). Nearby, in Little Salt Springs, at a depth of about 26 m in the sinkhole, two wooden stakes were found embedded between the plates of a tortoise shell (Clausen et al. 1979).

Preservation of human cultural remains under water depends on a number of factors. If artifacts are exposed on the seafloor, then they most likely will be lithic in nature. The in situ presence of stone mortars has been documented off southern California (Masters 1983). These mortars (and hundreds of other mortars previously discovered) lie in shallow water depths off La Jolla Shores. Other lithic artifacts from this site have been documented, including pestles, scrapers, and projectile points (Masters 1983). Clearly, as made evident here, stone artifacts can survive inundation. A submerged stone wall, part of a dwelling structure, which became submerged following the Holocene transgression has been documented in the Aegean Sea (Flemming 1969). Both of these examples reveal the preservation of stone artifacts that were indisputably manipulated by humans. Nonlithic artifacts would be much less likely to survive in underwater environments unless they are buried in the shallow sediment. An example is the degradation of organic remains due to inundation. A lakeside dwelling built primarily of organic material (such as wood and reeds) inundated by rising water levels would not survive very well. The process would erode away nearly the entire dwelling, leaving behind only what becomes buried, such as the wooden structural posts.

For submerged sites of human occupation on the continental shelf, preservation is strongly dependent on the rate of inundation and the slope of the shelf (Stright 1995). Bursts of rapid sea level rise during the late Quaternary transgression could have preserved ancient sites on the continental shelf. A surge around 14,000 BP raised sea level at a rate of about 3.7 m/100 years; and a second surge around 11,000 BP raised sea level at a rate of about 2.5 m/100 years (Bard et al. 1990). Such bursts in sea level rise at these times are important to the preservation of inundated sites, and these times represent significant periods in human history, especially in North and South America. The discovery of inundated prehistoric sites on the continental shelf has eluded researchers and explorers. Only now are we attempting to look.

References

Adams, J., M. Maslin, and E. Thomas (1999). Sudden climate transitions during the Quaternary. *Progress in Physical Geography* 23:1–36.

Adovasio, J. M., J. Donahue, J. E. Guilday, R. Stuckenrath, J. D. Gunn, and W. C. Johnson (1983). Meadowcroft Rockshelter and the peopling of the New World. In

Quaternary Coastlines and Marine Archaeology, ed. P. M. Masters and N. C. Flemming, 413–39. London: Academic Press.

Bard, E., B. Hamelin, and R. G. Fairbanks (1990). U-Th ages obtained by mass spectrometry in corals from Barbados; sea level during the past 130,000 years. *Nature* 346:456–8.

Barron, E. J. (1992). Paleoclimatology. In *Understanding the Earth*, ed. G. C. Brown, C. J. Hawkesworth, and R.C.L. Wilson, 485–505. Cambridge, UK: Cambridge University Press.

Bass, G. F. (1975). *Archaeology Beneath the Sea*. New York: Walker.

Belknap, D. F., and J. C. Kraft (1981). Preservation potential of transgressive coastal lithosomes on the U.S. Atlantic shelf. *Marine Geology* 42:429–42.

Blot, J-. Y. (1996). *Underwater Archaeology—Exploring the World Beneath the Sea*. Translated by Alexandra Campbell. London: Thames & Hudson.

Boothroyd, J. C., N. E. Friedrich, and S. R. McGinn (1985). Geology of microtidal coastal lagoons: Rhode Island. *Marine Geology* 63:35–76.

Broecker, W. S. (1989). Routing of meltwater from the Laurentide Ice Sheet during the Younger Dryas cold episode. *Nature* 341:318–21.

Butzer, K. W. (1971). *Environment and Archeology: An Ecological Approach to Prehistory*. Chicago: Aldine Atherton.

Center for International Earth Science Information Network (CIESIN), Columbia University. (2002). *National Aggregates of Geospatial Data: Population, Landscape and Climate Estimates (PLACE)*. Available at: http://sedac.ciesin.columbia.edu/plue/nagd/place. Accessed in September 2003. Palisades, NY: CIESIN, Columbia University.

Chappell, J., and N. J. Shackleton (1986). Oxygen isotopes and sea level. *Nature* 324: 137–40.

Clausen, C. J. (1975). The early man site at Warm Mineral Springs. *Journal of Field Archaeology* 2:191–213.

Clausen, C. J., A. D. Cohen, C. Emiliani, J. A. Holman, and J. J. Stipp (1979). Little Salt Spring, Florida: A Unique Underwater Site. *Science* 203:609–14.

Clottes, J., and J. Courtin (1996). *The Cave Beneath the Sea: Paleolithic Images at Cosquer*. New York: Times Mirror.

Culliton, T. J., M. A. Warren, T. R. Goodspeed, D. G. Remer, C. M. Blackwell, and J. J. McDonough (1990). *50 Years of Population Change Along the Nation's Coasts*. Rockville, MD: National Oceanic and Atmospheric Administration.

Dikov, N. N. (1983). The stages and routes of human occupation of the Beringian land bridge based on archaeological data. In *Quaternary Coastlines and Marine Archaeology*, ed. P. M. Masters and N. C. Flemming, 347–64. London: Academic Press.

Dunbar, J. S., and S. D. Webb (1996). Bone and ivory tools from submerged Paleoindian sites in Florida. In *The Paleoindian and Early Archaic Southeast*, ed. D. G. Anderson and K. E. Sassaman. Tuscaloosa, AL: University of Alabama Press.

Dyke, A. S., J. T. Andrews, P. U. Clark, J. H. England, G. H. Miller, J. Shaw, and J. J. Veillette (2002). The Laurentide and Innuitian ice sheets during the Last Glacial Maximum. *Quaternary Science Reviews* 21:9–31.

Edwards, R. L., and K. O. Emery (1977). Man on the Continental Shelf. *Annals of the New York Academy of Sciences* 288:245–56.

Erickson, J. (1990). *Ice Ages: Past and Future*. Blue Ridge Summit, PA: TAB Books.

Fairbanks, R. G. (1989). A 17,000-year glacio-eustatic sea level record: influence of glacial melting rates on the Younger Dryas event and deep-ocean circulation. *Nature* 342:637–42.

Fedje, D. W., and H. Josenhans (2000). Drowned forests and archaeology on the continental shelf of British Columbia, Canada. *Geology* 28:99–102.

Flemming, N. C. (1969). Archaeological evidence for eustatic change of sea level and earth movements in the Western Mediterranean during the last 2000 years. Boulder, CO: Geological Society of America Special Paper 109.

Flemming, N. C. (1983). Survival of submerged lithic and Bronze Age artifact sites: a review of case histories. In *Quaternary Coastlines and Marine Archaeology*, ed. P. M. Masters and N. C. Flemming, 135–73. London: Academic Press.

Flemming, N. C. (1985). Ice ages and human occupation of the continental shelf. *Oceanus* 28:18–25.

Goddio, F. (1998). *Alexandria—The Submerged Royal Quarters*. London: Periplus.

Harris, S. L. (2000). Archaeology and volcanism. In *Encyclopedia of Volcanoes*, ed. H. Sigurdsson, 1301–14. San Diego: Academic Press.

Haviland, W. A. (1997). *Human Evolution and Prehistory*. Fort Worth, TX: Harcourt Brace.

Josenhans, H. W., D. W. Fedje, K. W. Conway, and J. V. Barrie (1995). Post glacial sea levels on the Western Canadian continental shelf: evidence for rapid change, extensive subaerial exposure, and early human habitation. *Marine Geology* 125:73–94.

Mann, M. E., R. S. Bradley, and M. K. Hughes (1999). Northern Hemisphere temperatures during the past millennium: inferences, uncertainties, and limitations. *Geophysical Research Letters* 26:759–62.

Masters, P. M. (1983). Detection and assessment of prehistoric artifact sites off the coast of southern California. In *Quaternary Coastlines and Marine Archaeology*, ed. P. M. Masters and N. C. Flemming, 189–213. London: Academic Press.

Mather, I. R., and G. P. Watts, Jr. (2002). Ethics and underwater archaeology. In *International Handbook of Underwater Archaeology*, ed. C. V. Ruppe and J. F. Barstad, 593–607. New York: Kluwer Academic/Plenum Publishers.

Mayer, L. A., and L. R. LeBlanc (1983). The chirp sonar: a new quantitative high-resolution profiling system. In *Acoustics and the Seabed*, ed. N. G. Pace. Bath, UK: Bath University Press.

Miller, G. T. (1988). *Living in the Environment*. Belmont, CA: Wadsworth.

Milliman, J. D., and K. O. Emery (1968). Sea levels during the past 35,000 years. *Science* 162:1121–3.

Mörner, N.-A. (1980). Eustasy and geoid changes as a function of core/mantle changes. In *Earth Rheology, Isostasy, and Eustasy*, ed. N.-A. Mörner. New York: Wiley.

Morse, D. F., D. G. Anderson, and A. C. Goodyear (1996). The Pleistocene-Holocene transition in the eastern United States. In *Humans at the End of the Ice Age: The Archaeology of the Pleistocene-Holocene Transition*, ed. L. G. Straus, B. V. Eriksen, J. M. Erlandson, and D. R. Yesner, 319–38. New York: Plenum.

Muckelroy, K. (1978). *Maritime Archaeology*. New York: Cambridge University Press.

Peltier, W. R. (2002). On eustatic sea level history: Last Glacial Maximum to Holocene. *Quaternary Science Reviews* 21:377–96.

Penland, S., R. Boyd, and J. R. Suter (1988). Transgressive depositional systems of the Mississippi delta plain: a model for barrier shoreline and shelf sand development. *Journal of Sedimentary Petrology* 58:932–49.

Reeves, B.O.K. (1983). Bergs, barriers, and Beringia: reflections on the peopling of the New World. In *Quaternary Coastlines and Marine Archaeology*, ed. P. M. Masters and N. C. Flemming, 413–39. London: Academic Press.

Reich, D. E., and D. B. Goldstein (1998). Genetic evidence for a Paleolithic human population expansion in Africa. *Proceedings of the National Academy of Sciences, USA* 95:8119–23.

Renfrew, C. (1980). Ancient Bulgaria's golden treasures. *National Geographic Magazine* 158:112–29.

Renfrew, C., and P. Bahn (2004). *Archaeology—Theories, Methods, and Practice*, 4th ed. London: Thames & Hudson.
Snow, G. (1980). *The Archaeology of New England.* New World Archaeological Record. New York: Academic Press, 1980.
Stright, M. J. (1990). Archaeological sites on the North American continental shelf. In *Archaeological Geology of North America*, ed. N. P. Lasca and J. Donahue, 439–65. Boulder, CO: Geological Society of America, Centennial Special Volume 4.
Stright, M. J. (1995). Archaic period sites on the continental shelf of North America: the effect of relative sea-level changes on archaeological site locations and preservation. In *Archaeological Geology of the Archaic Period in North America*, ed. E. A. Bettis III, 131–47. Boulder, CO: Geological Society of America Special Paper 297.
Uchupi, E., N. Driscoll, R. D. Ballard, and S. T. Bolmer (2001). Drainage of late Wisconsin glacial lakes and the morphology and late quaternary stratigraphy of the New Jersey–southern New England continental shelf and slope. *Marine Geology* 172:117–45.
van Andel, T. H. (1989). Late Quaternary sea-level change and archaeology. *Antiquity* 63:733–45.
van Andel, T. H. (1994). *New Views on an Old Planet: A History of Global Change.* Cambridge, UK: Cambridge University Press.
van Andel, T. H., and P. C. Tzedakis (1996). Paleolithic landscapes of Europe and environs, 150,000–25,000 years ago: an overview. *Quaternary Science Reviews* 15: 481–500.
Verhart, L.B.M. (1995). Fishing for the Mesolithic. The North Sea: a submerged Mesolithic landscape. In *Man and Sea in the Mesolithic*, ed. A. Fischer, 291–302. Oxford, UK: Oxbow Books.
Webb, S. D., J. T. Milanich, R. Alexon, and J. S. Dunbar (1984). A *Bison Antiquus* kill site: Wacissa River, Jefferson County, Florida. *American Antiquity* 49:384–92.
Wells, S. (2002). *The Journey of Man: A Genetic Odyssey.* Princeton, NJ: Princeton University Press.
Whitmore, F. C., Jr., K. O. Emery, H.B.S. Cooke, and D.J.P. Swift (1967). Elephant teeth from the Atlantic continental shelf. *Science* 156:1477–81.
Whyte, I. D. (1995). *Climatic Change and Human Society*. London: Arnold.
Yesner, D. R. (1980). Maritime hunter-gatherers: ecology and prehistory. *Current Anthropology* 21:727–50.

10

Underwater Prehistoric Archaeological Potential of the Southern New England Continental Shelf off Block Island

Dwight F. Coleman and Kevin McBride

During the Wisconsin glaciation in the late Pleistocene Epoch, about 21,000 BP, the Laurentide ice sheet reached its southern terminus. South of what is now New England, the ice margin stretched from east to west across Long Island, Block Island, Martha's Vineyard, and Nantucket (figure 10.1) (Uchupi et al. 2001). These islands represent part of the terminal moraine, an enormous deposit of glacial sediment pushed in place by the advancing ice sheet and built upon by outwash sediment (Sirkin 1996). At this time, sea level was more than 100 m below the present level and the shoreline was near the edge of the continental shelf. As the glacier retreated, meltwater became trapped landward of the moraine and formed numerous large glacial lakes, which occupied coastal lowlands that presently form the Hudson River valley, Long Island and Block Island Sounds, Narragansett Bay, Cape Cod Bay, Nantucket Sound, and elsewhere. Recent studies indicate that these lakes subsequently underwent catastrophic drainage, primarily through two regions—the Hudson and Block Island valleys (figure 10.2) (Uchupi et al. 2001). These processes were responsible for forming much of the present-day morphology of the continental shelf off southern New England. By 19,000 BP, enough ice melted to cause rapid retreat of the glacier and by 12,000 BP, the ice margin was north of the St. Lawrence Seaway (Uchupi et al. 2001). The terrestrial landscape that is observed today throughout this region was sculpted by glacial and postglacial processes, and few of the major morphologic features on the submerged outer shelf have changed since that time. One of the more prominent features on this part of the continental shelf is the Block Island Valley, scoured away by the drainage of glacial lakes Connecticut and Block Island Sound through the Block Island Spillway (figure 10.2).

Block Island is a small island located about 20 km south of the southern Rhode Island coast and about 20 km east of the eastern tip of Long Island (see

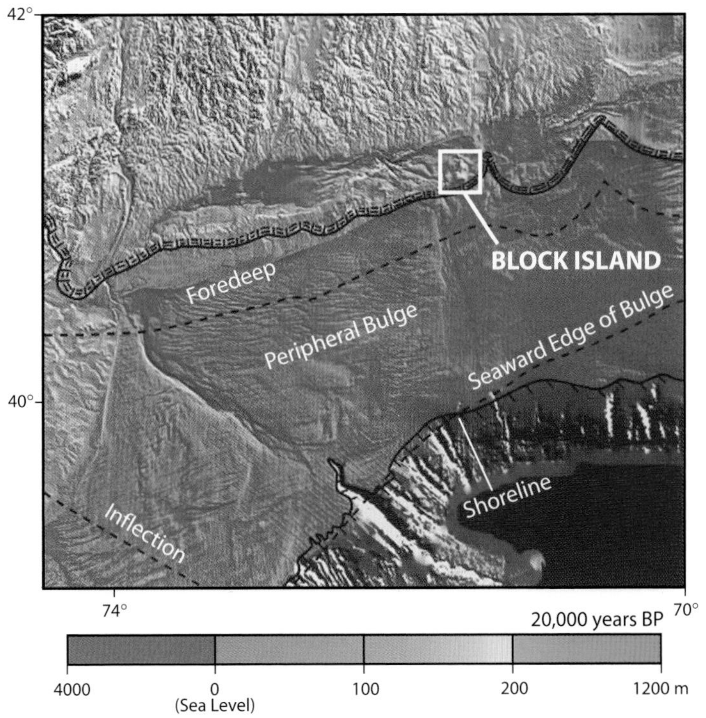

Figure 10.1. Topographic map of southern New England and adjacent continental shelf region to the south (after Uchupi et al. 2001). Southern extent of the Wisconsin glacial lobes is indicated by heavy black line. A 20,000 BP shoreline is indicated near the break of the continental shelf. Block Island is also identified. Color scale represents topographic elevations in meters.

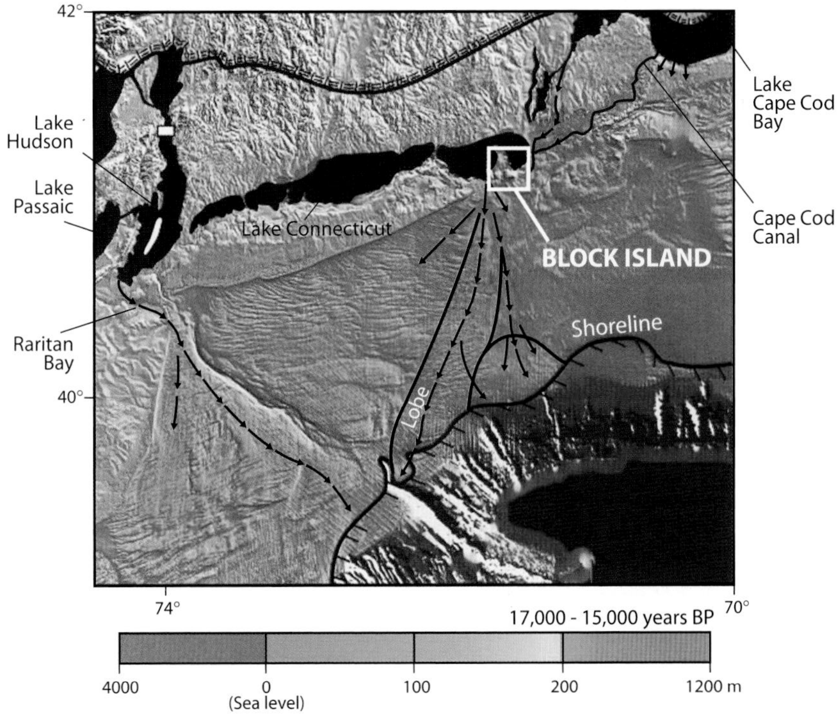

Figure 10.2. Topographic map as in figure 10.1 for the time period 17,000–15,000 BP (also after Uchupi et al. 2001). Large glacial lakes formed in what is now Long Island and Block Island Sounds, and these drained onto the outer shelf as indicated. Shoreline position retreated landward due to rising sea level. Color scale as in figure 10.1.

box in figure 10.1). It is situated on top of the aforementioned terminal moraine and is composed primarily of Pleistocene glacial drift that rests on an unconformity above Late Cretaceous shelf sediment (Sirkin 1976). Block Island is also situated in the foredeep isostatic zone, a zone of subsidence due to glacial loading that subsequently rebounded, north of the peripheral bulge, a zone that had been uplifted around the glacier, then subsequently subsided (figure 10.1) (Uchupi et al. 2001). Within this zone, subsidence occurred due to the land being depressed by the load of the glacier. After deglaciation, the zone was uplifted. Following the drainage of the glacial lakes through the spillway about 16,000 BP, sea level rose at a faster rate than the foredeep region was uplifted, resulting in inundation of the terrestrial landscape. Eventually, marine water flooded through the spillway and isolated the landmass that included Block Island.

The origin, timing, adaptation, and method of arrival of the first humans to enter the western hemisphere remains highly controversial (Dillehay and Meltzer 1991; Jablonski 2002; Powell 2005). Clovis-First theorists postulate that the first people to arrive in North America originated in Northeast Asia and entered northwestern North America across the Bering Strait land bridge between what is now Siberia and Alaska (refer to Chapter 9 for discussion of this and other land bridges). Dispersal through the rest of North America, and eventually into South America, would have been through an ice-free corridor between the Laurentide and Cordilleran ice sheets (Stanford and Bradley 2002).

An alternative theory to explain human entry into the Western Hemisphere, which does not preclude elements of the Clovis-First theory, argues that coastal migrations were involved in the initial colonization of the Americas by people utilizing boats and following the coastlines along the Pacific Rim (Erlandson 2002). In this scenario, the first inhabitants of North America may have arrived thousands of years before Paleo-Indians (i.e., Clovis people). These early colonizers, sometimes referred to as Paleo-Americans, may have arrived in the western hemisphere by a coastal route, down the west coast of North America, before they eventually turned inland. The evidence most often cited to support this scenario (or at least against a Clovis-First scenario) is (1) the widely accepted radiocarbon dates of ca. 13,000 BP for the Monte Verde site in Chile (Dillehay 1997) and ca. 14,000–16,000 BP for the Meadowcroft Rockshelter in southwestern Pennsylvania (Adovasio et al. 1990), which are both at least 1000 years before Clovis, and (2) the fact that recent research suggests that the ice-free corridor opened too late to allow entry into the continental United States and was therefore not a feasible route for the early Americans (Jackson et al. 1997).

Although their migration routes throughout North America, especially along the eastern seaboard, remain controversial, it is widely accepted that the first humans appeared in the Northeast by 11,000 BP (Bonnichsen and Will 1999). It is also theorized that Paleo-Indians in the New World settled in nearshore environments to take advantage of exploitable resources—shellfish and waterfowl for subsistence, and waterways for ease of transportation (Jones 1998, 2000; Snow 1980; Yesner 1980). The Paleo-Indian period lasted from this time until about 10,000 BP, the very end of the Pleistocene epoch. During this time period, a wide expanse of the continental shelf off southern New England was exposed. Megafaunal remains on the shelf attest to the fact that this region was inhabitable (figure 10.3) (Snow 1980), but that does not necessarily mean that

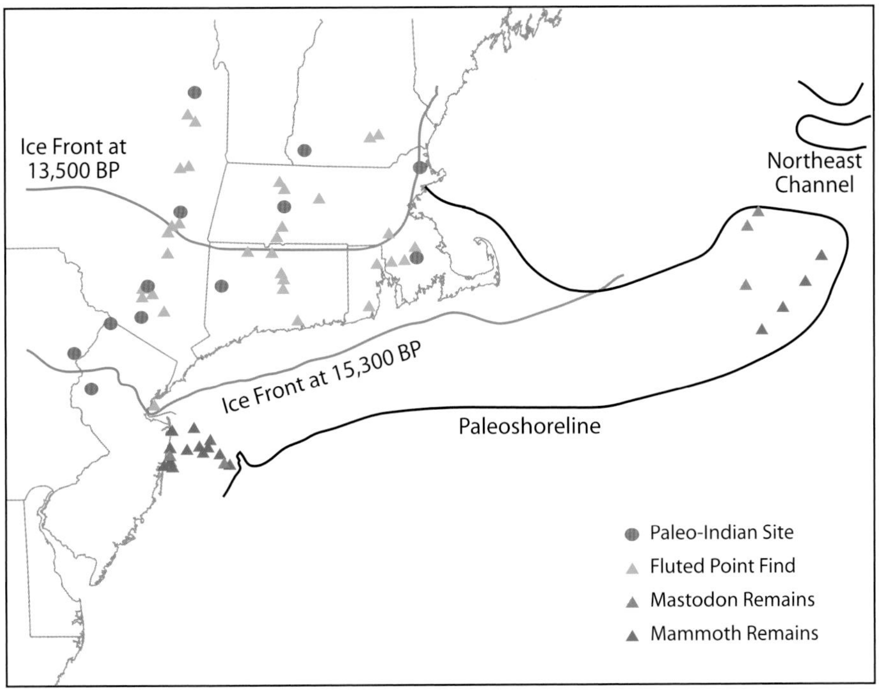

Figure 10.3. Distribution of mammoth and mastodon remains on the continental shelf and distribution of Paleo-Indian sites and fluted point finds in eastern New York, northern New Jersey, and southern New England (after Snow 1980). Positions of the ice front are accurate, but the dates are not. These are based on older data. Refer to figures 10.1 and 10.2 for more accurate dates.

it was occupied by humans, and no evidence of human occupation has been discovered or documented on the outer continental shelf, nor is there any direct evidence that Paleo-Indian people made use of coastal resources (Edwards and Emery 1977). Nonetheless, recent research on late Pleistocene and early Holocene hunter–gatherers has increasingly focused on submerged coastlines to identify evidence of early human occupation and/or to assess the timing and nature of coastal adaptations by Native people in the Northeast.

Lithic Paleo-Indian artifact finds, especially fluted points, are limited in number but widespread throughout the terrestrial northeastern United States (figure 10.3) (Snow 1980). Fluted points such as those illustrated in figure 10.4 are indicative of the Paleo-Indian culture and represent only this particular time period. Whereas lithic Paleo-Indian artifacts have not been found offshore, megafaunal remains have been found, as indicated by fossils that were recovered at depth then brought to the surface (Whitmore et al. 1967). These remains, especially teeth from mammoth and mastodon (figure 10.4), have been dredged up by modern scallop and sea clam fishermen, and the locations of these finds have been fairly well documented (figure 10.3) (Snow 1980; Whitmore et al. 1967); however, the origin of these remains is somewhat controversial. Uchupi et al. (2001) argue that the teeth and any other associated bones were probably carried there by enormous floods resulting from the catastrophic drainage of large glacial lakes as they spilled across the continental shelf from their original positions behind the terminal moraine. An alternative point of view is that these

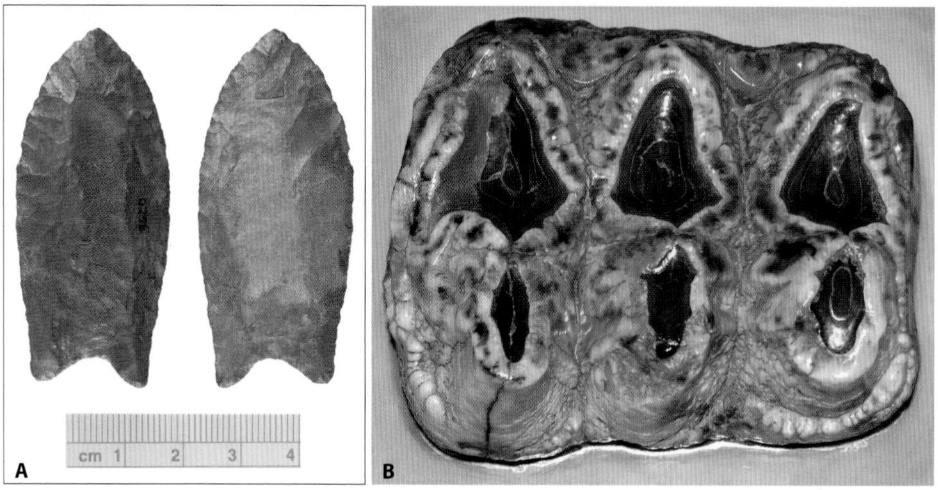

Figure 10.4. Lithic artifacts and faunal remains from Paleo-Indian times. (*Left*) An example of fluted points, typically produced by Paleo-Indians of eastern North America like those found at sites indicated in figure 10.3. (*Right*) Photograph of a mastodon tooth such as those found by fishermen on the continental shelf. (Photographs courtesy of Mashantucket Pequot Museum and Research Center)

late Pleistocene faunal remains could exist today at the location where the animal perished, more or less in situ, with some slight redistribution due to modern postdepositional processes. Because the teeth do not appear to be heavily reworked, the authors believe this latter scenario is more likely. Either way, their preservation on the shelf encourages the possibility that other remains (possibly human) or lithic artifacts may also be preserved in offshore environments.

Evidence for human association with Pleistocene fauna and megafauna is sparse for northeastern North America. Elsewhere in the Southeast and especially the Southwest and Great Plains, however, numerous dated finds provide indisputable evidence that indicate Paleo-Indians did hunt, kill, and butcher these animals (Morse et al. 1996). Whether or not people played a role in their extinction, however, remains controversial. A few examples of these finds are given here, all from Florida (Morse et al. 1996): (1) At Little Salt Spring, Clausen et al. (1979) reported on a giant land tortoise with an embedded wooden stake that clearly had been sharpened; (2) in the Wacissa River (panhandle region), Webb and others (1984) reported on a bison skull with an embedded projectile point; and (3) in several different localities, mastodon and mammoth remains have been recovered showing clear signs of being worked with tools (Dunbar and Webb 1996). One example from the Northeast also appears authentic and was found under water in Narragansett Bay, Rhode Island: A late Paleo-Indian lanceolate point that had been embedded in part of a femur from an immature bovid animal and was recovered by a shellfisherman (Turnbaugh 2002, personal communication). In addition, a number of finds of lithic artifacts from off La Jolla Shores, California (Masters 1983) and, more recently, a flake tool discovered off British Columbia in 53 m water depth (Fedje and Josenhans 2000) provide evidence that humans lived on the presently submerged shelf off western North America.

Due to the preservation of faunal remains on the southern New England continental shelf, and the potential for preservation of human remains and artifacts based on known associations, we identified several offshore areas for exploration and sampling on the shallow submerged shelf off Block Island that had been previously identified as containing preserved relict coastal features (McMaster and Garrison 1967). The time interval between 10,000 and 8000 BP and the location off Block Island were identified as being significant for a number of reasons: (1) there is evidence of early Archaic habitation on Block Island, so we wanted explore for evidence from an earlier time proximal to where sites are known to exist; (2) McMaster and Garrison (1967) identified a preserved paleoshoreline from this time period south of Block Island; (3) no submerged cultural remains from this time period have been identified anywhere throughout the region, so any discovery would potentially be significant; and (4) off the south coast of Block Island, a region that was exposed during this time interval, is a location where we did not expect to find evidence of heavy bottom-trawling fishing activity that would have disturbed the seabed.

Through close collaboration between the Institute for Exploration (Mystic, CT) and the Mashantucket Pequot Museum and Research Center (Mashantucket, CT), a research program commenced in 1997 to investigate the continental shelf off southern New England for evidence of Paleo-Indian occupation. Phase I of the program took place in 1998, when the U.S. Navy's nuclear research submarine *NR-1* (out of Groton, CT) was used to conduct a series of dives starting in the vicinity of Hudson Canyon and traversing toward the Long Island platform (Uchupi et al. 2001). The location for the transect beginning was chosen to coincide with the intersection of a paleoshoreline, the Fortune Shoreline, with the canyon wall (Uchupi et al. 2001), and near where megafaunal remains have been discovered (Whitmore et al. 1967; Edwards and Emery 1977). The transect crossed the Fortune Shoreline, and numerous visual observations were made. The paleoshoreline, which follows closely to the 68-m depth contour, is situated within an undulating topography with sediment waves aligned parallel to the strike of the shelf (Uchupi et al. 2001). Other paleoshorelines were crossed along the transect, including the Franklin and Atlantis Shorelines, and relict features such as oyster shells and glacial erratics were observed (Uchupi et al. 2001). The preservation of these relict shoreline features on the outer shelf with very little sediment overburden, combined with the occasional presence of megafaunal remains, lends credence to the notion that human artifacts could also be preserved in this environmental setting. Phase II of the program, to map the submerged environmental setting off Block Island, took place in the spring and summer of 2000, and the methods and results from this survey are presented here.

Methodology

In May 2000, a high-resolution geophysical survey of the shallow shelf off Block Island, Rhode Island, was conducted. Following the geophysical survey and data processing, in July 2000, sediment cores were collected to aid in the paleo-geographic reconstruction of Block Island as it existed between 8000 and

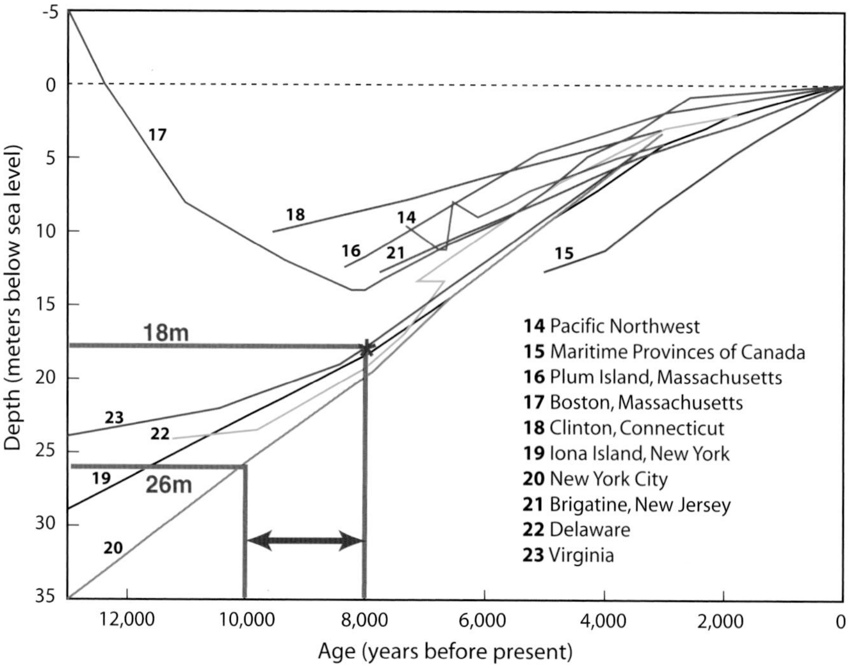

Figure 10.5. Relative sea level rise curves for the northeastern United States and maritime Canada based on a number of different studies (after Stright 1995). For nearly all curves, the 8000 to 10,000 BP paleoshoreline would be expected to exist between 18 and 26 meters water depth, with the absence of significant Holocene sedimentary overburden.

10,000 BP. Published relative sea level rise curves for the northeastern United States and maritime Canada were first examined to interpret the position of shorelines during this time period. For nearly all curves, it was determined that the 8000 to 10,000 BP shorelines could be found submerged between 18 and 26 m water depth (figure 10.5) (Stright 1995). So this was our starting point, and a bathymetric contour map was generated with contours colored to highlight this region of interest (figure 10.6). Then a different map was created that showed only Block Island inside the 26-m contour (figure 10.7). This map essentially illustrates how a "Paleo"-Block Island would have appeared with sea level 26 m lower than at present. From this map, regions were identified for more detailed surveying, including several promising areas to the west and south of Block Island. To the south, the shelf is shallow, flat, and broad and contains interesting postglacial landforms, as made evident by the side-scan mosaic (figure 10.8). To the west, the Block Island Spillway cuts a deep channel that is flanked by what appears to be submerged levees, bars, and spits. The focus for this discussion is only on the region to the south of Block Island.

Using the University of Connecticut's ship, R/V *Connecticut,* out of the Avery Point Campus in Groton, Connecticut, we surveyed a roughly 15-square-nautical-mile region to the south and west of Block Island using a Chirp side-scan sonar/subbottom profiling system (figure 10.8). We conducted the survey for three days, working 24 hours per day. There were several regions where we wanted to survey but could not due to the presence of fishing gear, such as traps

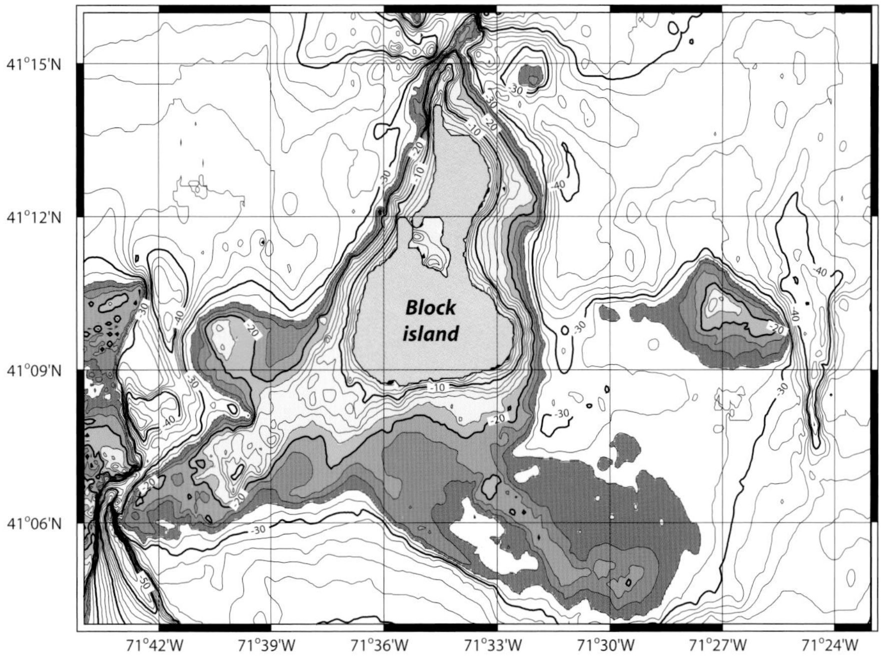

Figure 10.6. Bathymetric contour map surrounding Block Island, Rhode Island based on digital gridded soundings from the National Geophysical Data Center Coastal Relief CD-ROMs. Contours between 18 and 26 m depth are indicated by shades of green and represent the region containing the 8000- to 10,000-year-old shorelines, as suggested by figure 10.5.

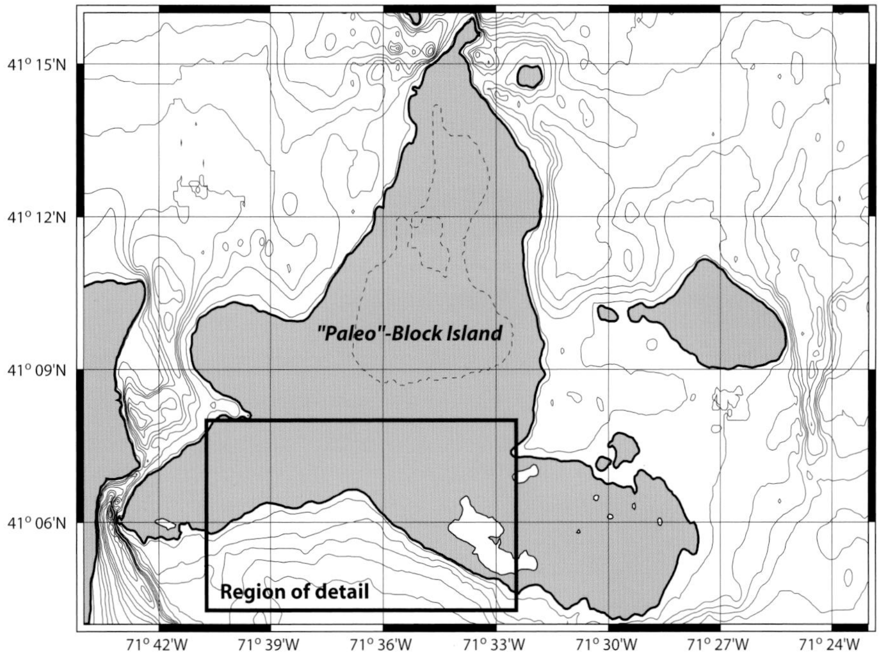

Figure 10.7. Bathymetric contour map with the entire region within the 26-m contour shaded to depict a "Paleo"-Block Island. The shape represents the shoreline positions about 10,000 years ago. The lower sea level exposes an island 3–4 times the size of present-day Block Island. Box indicates region depicted in figure 10.8.

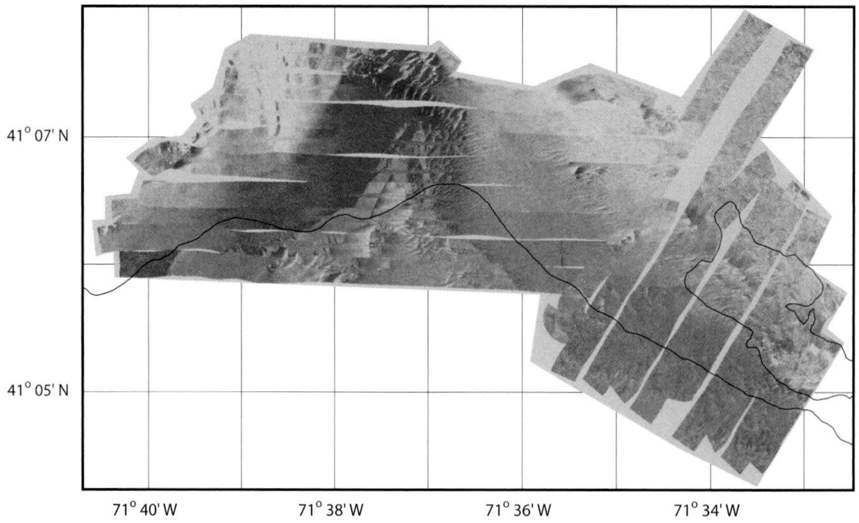

Figure 10.8. Side-scan sonar mosaic for a region south of Block Island (boxed region in figure 10.7), revealing details of the seafloor between the 8000 and 10,000 BP paleoshorelines. Lighter gray regions represent regions of low acoustic reflectivity, such as soft mud and clay. Darker gray regions represent regions of higher acoustic reflectivity, such as coarse sand and cobbles.

and nets. One area in particular, Southwest Ledge off southwestern Block Island, contained a dense array of lobster traps and gill nets, making it too risky for towing geophysical equipment, so this survey region had to be avoided. The primary survey tool was the Datasonics (Benthos) SIS-1000 seafloor imaging system. The SIS-1000 is a 100-kHz side-scan sonar towfish with an integrated Chirp subbottom profiler. A backup Datasonics Chirp II system was a secondary survey tool, but this did not collect side-scan imagery, only subbottom profiles. Position information was collected using a differential GPS receiver that was integrated to navigational software. The position and time for each sonar/subbottom ping was also written to the digital data. A Furuno echosounder was used to collect single-beam bathymetric information. All navigational and geophysical data were stored and backed up on computer disks and hard drives.

On several occasions during the survey geological sampling was attempted. A gravity corer was used to try to penetrate the sediment and collect a core sample, but these attempts failed because the sandy and rocky seafloor was too hard. On several occasions, a rock dredge was used to collect seafloor samples. This technique worked fine, but precision sampling could not be performed, just poorly located dredge hauls of the uppermost sediment along a several-hundred-meter track.

The geophysical data were processed at the U. S. Geological Survey in Woods Hole, Massachusetts. The side-scan sonar data were processed on a Unix workstation and displayed using GIS software (figure 10.8). Multiple tracklines were merged together to create a mosaic of side-scan imagery, resulting in an acoustic picture of the seafloor. Corrections had to be made for slant range, speed variations, and navigational errors. The subbottom profiler data were extracted using the Seismic Unix processing software. Subbottom tracklines were

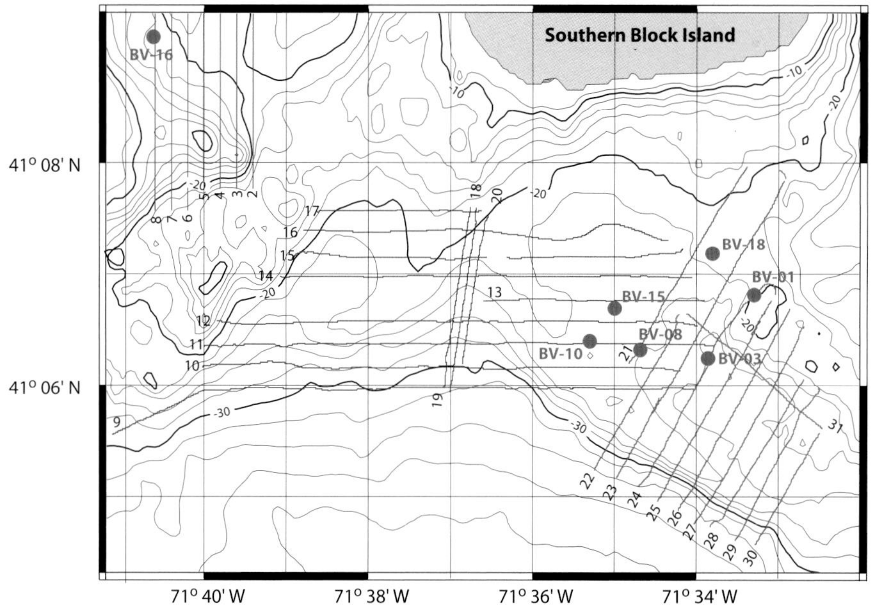

Figure 10.9. Location of selected side-scan sonar and subbottom profiler tracklines and coring sites south and west of Block Island. Subbottom tracklines are annotated every 1000 pings, except for lines 2–8, where poor navigation data was recorded.

plotted with every 1000 pings annotated so that similar subbottom features could be correlated between the cross sections and identified in plan view on the trackline chart (figure 10.9). Figures 10.10 through 10.13 show the subbottom profiles. These profiles would be better viewed on large-scale plots, but are presented here for illustration purposes. The subbottom profiles were used to interpret the subsurface geology and to identify regions for coring.

During a 4-day period in July 2000, using the M/V *Beavertail* out of Jamestown, Rhode Island, we collected nine sediment cores from selected locations identified on the subbottom profiles. Core locations are presented in table 10.1. These locations were chosen based on a preliminary interpretation of the subbottom profiles. The target was to core in specific paleo-geographic environments, including a possible paleo-river channel and paleo-barrier beach and lagoon. For each core location, a three-point anchor had to be set to secure the vessel, and then a vibracoring system was used to collect the samples (figure 10.14). As illustrated in figure 10.14, after the coring device was deployed and secured, the pneumatic vibracoring head drove the core barrel into the sediment, and then it was slowly pulled out. The core liner could be extracted from the barrel after it was decoupled from the head. Several good quality cores were collected, including within regions of coarse cobbles. Poor weather prevented us from collecting more cores, but we did successfully sample several key sites.

The cores were returned to the Mashantucket Pequot Museum and Research Center for analysis, where they were split, described, photographed, subsampled, and stored. Photographs from six of the cores are shown in figures 10.15 through 10.18. Four shells were selected from four different cores for radiocarbon dating (the only cores that contained shell material). The samples were

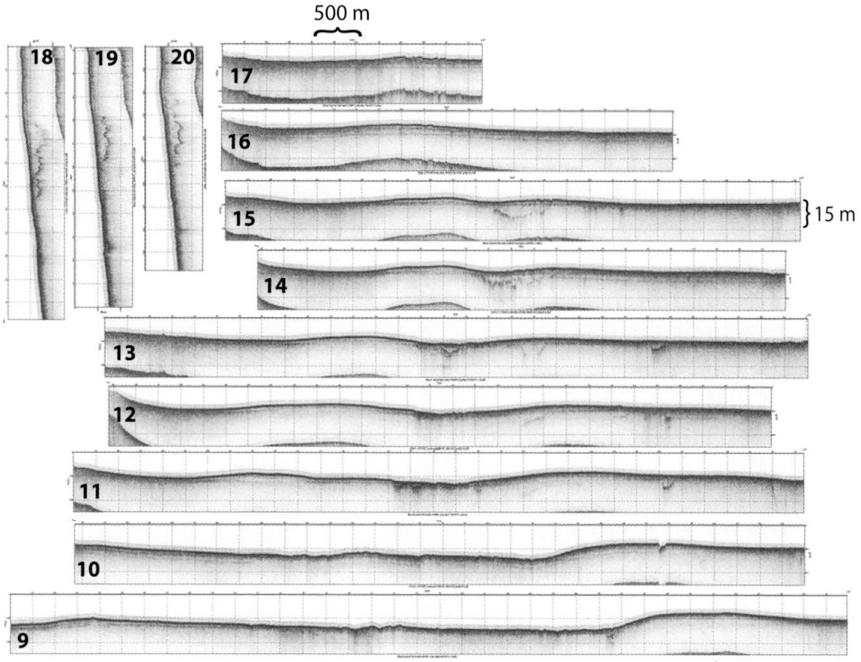

Figure 10.10. Series of east–west (and three north–south) trending parallel subbottom profiles from south of Block Island. Refer to figure 10.9 for location and figure 10.11 for more detail. Approximate vertical and horizontal scales are given. Profiles have about 20x vertical exaggeration.

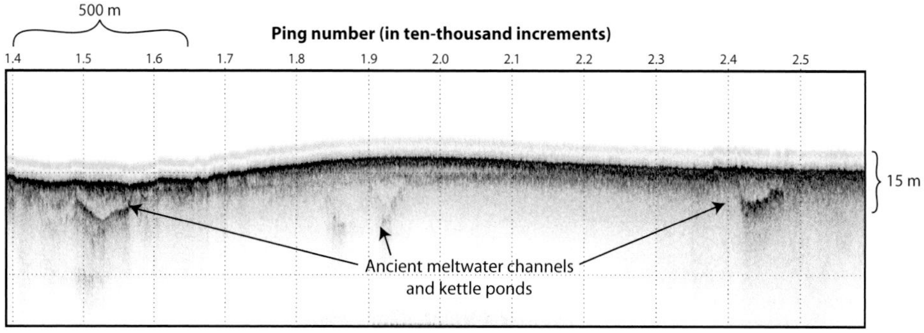

Figure 10.11. Subbottom profile section from line 13 revealing geological features buried beneath the seafloor. For scale, each dotted rectangle represents about 200 m horizontally and 15 m vertically. Profile has about 20x vertical exaggeration.

submitted to Beta Laboratories for AMS processing. The dating results are presented in table 10.2. Finally, the geophysical and sediment sampling data were combined and analyzed to generate an interpretation of the paleo-geographic setting for offshore southern Block Island. This interpretation can be used to identify specific regions for future detailed archaeological investigations.

Results and Interpretation

The side-scan sonar mosaics that were generated by splicing individual tracklines together help to reveal the continuity and spatiality of geologic features on the seafloor. As illustrated in figure 10.8, diverse patterns can be identified on the seabed, indicating different sediment types. The patterns noted by varying

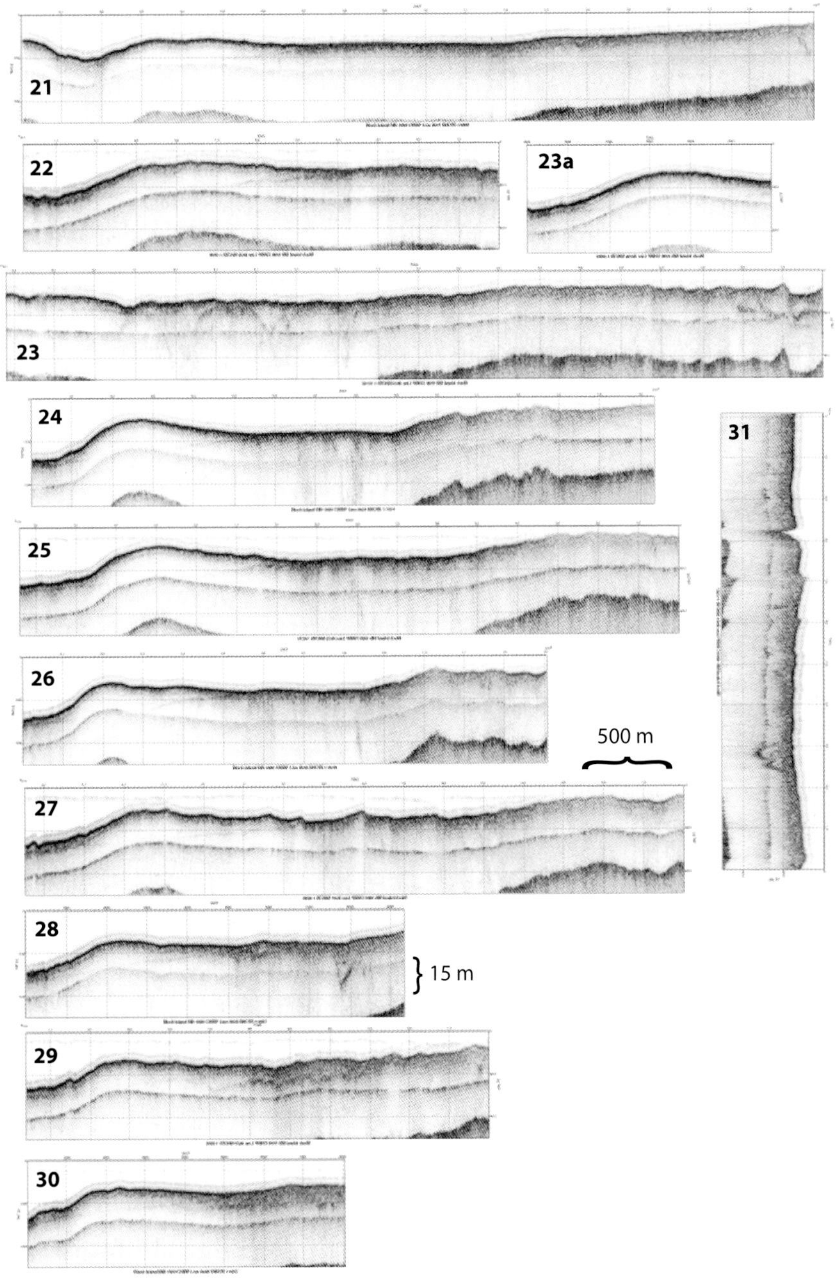

Figure 10.12. Series of northeast–southwest (and one northwest–southeast) trending parallel subbottom profiles from south of Block Island. Refer to figure 10.9 for location and figure 10.13 for more detail. Approximate vertical and horizontal scales are given. Profiles have about 20x vertical exaggeration.

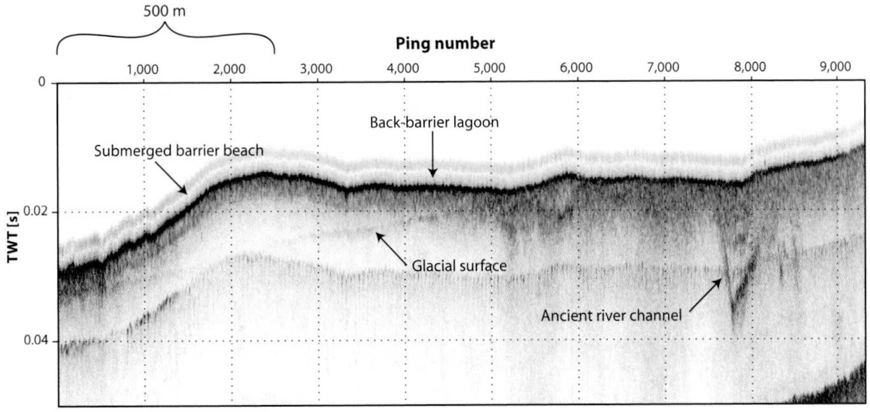

Figure 10.13. Subbottom profile of line 28 revealing an interpretation of the seafloor morphology and some subtle geological features buried beneath the seafloor. For scale, each dotted rectangle represents about 200 m horizontally and 15 m vertically. Profile has about 20x vertical exaggeration.

TABLE 10.1.
LOCATIONS OF CORES COLLECTED OFF BLOCK ISLAND

Core #	*Date*	*Time*	*Core length (m)*	*Latitude*	*Longitude*	*Water depth (m)*
BV-01	7/19/2000	1251	2.59	41° 06.844'	–71° 33.293'	20
BV-15	7/19/2000	1549	4.12	41° 06.725'	–71° 34.994'	23
BV-GSP-1	7/20/2000	1158	3.96	41° 11.079'	–71° 34.929'	06
BV-GSP-2	7/20/2000	1507	5.56	41° 11.079'	–71° 34.929'	06
BV-10	7/21/2000	0745	4.21	41° 06.433'	–71° 35.300'	24
BV-08	7/21/2000	1020	4.06	41° 06.355'	–71° 34.681'	25
BV-03	7/21/2000	1215	1.30	41° 06.277'	–71° 33.857'	28
BV-18	7/21/2000	1518	3.11	41° 07.217'	–71° 33.805'	22
BV-16	7/22/2000	0940	3.54	41° 09.271'	–71° 40.632'	23

NOTE: Times are local.

shades of gray, mostly either light or dark, correspond to regions of high or low acoustic reflectivity. Darker regions represent sediment on the seabed that are more reflective, indicating the presence of coarser material such as sand, till, or bedrock. Lighter-colored regions represent sediment on the seabed that absorbed more acoustic energy, indicating the presence of finer-grained material such as silt, mud, and clay. Very stiff, dense clay, with low water content, could produce strong reflections, however. The patterns observed in the mosaic presented in figure 10.8 are also influenced by the seafloor morphology. Large isolated features, such as large boulders, stand out as small reflective targets with an associated acoustic "shadow" that is created when the target impedes the acoustic pulses from returning off the seabed behind the target. Other depressions and prominent features are represented by different sonar patterns. Large sand waves on the seafloor are made evident in the mosaic by alternating wavy bands of light and dark gray. One significant result from interpreting the sonar data is the lack of evidence for bottom trawling, indicating that this region is not heavily fished, and therefore anything remaining on the seafloor has a higher chance for preservation.

The subbottom profiler data (figures 10.10–10.13) reveal a number of diverse relict geomorphologic features buried beneath the seafloor. The Chirp system is relatively high frequency compared to other subbottom imaging systems. Due to the presence of coarse deposits in the area we mapped, the frequency of this system was for the most part inappropriate—too high for sufficient penetration beneath the uppermost layers. Despite this fact, several prominent features can

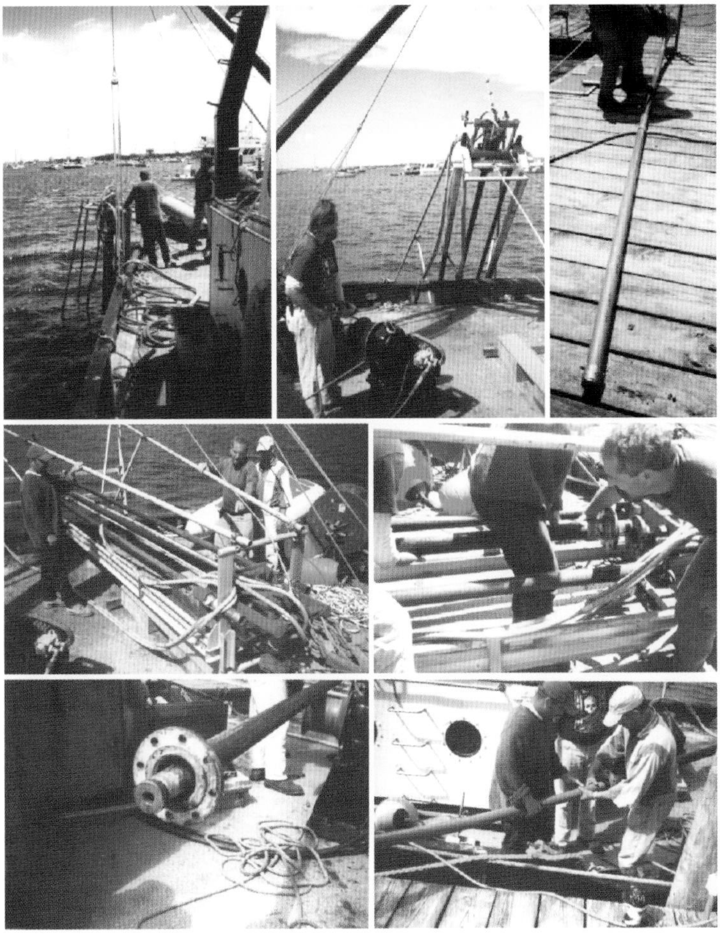

Figure 10.14. Vibracoring operation. The device is deployed over the rail of the vessel. Compressed air provides the power to the vibrating head, which drives the core barrel down into the sediments. The barrel is then decoupled from the head, and the core tube inside the barrel is extruded.

easily be discerned. Two distinct regions are represented here. The first is a series of east–west-trending, contour-parallel profiles collected over a relatively flat portion of the shelf south of Block Island (figures 10.10, 10.11). The second is a series of northeast–southwest-trending, contour-normal profiles collected over a slightly hummocky portion of the shelf immediately adjacent to the southeast of Block Island (figures 10.12, 10.13).

Figure 10.10 presents all of the lines from the first region at a small scale, but arranged more or less accurately relative to each other so features can be correlated. Three north–south-trending lines (lines 18–20) were run nearly perpendicular to the east–west-trending lines (lines 9–17) and are plotted sideways in the upper left portion of the figure. Throughout most of these profiles, prominent subbottom channel-like features can be identified and quantitatively mapped. On the right half of lines 9–15, a meandering channel is observed that at its widest and deepest point along line 11 is about 8 m deep and 100 m wide. Its shape, as indicated in this profile, is asymmetrical, typical of what would be expected in a stream meander cross section. This channel was likely scoured by runoff from the melting glacier during times just following its retreat. The channel probably remained as a low-lying riverbed or dry channel for several

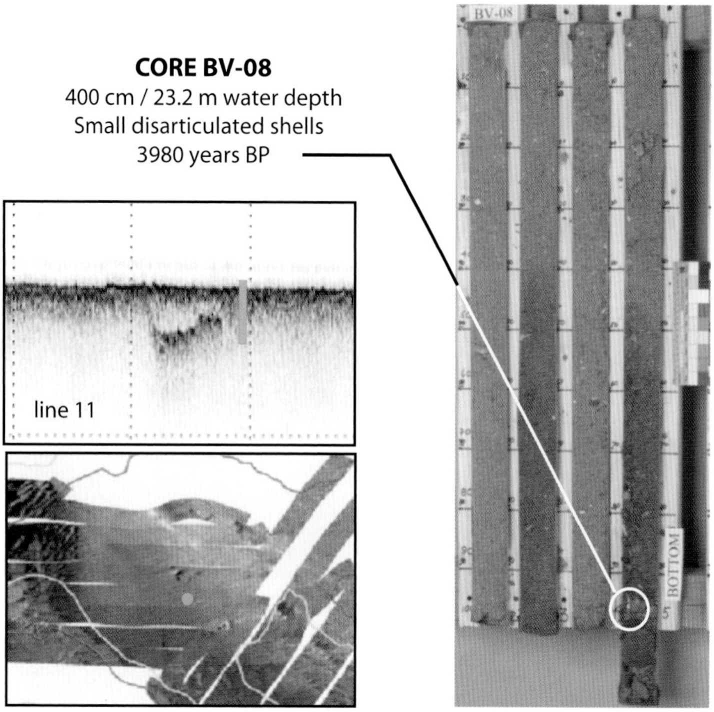

Figure 10.15. Core BV-08, collected immediately to the east of the meltwater channel indicated in the subbottom profile for line 11. Refer to figure 10.9 for location. The core primarily consists of reworked coarse beach sand. Some coarse glacial till was present at 4 m below the seafloor. A radiocarbon date from a disarticulated shell just above this till layer is 3980 BP.

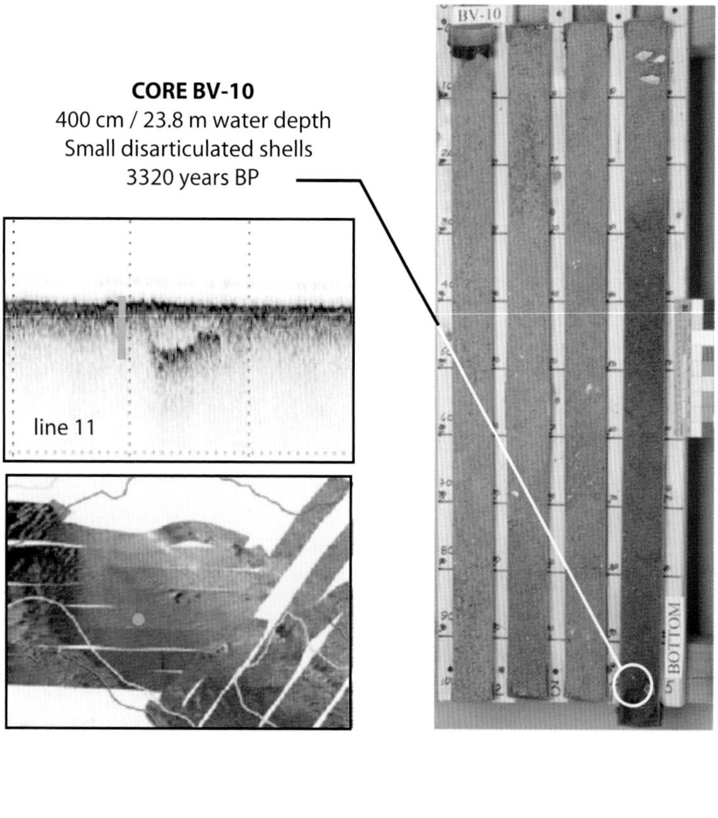

Figure 10.16. Core BV-10, collected immediately west of the same channel indicated in line 11. Refer to figure 10.9 for location. The same glacial till layer was encountered at 4 m below the seafloor. Another disarticulated shell from just above this layer dated to 3320 BP.

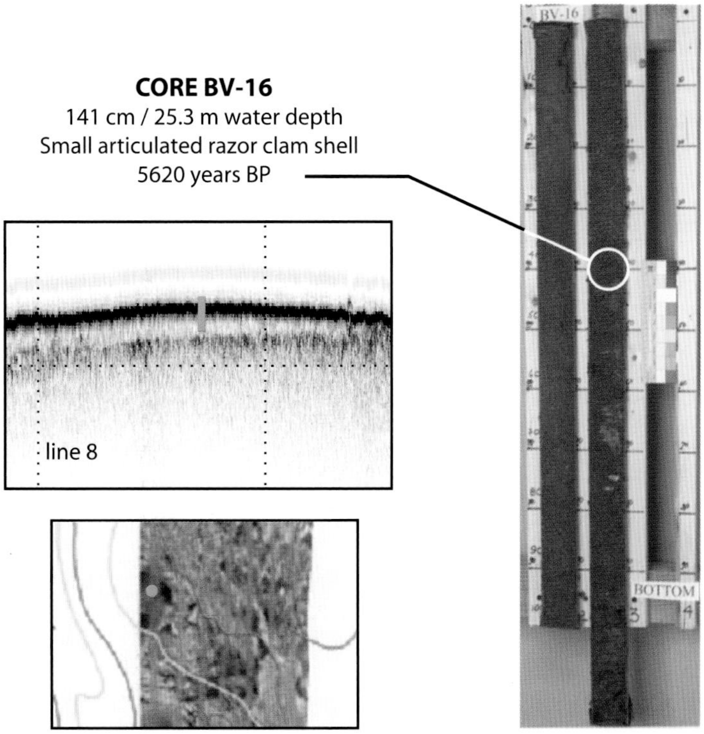

Figure 10.17. Core BV-16, collected west of Block Island. Refer to figure 10.9 for location. The high attenuation of the acoustic signal indicated in the upper layer just below the surface in the subbottom profile was interpreted to be an organic-rich layer, possibly peat. Core penetration was insufficient to test this theory, but a radiocarbon date from an articulated razor clam shell revealed the deposit was 5620 years old at 140 cm below the surface, indicating the preservation of fragile coastal features by a thin layer of sediment.

millennia following deglaciation, until inundation occurred some time after 8000 BP. Just prior to inundation, after the island became isolated, this channel was likely to be a small freshwater tributary that evolved into a tidal inlet with rising sea level. Other features can be identified as well, which do not have the same spatial continuity as this channel. Toward the center of line 13 (figure 10.11), a strong reflector can be identified that could be interpreted as part of an ancient tidal inlet that connects to a paleo-lagoon landward of this feature. This possible lagoon feature can be identified in lines 18–20 and on the east–west lines north of line 13. Other small features could be kettle holes that survived as ponds until they became inundated. All of these features became buried following inundation. The marine transgression submerged the landscape, and modern sediments filled the relict channels.

Figure 10.12 presents all of the lines (lines 21–30) from the second region; plotted so similar features can be correlated, with one northwest–southeast-trending line (line 31) that ran perpendicular to the others plotted sideways. For these profiles, coherent subbottom features are difficult to distinguish, but the general surface morphologic features can be correlated. One possible interpretation, based on shape alone, is that this represents a submerged relict barrier beach and lagoon system as originally identified by McMaster and Garrison (1967). This interpretation is also given for line 28 (figure 10.13). In this figure, the horizontal axis is roughly 3 km long and the vertical axis is roughly 40 m deep, and water depth is about 26 m. The left side of figure 10.13 represents the ancient shoreline to the southwest. A weak subbottom reflector exists below the sloping shoreline (barrier beach), and this can be correlated to adjacent profiles. This reflector slopes gently away to the southwest, toward the open ocean, and could represent the original glaciated land surface on which the barrier beach was built. Working across the profile from left to right (southwest to northeast), the seafloor slopes back down into an area interpreted as a relict back-barrier lagoon. Other interpretations are possible for this geomorphologic feature, however. It could represent an area that has been scoured by longshore currents, or it could just be a depression into a relict kettle pond. Continuing right, the seafloor then starts to slope back up to what is interpreted as a shoreface behind the lagoon.

Figure 10.18. Photographs of cores BV-15, BV-03, and BV-18. Refer to figure 10.9 for location. BV-15 was collected from inside the meltwater channel between, but north of BV-08 (figure 10.13) and BV-10 (figure 10.14). BV-03 and BV-18 were collected along line 24, which looks similar to line 28 (figure 10.11). Refer to text for interpretation and discussion.

TABLE 10.2.
RESULTS FROM RADIOCARBON DATING OF SHELLS FROM WITHIN SELECTED CORES

Core #	*Depth in core (cm)*	*Shell description*	*Age*
BV-GSP-1	105	*Crepidula fornicata,* articulated	109
BV-08	405	Small disarticulated clam shell	3980
BV-10	402	Small disarticulated clam shell	3320
BV-16	141	Small articulated razor clam shell	5620

NOTE: Age in radiocarbon years BP.

Figures 10.15–10.18 show photographs of selected cores, including their locations on the side-scan sonar and subbottom imagery. (Refer to table 10.1 for core locations and table 10.2 for the results of radiocarbon dating of shells.) Core BV-08 was collected immediately east of the channel (figure 10.15). It consists primarily of reworked coarse beach sand containing broken shell fragments and rounded beach cobbles. At a depth of about 4 m in the core, very coarse glacial till was encountered, probably representing a terrestrial paleo-land surface that was scoured by the glacier. Just above this layer, a shell was collected and submitted for radiocarbon dating. This shell dated to 3980 BP. Because the shell was disarticulated, it was probably not collected in situ, but had been transported there by the reworking and redistribution of sand. Similarly, in core BV-10, which was collected immediately west of the channel (figure 10.16), a disarticulated shell fragment from about 4 m deep in the core dated to 3320 BP. This core also contained a thick deposit of reworked coarse beach sand, but no till layer was encountered.

One core was collected from a region not mentioned in detail here, but it is worth noting. Core BV-16 was collected to the west of Block Island in an area that was interpreted in the subbottom profile as possibly containing peat, a sediment deposit with high organic content (figure 10.17). In this figure, as revealed by the subbottom profile, a layer of no reflections, or high attenuation of the Chirp signal, is evident. This could be due to a number of factors, but one that is likely is that it represents an organic-rich layer of sediment. The core did not penetrate deep enough to confirm this, but an articulated razor clam sampled from 1.4 m deep in the core gave a date of 5620 radiocarbon years BP. This deposit had not been reworked extensively like the deposits south of Block Island, based on the interpretation of cores BV-08 and BV-10. In addition, based on the radiocarbon date, the rate of sediment deposition is significantly slower in this region compared to the south, resulting in a much thinner layer of Holocene sedimentary overburden.

The next series of cores did not contain any shell material, so no radiocarbon dates were obtained (cores BV-15, BV-03, and BV-18, left to right, figure 10.18). Core BV-15 was collected from inside the channel, located between but north of cores BV-08 and BV-10. It contains a very thick deposit of reworked beach sand. Unlike the flanking cores, this core contained many more rounded cobbles mixed in with the coarse sand. Also unlike the flanking cores, no glacial till was encountered at the bottom. Cores BV-03 and BV-18 were collected in the second region, to the southeast. Only about 1.25 m of penetration was achieved in both of these cores (figure 10.18). Core BV-03 was collected from slightly deeper water from the region that was interpreted in the subbottom profiles as a back-barrier lagoon. Below a thin layer of coarse sand, a thick deposit of dark gray marine clay dominates the rest of the core. Core BV-18 was collected from shallower water from the region interpreted in the subbottom profiles as the terrestrial shoreface landward of the back-barrier lagoon. This core encountered coarse glacial till at only 1 m depth. Above the till layer, a tan-colored coarse beach sand layer was found at the top, which graded downward to a dark brown finer-grained, silty sand (figure 10.18).

An interpretive fence diagram illustrating the relative vertical elevation and relationships among all of the aforementioned cores, except core BV-16, is presented in figure 10.19. The east-to-west portion of the diagram was chosen to fit between cores BV-10, BV-15, and BV-08, so their actual locations are projected onto this line. Then the diagram hinges to run through cores BV-03 and BV-18. The horizontal scale is arbitrary, but the vertical scale depicts true relative positions of the cores. Three primary lithostratigraphic units are present: marine clay, coarse reworked sand, and glacial till. Not all cores in the fence diagram contain each unit, but a correlation can still be made, as indicated in the figure. From east to west, the region south of Block Island represents primarily a depositional sequence of coarse reworked beach sand that is redistributed by littoral drift caused by longshore currents. This environment is similar to an environment mapped off Fire Island (central Long Island) that also contained paleo-channels and reworked coarse sand (Schwab et al. 1999), although the deposits described here are generally younger. From north to south, and further to the east, the diagram illustrates a different environment. A very thin layer of reworked coarse sand overlies a sequence of marine clay. The clay was deposited in a depression that may represent a relict coastal lagoon, as described above, although no dates were obtained (figure 10.19). If the thin layer of sand correlates with the thicker sequence to the north (at core BV-18), then it is possible that the clay sequence would lie above the Pleistocene glacial till representing an ancient terrestrial surface, but this sequence is not present in shallower depths. Nevertheless, it can be confirmed that the environment represented here has significantly thinner overburden covering the ancient surfaces.

Two additional cores were collected from inside Great Salt Pond, where New Harbor is located on Block Island, while the vessel was tied alongside the pier (figure 10.14). One of these cores (BV-GSP-2) was more than 5 m long, and revealed some details of the geologic history of the harbor. At the base of the core, we found glacial till, which represents the land surface just after retreat of the glacier. Since this was a low-lying area, it filled with water to form a lake. The next sequence of sediment is very thick varved clays, which represent finely laminated mud deposited in a pond that froze over seasonally. Then a dramatic change occurred, when sea level rose and marine water inundated the lake, forming the harbor. On top of the varved clays is a dark, organic-rich marine mud, complete with clam and snail shells, representing this transition from a lacustrine to a marine environment.

Although a few data were collected, there is enough information to formulate a possible interpretation of the paleo-geography and transgression history for the region south of Block Island. This interpretation is presented in figure

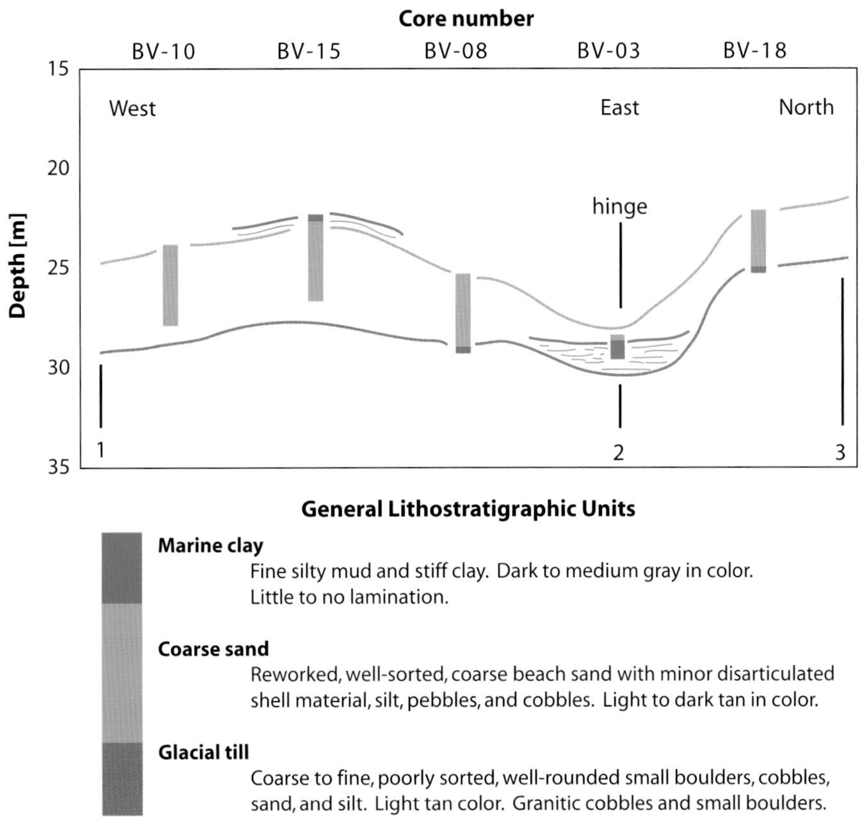

Figure 10.19. Fence diagram showing general details of five cores collected south of Block Island. Refer to figure 10.9 for core locations. Cross-section line 1-2-3 indicated in figure 10.20. Refer to text for discussion.

10.20, which illustrates the environmental conditions prior to about 10,000 years ago, with the Holocene sedimentary overburden stripped away, leaving behind the primary geomorphologic features identified in the subbottom profiles. A low elevation, broad, flat coastal plain existed that contained a number of depressions and channel-like features. These low-lying regions that probably once contained shallow coastal ponds were the first regions to become inundated by rising sea level. A small tidal inlet formed that connected the ponds to the ocean, by cutting through the beach shoreface or barrier, creating a spit and inundating the pond with marine water. For these features to be preserved, the shoreline probably stabilized for a time period around the Younger Dryas (approximately 11,000 BP; Fairbanks 1989) when the rate of sea level rise slowed. A transgressive shoreline sequence developed stratigraphically above the inundated landscape. This sequence, which lies above the glaciated land surface and till layer, then retreated inland as sea level rose to higher elevations. After sea level reached a point within about 10 m of the present-day level, about 5000 BP, this transgressive shoreline deposit became reworked and redeposited over the shallow shelf, creating the geology that is observed today. The thickness of the reworked deposit varies from place to place, primarily governed by the subsurface morphology.

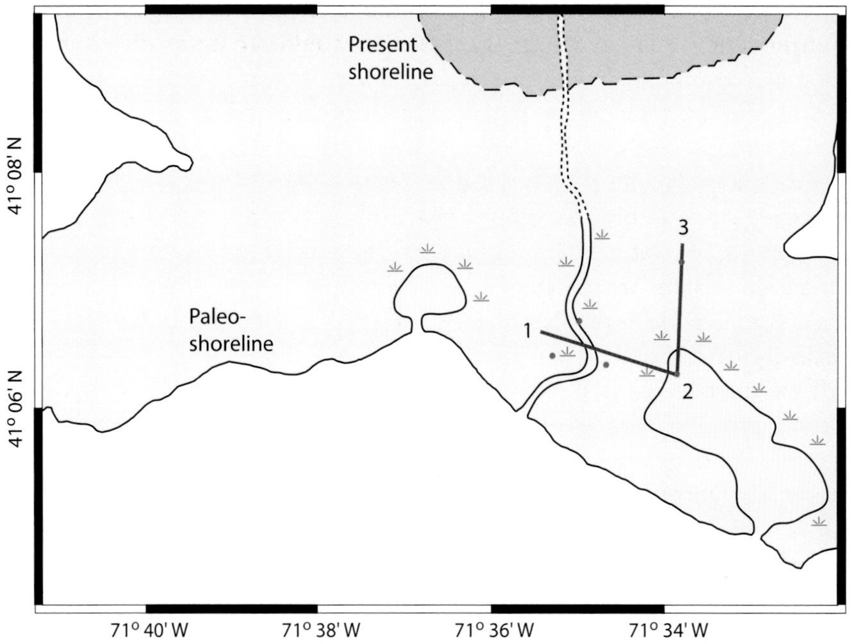

Figure 10.20. Paleogeographic reconstruction off southern Block Island for the approximate time period between 8000 and 10,000 BP based on interpretation of subbottom profiles. Areas shaded blue represent relict geomorphologic features, interpreted here as the remnants of ancient lagoons and river channels. Cross-section line 1-2-3, indicated by blue line, is shown in detail in figure 10.19.

Discussion and Conclusions

This project represents a first approach to characterizing the paleo-environmental setting of a region south of Block Island, Rhode Island, as it existed approximately between 8000 and 10,000 BP. The purpose for this reconstruction is to identify potential geographic settings that may have been favored by Paleo-Indians. The survey and sampling strategy discussed here proved effective for characterizing geomorphologic features for specific regions. High-resolution geophysical data and sediment cores provide the baseline for identifying future sites for more detailed archaeological investigations. At the same time, more information was learned about the sedimentological history and depositional context for this particular shelf region. However, the presence of thick sequences of Holocene sediment creates an overburden, burying potential artifacts and other evidence of human occupation such as shell middens, hearths and living floors, or burial grounds. The preservation of shoreline features such as the ones described here has been investigated by others (Belknap and Kraft 1981; Josenhans et al. 1995), and the main conclusion is that features on the outer shelf have a much better chance at being well preserved than features on the inner shelf. This has been confirmed by more recent work (Uchupi et al. 2001), where significant relict features on the outer shelf have been visually identified. Deeper regions on the outer shelf became inundated during a rapid rise in sea level and their existence far offshore has kept them protected from the influence of tidal currents and near shore sedimentary processes that could destroy and/or bury these regions.

Belknap and Kraft (1981) have identified significant variables that influence the preservation of paleo-coastal features. These are subsurface topography, erosion depth and resistance, waves, tides, sediment supply, and relative sea level rise. Nearshore Block Island, which is on the inner shelf in a region with large sediment supply and significant wave energy, would not be predicted to have high preservation potential for fragile coastal features, and this does appear to be the case. However, during the early part of the Holocene in this region, sea level rise was comparatively rapid (Belknap and Kraft 1981), increasing the chance for preservation further offshore. Penland et al. (1988) described the evolution of barrier systems during transgressions and noted that the uppermost sedimentary sequences do not survive transgressions. These features (dunes and beaches) become inundated, eroded, and finally formed into low-relief offshore sandy shoals. If this model is correct, then there would be little chance for survival of archaeological material.

If the coastal features interpreted here (lagoons, inlets, shorelines, and channels) are preserved, then paleo-geographic regions can be accurately identified for future sampling. The interpretation presented in figure 10.20 reveals features that can be compared to modern-day coastal features for the Rhode Island shoreline. A semiqualitative comparison between these features and those that exist today along the southern Rhode Island coast is that the size, relief, and depth of the inlets and coastal lagoons are very similar. Boothroyd et al. (1985) presented an overview of the geology of present-day microtidal coastal lagoons in Rhode Island. On average, these lagoons are several kilometers long by a few kilometers wide, with narrow and shallow tidal inlets connecting them to the ocean. These lagoons probably were nearshore kettle holes containing freshwater wetlands that were inundated by rising marine waters (Boothroyd et al. 1985). The model presented here based on the subbottom profiles is similar. Unfortunately, however, the amount of core data collected are insufficient to establish the paleo-environment completely. If freshwater and saltwater peat deposits had been recovered in the cores, then this would have further reinforced the paleogeographic interpretation.

Based on the interpretations presented here, the paleo-environments south of Block Island are not ideally suited for archaeological sampling, although more work needs to be done, particularly in the region that we have interpreted as remnant of a back-barrier lagoon. Due to the amount of reworked sediment immediately south of the island (more than 4 m in 4000 years), further investigation or sampling in this area would probably not yield any evidence of Paleo-Indian occupation. The region further to the southeast, however, does have less overburden, and could possibly contain relict coastal features that could have been attractive to humans for hunting and shellfishing. The region of Southwest Ledge, where we were not able to survey or sample, could be promising for future work. To the west of Block Island, where core BV-16 was collected, there appears to be better preservation of older material not very far beneath the surface. This region is close to the Block Island Spillway, which currently is an erosional environment, but with preserved glacial cobbles nearby, indicating very little modern deposition. In addition, the subbottom profiles seem to indicate the potential presence of organic mud and/or peat, so this represents an area we have identified for future work. Elsewhere, no data were collected north or east of Block Island. Bathymetric contours in these regions are much closer together, indicating steeper topography (figure 10.6) than to the south and east, however, so these environments did not have a broad coastal plain 10,000 years ago, but they could still be favorable areas to search because the wave energy and sediment reworking would be reduced.

In conclusion, we feel that the reconstructions presented here represent a good start to understanding the paleo-environement in this region. In terrestrial archaeological surveys, a first approach to understanding a site is to gain an understanding of the present day natural environment and geological context of the site. The approach then applies data and clues collected from the site to reconstruct the past. Therefore the approach to understanding the submerged sites off Block Island that are presented here can be applied to other regions of archaeological interest and the collection of geological and geophysical data to enable paleo-geographic reconstructions must be included.

References

Adovasio, J.M.J., J. Donahue, and R. Stuckenrath (1990). The Meadowcroft Rockshelter radiocarbon chronology 1975–1990. *American Antiquity* 55:348–54.

Belknap, D. F., and J. C. Kraft (1981). Preservation potential of transgressive coastal lithosomes on the U.S. Atlantic shelf. *Marine Geology* 42:429–42.

Bonnichsen, R., and R. Will (1999). Radiocarbon Chronology of Northeastern Peleo-American sites: discriminating natural and Human Burn Features. In *Ice Age People of North America: Environments, Origins, and Adaptations*, ed. R. Bonnichsen and K. Turnmire, 395–415. Corvallis: Center for the Study of the First Americans, Oregon State University.

Boothroyd, J. C., N. E. Friedrich, and S. R. McGinn (1985). Geology of microtidal coastal lagoons: Rhode Island. *Marine Geology* 63:35–76.

Clausen, C. J., A. D. Cohen, C. Emiliani, J. A. Holman, and J. J. Stipp (1979). Little Salt Spring, Florida: a unique underwater site. *Science* 203:609–14.

Dillehay, T. D. (1997). *Monte Verde*, Vol. 2. Washington, DC: Smithsonian Institution Press.

Dillehay, T. D., and D. J. Meltzer, eds. (1991). *The First Americans: Search and Research.* Boston: CRC Press.

Dunbar, J. S., and S. D. Webb (1996). Bone and ivory tools from submerged Paleoindian sites in Florida. In *The Paleoindian and Early Archaic Southeast*, ed. D. G. Anderson and K. E. Sassaman. Tuscaloosa, AL: University of Alabama Press.

Edwards, R. L., and K. O. Emery (1977). Man on the Continental Shelf. *Annals of the New York Academy of Sciences* 288:245–56.

Erlandson, J. M. (2002). Anatomically Modern Humans, maritime Voyaging, and the Pleistocene of the Americas. In *The First Americans: The Pleistocene Colonization of the New World*, ed. N. G. Jablonski, 59–92. Memoirs of the California Academy of Sciences, Number 27.

Fairbanks, R. G. (1989). A 17,000-year glacio-eustatic sea level record: influence of glacial melting rates on the Younger Dryas event and deep-ocean circulation. *Nature* 342:637–42.

Fedje, D. W., and H. Josenhans (2000). Drowned forests and archaeology on the continental shelf of British Columbia, Canada. *Geology* 28:99–102.

Jablonski, Nina G., ed. (2002). The First Americans: The Pleistocene Colonization of the New World. *Memoirs of the California Academy of Sciences, No. 27.* San Francisco: Allen Press.

Jackson, L. E. Jr., F. M. Phillips, K. Shimamura, and E. C. Little (1997). Sosmogenic 36CI fdating of the Foothills Erratics Train, Alberta, Canada. *Geology*, 25:195–98.

Jones, B. D. (1998). Human Adaptation to the Changing Northeastern Environment at the End of the Pleistocene. Unpublished PhD Dissertation, University of Connecticut. University Microfilms No. 9906705, UMI, Ann Arbor.

———. (2000). Constraints and Assumptions for Modeling the PaleoIndian Colonization of new England. Paper Presented at the Conference on new England Archaeology, Sturbridge, MA.

Masters, P. M. (1983). Detection and assessment of prehistoric artifact sites off the coast of southern California. In *Quaternary Coastlines and Marine Archaeology*, ed. Masters, P. M., and N. C. Flemming, 189–213. London: Academic Press.

McMaster, R. L., and L. E. Garrison (1967). A submerged Holocene shoreline near Block Island, Rhode Island. *Journal of Geology* 75:335–40.

Morse, D. F., D. G. Anderson, and A. C. Goodyear (1996). The Pleistocene–Holocene transition in the eastern United States. In *Humans at the End of the Ice Age: The Archaeology of the Pleistocene–Holocene Transition*, ed. L. G. Straus, B. V. Eriksen, J. M. Erlandson, and D. R. Yesner, 319–38. New York: Plenum Press.

Pelletier, B. G., and B. S. Robinson (2005). Tundra, ice and a Pleistocene cape on the Gulf of Maine: a case of PaleoIndian transhumance. *Archaeology of Eastern North America* 33:163–76.

Penland, S., R. Boyd, and J. R. Suter (1988). Transgressive depositional systems of the Mississippi delta plain; a model for barrier shoreline and shelf sand development. *Journal of Sedimentary Petrology* 58:932–49.

Powell, J. F., ed. (2005). *The First Americans: Race, Evolution, and the Origin of Native Americans*. Cambridge, UK: Cambridge University Press.

Schwab, W. C., E. R. Thieler, J. S. Allen, D. S. Foster, B. A. Swift, J. F. Denny, and W. W. Danforth (1999). Geologic mapping of the nearshore area offshore Fire Island, New York. In *Proceedings of Coastal Sediments '99*, 1552–67.

Sirkin, L. (1976). Block Island, Rhode Island: evidence of fluctuation of the late Pleistocene ice margin. *Geological Society of America Bulletin* 87:574–80.

Sirkin, L. (1996). *Block Island Geology*. Watch Hill, RI: Book and Tackle Shop.

Snow, G. (1980). *The Archaeology of New England*. New World Archaeological Record. New York: Academic Press.

Stanford, D., and B. Bradley (2002). Ocean trails and prairie paths? Thoughts about Clovis origins. In *The First Americans: The Pleistocene Colonization of the New World*, ed. N. G. Jablonski, 255–71. Memopirs of the California Academy of Sciences, No. 27. San Francisco: Allen Press.

Stright, M. J. (1995). Archaic period sites on the continental shelf of North America: the effect of relative sea-level changes on archaeological site locations and preservation. In *Archaeological Geology of the Archaic Period in North America*, ed. E. A. Bettis, III, 131–47. Boulder, CO: Geological Society of America Special Paper 297.

Uchupi, E., N. Driscoll, R. D. Ballard, and S. T. Bolmer (2001). Drainage of late Wisconsin glacial lakes and the morphology and late quaternary stratigraphy of the New Jersey–southern New England continental shelf and slope. *Marine Geology* 172:117–45.

Webb, S. D., J. T. Milanich, R. Alexon, and J. S. Dunbar (1984). A *Bison Antiquus* kill site: Wacissa River, Jefferson County, Florida. *American Antiquity* 49:384–92.

Whitmore, F. C., Jr., K. O. Emery, H.B.S. Cooke, and D.J.P. Swift (1967). Elephant teeth from the Atlantic continental shelf. *Science* 156:1477–81.

Yesner, D. R. (1980). Maritime hunter-gatherers: ecology and prehistory. *Current Anthropology* 21:727–50.

11

Sinkholes in Lake Huron and the Possibility for Early Human Occupation on the Submerged Great Lakes Shelf

Dwight F. Coleman

In 2000, the National Oceanic and Atmospheric Administration (NOAA) and the State of Michigan established the Thunder Bay National Marine Sanctuary and Underwater Preserve (TBNMS/UP, figure 11.1). TBNMS/UP, which is located in northeastern Lower Michigan off Alpena County in western Lake Huron, is the thirteenth designated National Marine Sanctuary in the United States and the first in the Great Lakes. The Sanctuary was established primarily for the protection, preservation, and long-term management of its submerged cultural resources, namely historically significant shipwrecks from the mid-19th century to the mid-20th century. As part of the management plan for TBNMS/UP, NOAA promotes the Sanctuary and its cultural resources through public outreach, educational programs, and scientific research. To accomplish this goal, NOAA partnered with the Institute for Exploration (IFE) in Mystic, Connecticut, to start the first phases of research and outreach involving the shipwrecks of Thunder Bay.

The Sanctuary was suspected to contain more than 100 shipwrecks, from wooden schooners and steamers to steel barges and freighters, ranging in size from less than 100 ft long to greater than 500 ft long. Several vessels are of potential national historic significance, including the *New Orleans, Grecian,* and *Isaac M. Scott* (U.S. Department of Commerce, 1999). Of the 30 or so shipwrecks that are known throughout the Sanctuary, about half rest in shallow water (<15 m). One vessel, the *Nordmeer*, which ran aground on some shoals northeast of Thunder Bay Island in 1966 (U.S. Department of Commerce, 1999), has a rapidly deteriorating hull still partially exposed above water. The remaining shipwrecks found within the deeper parts of the Sanctuary were primarily lost due to collisions in the fog. Many of these known shipwrecks are popular scuba diving attractions and contain buoyed markers; some are less popular and poorly located, but occasionally visited by advanced divers. More than 70 other ships are thought to be lost in the same region.

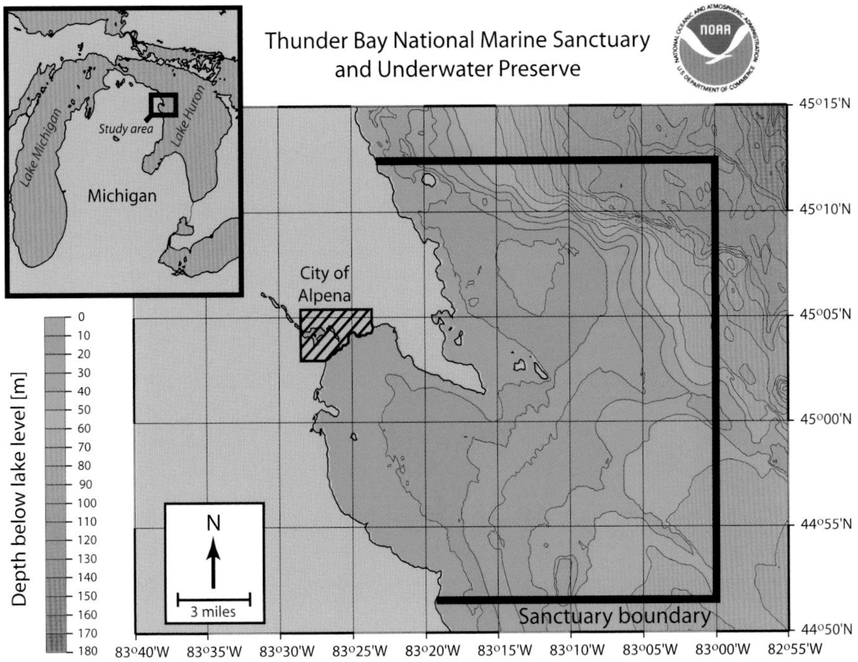

Figure 11.1. Location map of Thunder Bay National Marine Sanctuary and Underwater Preserve. Contours represent depths below lake level in meters.

During a 15-day expedition in June 2001, IFE conducted a side-scan sonar survey throughout the deep-water portions of the Sanctuary to determine the extent and character of cultural resources on the lakebed. The primary goal of this first phase of research was to establish an archaeological baseline for the Sanctuary from which more intensive archaeological survey work could develop. A secondary goal was to explore for shipwrecks that were lost in the vicinity of the Sanctuary, but never found. A tertiary goal was to explore for potential sites of human occupation and artifacts on the lakebed, remnants from when the lake was at a lower level following the last Ice Age. The goal of the second phase of the project was to augment the acoustic survey with a detailed visual survey. Shipwrecks, side-scan sonar targets, and other features on the lakebed identified during the 2001 survey were inspected and imaged using an advanced remotely operated vehicle (ROV) system during a two-week period in August and September 2002. In addition, several sediment cores were collected from several deep basins in the region and from one of a large number of sinkholes that was discovered in the northeastern corner of the Sanctuary. To achieve the aforementioned tertiary goal of identifying submerged prehistoric sites, the bathymetry of the Sanctuary was examined to map paleoshorelines according to published land and lake level fluctuation data, which varied considerably following the retreat of the Laurentide ice sheet.

Lake Huron Levels

The elevation of Lake Huron has fluctuated dramatically since the Laurentide ice sheet, which covered all of the Great Lakes 18,000 years ago, retreated (figure 11.2) (Anderson and Lewis 2002). Lake Huron's elevation history has been

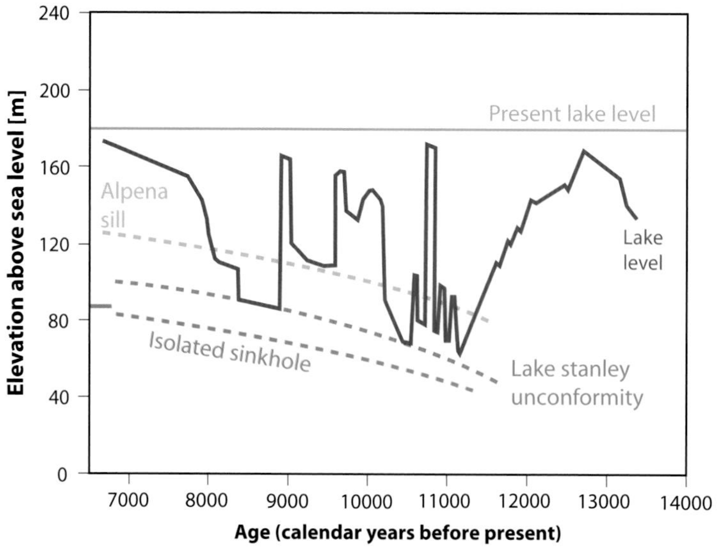

Figure 11.2. Variation in isostatically corrected lake level curve in the Lake Huron Basin during the time period 14,000 to 7000 BP (after Lewis and Anderson 1989; Anderson and Lewis 2002). Also indicated are the isostatic uplift curves for notable geologic features and for the isolated sinkhole, refer to text for details.

well studied (Hough 1962; Eschman and Karrow 1985; Lewis and Anderson 1989; Rea et al. 1994; Anderson and Lewis 2002), and the lake level rise curve has evolved to the curve portrayed in figure 11.2 as new data became incorporated. From about 14,000 to 7000 BP, the elevation of paleo-Lake Huron fluctuated, depending on several factors, such as the position of the ice margin, glacial meltwater supply, and postglacial isostatic rebound. These factors governed the amount of water available to the Great Lakes basin and the route of water flow into and out of the basin. Postglacial isostatic adjustment raised and lowered flow-restricting horizons such as sills between smaller basins. In general, the amount of isostatic uplift increased from southwest to northeast across the entire Great Lakes basin (Lewis and Anderson 1989). The Main Algonquin shoreline, which represents an ancient high stand of the lakes that completely encircles the basin, has been dated to about 11,700 radiocarbon years BP, which converts to about 13,500 calendar years BP (figure 11.3a) (Lewis and Anderson 1989; Lewis et al. 1994). In its present position, the Algonquin paleoshoreline is found at elevations varying from about 164 m at the Port Huron outlet near the southern tip of Lake Huron to about 450 m at the North Bay outlet in eastern Lake Nippising (Lewis and Anderson 1989). This southwest-to-northeast uplift differential for the Algonquin paleoshoreline is nearly 300 m across a 450-km span. The magnitude of postglacial rebound must be accounted for when determining lake-level fluctuations through time.

The chronology of this time period is represented in figures 11.2 and 11.3, where the history of Lake Huron's elevation is depicted graphically in cross-section view and spatially in map view. The radiocarbon ages have been converted to calendar years in these figures according to the corrections given by Anderson and Lewis (2002). During the Younger Dryas cold episode, from about 11,000 to 10,000 radiocarbon years before present, the Laurentide ice sheet readvanced slightly across the Great Lakes basin (figure 11.3c and d) (Lewis et al. 1994). It

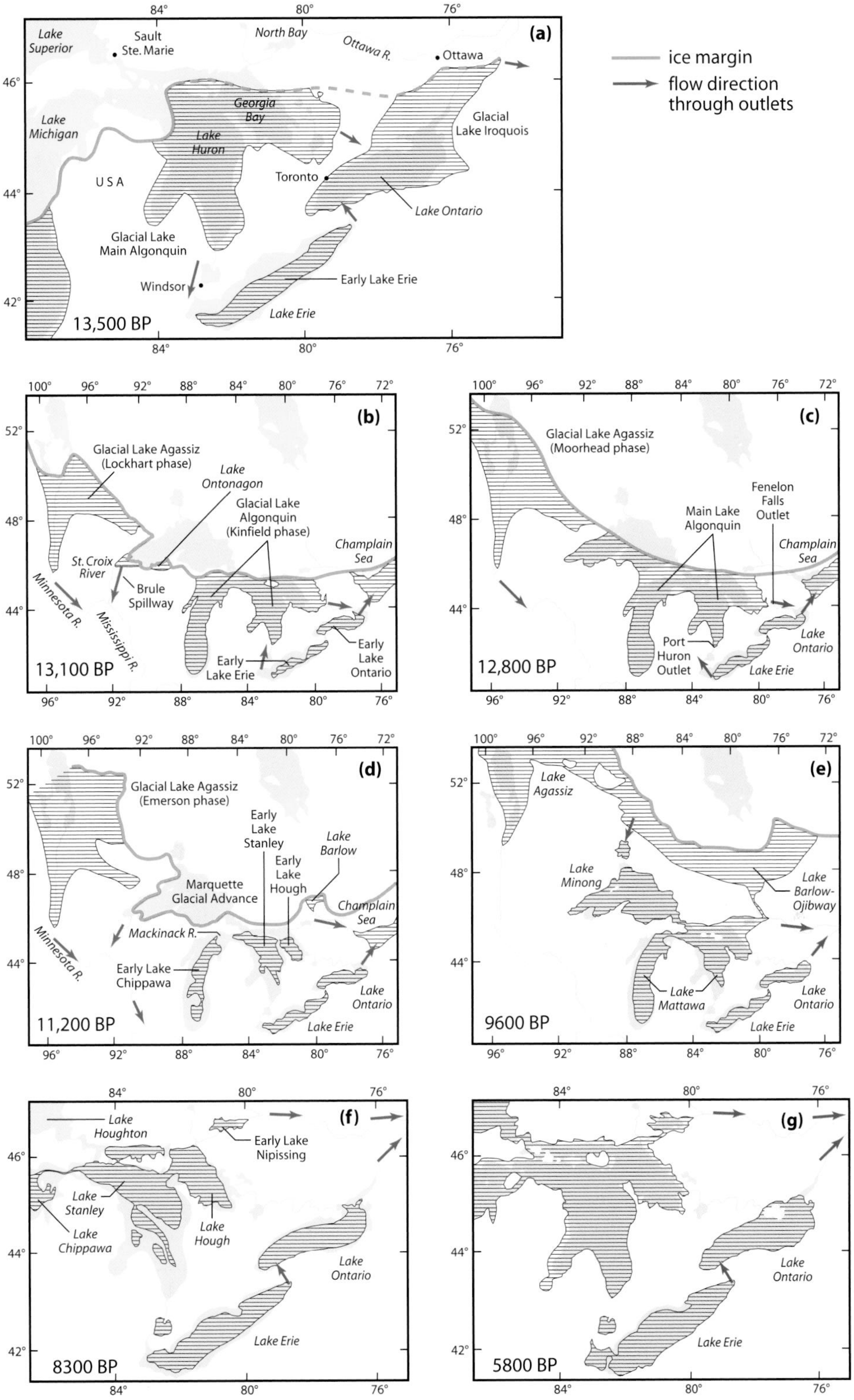

Figure 11.3. Evolution of the Great Lakes during selected postglacial times (after Lewis et al. 1994). Red lines indicate position of the ice margin, blue arrows indicate flow direction of lake water through outlets. Prior to 9600 BP, mid-continent glacial meltwater drained out of Glacial Lake Agassiz to the paleo-Mississippi River and ultimately to the Gulf of Mexico. After 9600 BP, the major discharge was to the Champlain Sea and paleo-St. Lawrence River. Ages are in calendar years before present, converted from radiocarbon years (Anderson and Lewis 2002). Refer to text for details.

has been postulated that this episode was initiated by a rerouting of glacial meltwater runoff from the Mississippi drainage system to the St. Lawrence drainage system via the Champlain Sea (Broecker et al. 1989; Lewis et al. 1994; Moore et al. 2000). There is strong supporting evidence to support a theory that this rerouting event increased production of North Atlantic Deep Water (NADW), which then led to a brief climate change (Broecker et al. 1989; Moore et al. 2000). The climate change caused Glacial Lake Algonquin to shrink substantially, drawing the levels in the Lake Huron basin down. During the following several thousand years, meltwater continued to discharge to the Champlain Sea and differential isostatic uplift, which affected the elevations of outlets and caused lake levels to fluctuate, but the overall long-term trend was lake lowering. This continued until about 8300 calendar years BP, when the Nippising transgression commenced and the basins filled again (Figure 11.2) (Lewis et al. 1994).

The Lake Stanley unconformity (figure 11.2), which occurs between layers of sediment in the deep parts of the lake, represents the lowest level of water in the Lake Huron basin (Hough 1962; Lewis and Anderson 1989). Glacial Lake Stanley existed when water levels had been drawn down to a minimum, and this occurred basically during three time periods, referred to as early, middle, and late Stanley low stands, with two of these periods represented in figure 11.3d and 11.3f) (Lewis et al. 1994). In figure 11.2, the unconformity elevation was corrected for postglacial rebound (Anderson and Lewis 2002). This feature separates deep-water lake clays from stratigraphically lower sands containing shallow-water fossils. The Alpena Sill (figure 11.2) was one of several sites actively controlling the outflow from the basin during this time period, until levels were either drawn down below the sill depth or raised much higher than the sill flanks (Lewis and Anderson 1989). In summary, according to the data in figure 11.2, during the Lake Stanley low stands, much of the lakebed of TBNMS/UP was exposed subaerially, creating a landscape that was suitable for human occupation.

Michigan Sinkholes

The underwater sinkholes in the Sanctuary, like those found in Misery Bay, off Middle Island, and in deeper water (figure 11.4), were of particular interest due to their potential for representing inundated prehistoric sites on the lakebed. The archaeological and paleontological significance of submerged sinkholes off Florida has been well established (Clausen et al. 1979), therefore the sinkholes off Michigan described here could be equally significant. Sinkholes and other karst features found throughout northeastern Lower Michigan, including Alpena County (the land immediately adjacent to TBNMS/UP), have been described and their distribution has been documented (Warthin and Cooper 1943; Black 1983). In this study region, karst formations are associated with dissolution of limestone and evaporite bedrock, causing cavities into which overlying bedrock and surface sediments collapsed (Black 1984). The surficial geology of the TBNMS/UP region is underlain by limestone rocks of the Traverse Group and evaporite rocks of the Detroit River Group, and the latter bedrock type shows evidence of karst formations (Black 1997). The cavities in the bedrock may be structurally controlled by tectonic faulting, as many of the sinkholes and other

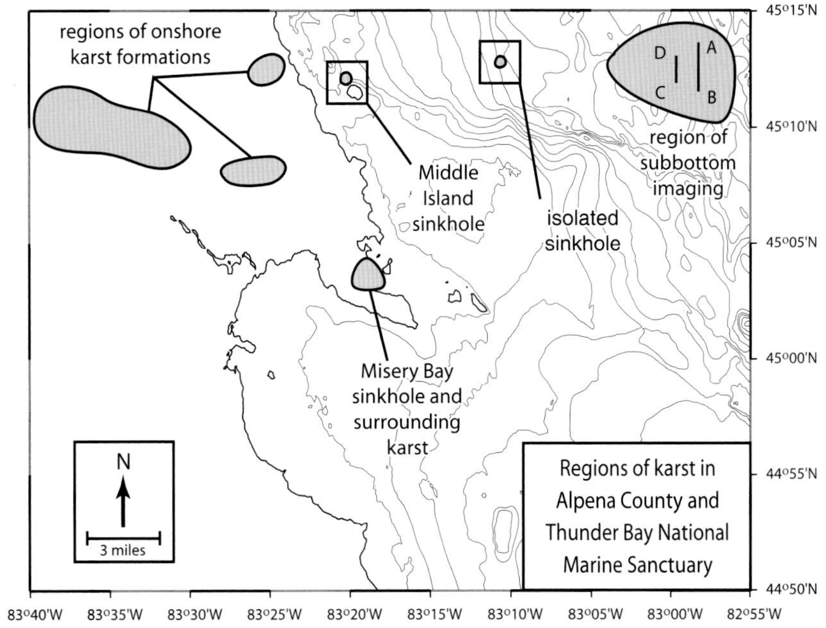

Figure 11.4. Regions of karst formations, as known to exist on land (after Black 1983), in shallow water (Middle Island, Misery Bay), and in deep water (from this investigation). Subbottom profile locations for figures 11.8–11.13 are shown in the upper right.

karst features exist along a line that follows the faulting trend (Black 1997). The sinkhole in Misery Bay, which bottoms at about 24 m water depth, is spring-fed, representing a site of groundwater discharge (Warthin and Cooper 1943; Black 1983), and this had been suspected as well in the Middle Island sinkhole. There has never been any mention in the literature of the archaeological significance of Michigan sinkholes, and, in general, little research has been done on the geological formation and evolution of northern Michigan sinkholes.

Paleo-Indian and Archaic Occupation

Sometime following the Last Glacial Maximum, humans entered North America. Many researchers believe people crossed into Alaska from Siberia across Beringia, a land bridge that spanned the Bering Strait created by the glacial lowering of sea level (Butzer 1971; Dikov 1983). As climate warmed, it was very possible that an ice-free corridor between the Cordilleran and Laurentide ice sheets opened, permitting people to migrate through central Alaska and northwestern Canada, east of the Rocky Mountains, and into the Great Plains region of the central United States (Reeves 1983). The coastal region from the Aleutian Islands to Puget Sound in the northwestern United States, which was covered by a thick ice sheet that calved into the Pacific Ocean, was probably not a favorable route for human migration (Butzer 1971). If humans followed the ice-free corridor route and entered the Great Plains region, they would have had little difficulty reaching the Great Lakes region. In fact, the presence of a well-studied site of early Paleo-Indian occupation at the Meadowcroft Rockshelter in western Pennsylvania, within 200 km of Lake Erie, confirms the existence of people in this part of North America, and is supporting evidence for the ice-free corridor migration theory (Adovasio et al. 1983).

Throughout the Great Lakes region, Paleo-Indian occupation sites are rare. However, several sites have been discovered and dated, indicating the presence of humans who occupied coastal regions surrounding Glacial Lake Algonquin around 12,000 BP (Cleland 1992). Several sites have been documented in southern and central Michigan where fluted points were found, some in association with butchered late Pleistocene faunal remains, such as mastodon, caribou, and other big-game species (Mason 1981). Around 10,600 BP (12,800 calendar years BP), Glacial Lake Algonquin, which had been held back by the Port Huron outlet (figure 11.3) (Lewis et al. 1994), began to drain. This drainage converted shallow portions of the main lakes into coastal freshwater wetlands and marshes, which would have attracted waterfowl and other game species, establishing habitats attractive to late Paleo-Indians (Mason 1981). The lithic artifacts left behind by Paleo-Indians, namely fluted points, disappeared by 10,000 BP, indicating the transition to the Archaic period, when a change in the manufacture of stone tools became evident in the archaeological record (Cleland 1992). During this time period water levels in the Huron basin approached their lowest elevation, forming the early Stanley lowstand and much of the lakebed of present-day Thunder Bay was exposed. A number of early Archaic sites have been discovered in central Lower Michigan and regions of Ontario near the shores of Glacial Lakes Stanley and Hough. Since these low stands persisted intermittently for the next several thousand years until the Nipissing transgression, which inundated much of the Sanctuary lakebed, Archaic human occupation sites undoubtedly exist beneath the present day Great Lakes (Mason 1981; Halsey 1991).

Methodology

For the 2001 expedition, the R/V *Lake Guardian*, a 187-ft research vessel operated throughout the Great Lakes by the U.S. Environmental Protection Agency, was used to tow Echo, a dual-frequency side-scan sonar towfish (figure 11.5). The ship's main laboratory space was converted to a sonar control room and data processing center for 24-hour survey operations (Coleman 2002).

The side-scan sonar survey was designed to cover as much of the lakebed as possible during a 15-day cruise, so an optimal lane spacing, swath width, and towing speed was chosen. For archaeological surveying, this represented a reconnaissance investigation simply designed to locate shipwrecks. Echo is a dual-frequency deep-towed Chirp side-scan sonar, with the high-frequency signal centering around 400 kHz and the low-frequency signal centering around 100 kHz. Echo has buoyancy that is slightly positive, so the towfish is used in tandem with a lead depressor weight. The depressor weighs about 650 kg and connects to the towfish using a 30-m-long fiber-optic tether (figure 11.5). A steel-armored fiber-optic cable, which mechanically terminates at the depressor weight, is mounted to a winch on the ship's deck, and carries the power to the vehicle and sends the sonar signal and telemetry data back to the surface.

The side-scan sonar data were collected on computers in the control lab, where the sonar signal was demultiplexed, displayed, and logged. Specialized computer software was used that displays and records all channels of the incoming side-scan sonar data, which includes navigation, telemetry, and auxiliary sensor data. The horizontal distance between the GPS antenna on the ship and the towfish position while undertow, or "layback," was entered into the acquisition software and constantly updated as the slant range changed by winching out more cable. The raw sonar data were processed to correct for heading and navigation errors, bottom tracking, and speed (ship accelerations). As the data were collected, acoustic targets and other interesting features on the lakebed were continuously logged. Shipwrecks stood out as strong sonar reflectors and could easily be identified based on their size, shape, and acoustic character. A large number of sinkholes and other geologic features on the lakebed were also easily identified on the sonar record and logged.

For the 2002 survey, the R/V *Connecticut* was used (figure 11.5), along with IFE's ROV system Argus and Little Hercules (figure 11.6). The R/V *Connecticut* had been recently outfitted with a dynamic positioning system. This system employs computer-driven thrusters to stabilize and maneuver the ship with very high precision. The main lab inside the vessel was converted to an ROV operations lab where the video and other data feeds were displayed on monitors and computers. The pilots and engineers operated the vehicle systems from this space, and a navigator had constant communications with the ship's bridge. The ROV system was successfully utilized to collect high-quality video images of 15 shipwrecks that had been mapped by the 2001 sonar survey. In addition, a number of dives were made in several different sinkholes. All the video imagery was recorded on digital tape media, including the high-definition format. The vehicles' telemetry data, such as depth, altitude, and heading, plus all scanning sonar data were logged by the primary data acquisition computer.

Figure 11.5. Research platforms for 2001 and 2002 surveys. R/V *Lake Guardian* (*upper left*), R/V *Connecticut* (*lower left*), and side-scan sonar towfish Echo (*right*), with its depressor weight. (Copyright Institute for Exploration)

In addition to the remotely operated vehicles, Echo was also used in 2002 to collect additional acoustic data from the sinkholes and other basins surrounding the Sanctuary. The towfish had been retrofitted with a Chirp subbottom profiler. The Chirp sonar technology, designed in the early 1980s by engineers at the University of Rhode Island, employs high-bandwidth frequency-sweeping acoustic transmissions that yield high-resolution subbottom images when focused toward the bottom with a narrow-width beam (Mayer and Leblanc 1983). The Echo towfish was equipped with a Chirp subbottom profiler that swept through frequencies between 2 and 7 kHz. The data collected by this instrument were used to identify sediment coring locations and to image the subsurface structure of the sinkholes.

Sediment cores were also collected during the 2002 expedition, using the R/V *Shenehon*, a vessel operated for TBNMS/UP during the summer of 2002 by NOAA's Great Lakes Environmental Research Laboratory. A coring team from the University of Rhode Island Graduate School of Oceanography (GSO) collected a number of 7- to 8-m-long piston cores from several deep basins within and surrounding the Sanctuary. The deep basin coring sites were chosen by examining the subbottom images to determine where the longest and most complete sedimentary record could be sampled with a 9-m core, and this was typically where the layers thinned near the basin margins. Two basins, Alpena and Manitoulin, were sampled, with several cores collected from each. The deep penetration of these cores allowed a sedimentary sequence to be recovered that spanned the Holocene and very late Pleistocene epochs, including the dry periods when the level of Lake Huron was drawn down significantly. In addition to the deep cores, one core was collected from one of the sinkholes. This core was collected last for fear of damaging the core barrel at the bottom of the sinkhole.

Figure 11.6. Underwater video still image of remotely operated vehicle Little Hercules and optical towsled Argus examining the propeller of a shipwreck in Thunder Bay. These vehicles, which are tethered together and work in tandem, were also deployed from the R/V *Connecticut* to examine the sinkholes. (Video still image courtesy of John Brooks)

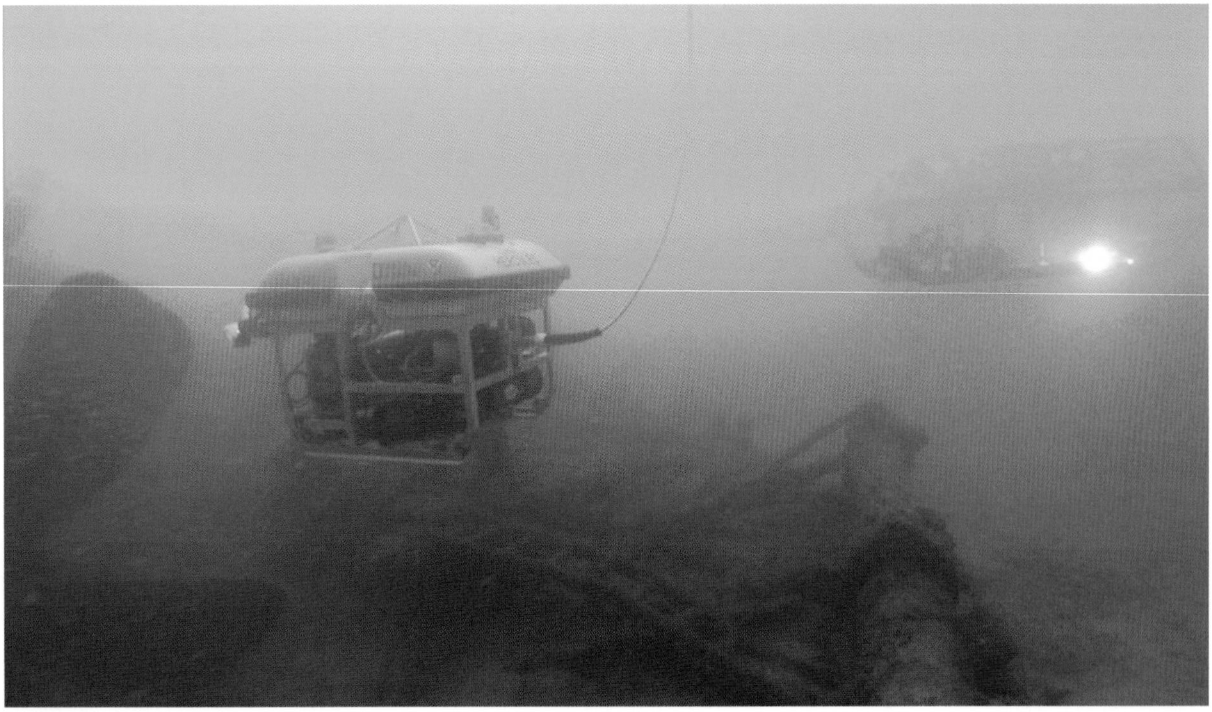

Also during the 2002 expedition, the sinkhole off Middle Island (figure 11.4) was examined using a conductivity–temperature–depth (CTD) sensor from onboard the R/V *Shenehon*. The CTD was lowered toward the bottom while drifting across the sinkhole. The sensor was held at about 1 m off the bottom while drifting, then slowly raised back toward the surface. The CTD data were stored internally and later downloaded, processed, and interpreted.

Results and Interpretation

During the 2001 shipwreck mapping survey more than 250 square kilometers of lakebed within the Sanctuary was acoustically mapped, representing more than half of the Sanctuary and covering all the deep-water portions greater than 15 m water depth. In addition, nearly 100 square kilometers of lakebed north of the Sanctuary boundary was surveyed. More than 50 tracklines of processed side-scan sonar images were combined in a complete mosaic of the Sanctuary lakebed (figure 11.7). In all, 17 shipwrecks were precisely located and imaged, most of which were previously known, especially by the technical diving community. In addition 2 previously unknown shipwrecks were found. Many other sonar targets were also imaged that appear to be man-made objects on the lakebed, not geological features, because they were isolated and not contiguous with the surrounding lakebed geology, or because they had a particularly unique-looking echo.

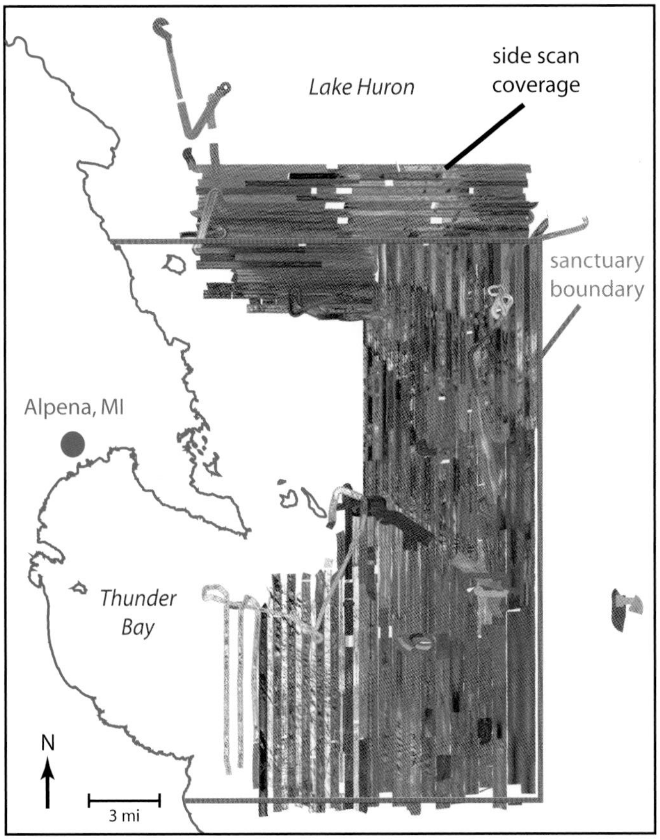

Figure 11.7. Raw mosaic of side-scan sonar trackline coverage, both inside the Sanctuary boundary and farther to the north. Shallower regions inside the sanctuary could not be surveyed primarily due to the maneuverability of the ship and towfish. Gaps in the coverage, or "holidays," were filled in after the primary survey was completed. (side-scan mosaic created by Jeremy Weirich)

In addition to the shipwrecks, a large number of underwater sinkholes and pockmark features were discovered on the lakebed and mapped during the 2001 survey. During the 2002 survey, Echo, which had been equipped with a subbottom profiler, imaged several of these sinkholes again, but in more detail with the added channel of data that imaged beneath the lakebed. The data collected from these sinkholes are portrayed in figures 11.8–11.11, which depict both plan views (side-scan) and cross-section views (subbottom). The first profile, line A-B (figures 11.8 and 11.9), crosses over two sinkholes. The southern sinkhole has a larger diameter and is deeper than the northern sinkhole. The second profile, line C-D (figures 11.10 and 11.11), crosses over a third sinkhole, which is about as deep and wide as the larger of the first two sinkholes imaged. Figure 11.4 reveals the location of both profiles and the regions within the Sanctuary occupied by karst features. For the most part, sinkholes and other karst features were found only in the northeastern corner of the Sanctuary, except for a few that were isolated. The dimensions of the southern sinkhole along line A-B are about 250 m long by 150 m wide by 10 m deep (figures 11.8 and 11.9). The

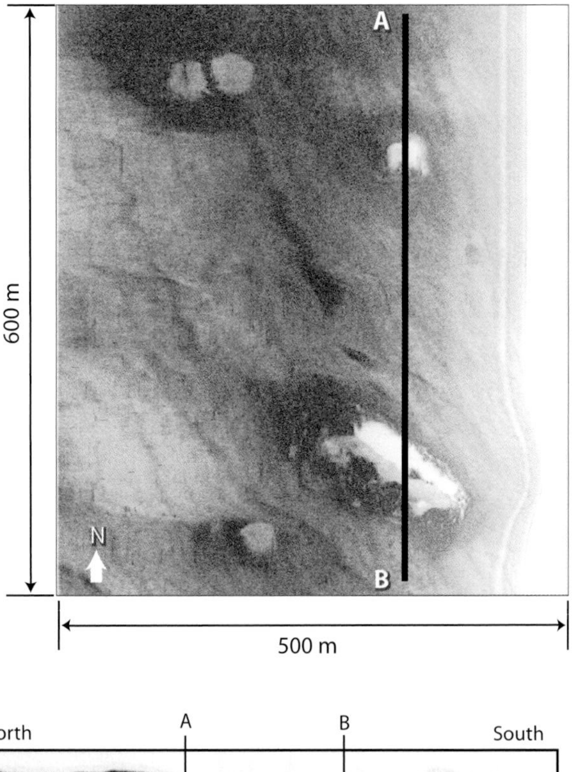

Figure 11.8. Side-scan sonar track revealing two prominent sinkholes in the port lobe. Image is not geocorrected. Location of subbottom profile in figure 11.9 is shown. Refer to figure 11.4 for location.

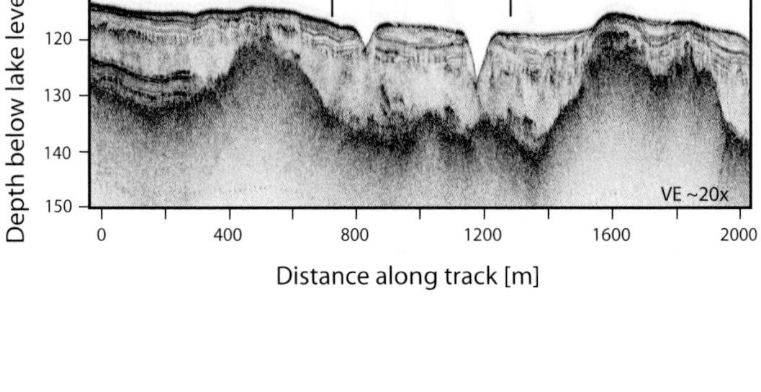

Figure 11.9. Subbottom profile across the two sinkholes in figure 11.8. Smaller sinkhole is about 5 m deep, larger sinkhole is about 10 m deep. Shown with substantial vertical exaggeration.

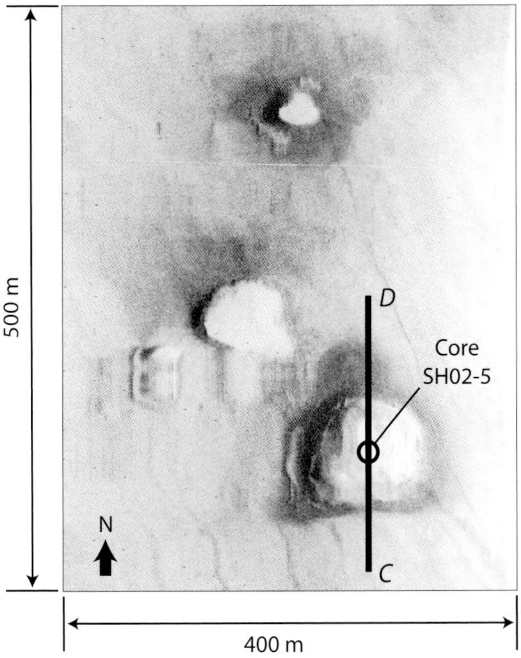

Figure 11.10. Side-scan sonar track revealing several prominent sinkholes in the port lobe. Image is not geocorrected. Location of subbottom profile in figure 11.11 is shown. Refer to figure 11.4 for location. Location of core shown in figure 11.15 is also depicted.

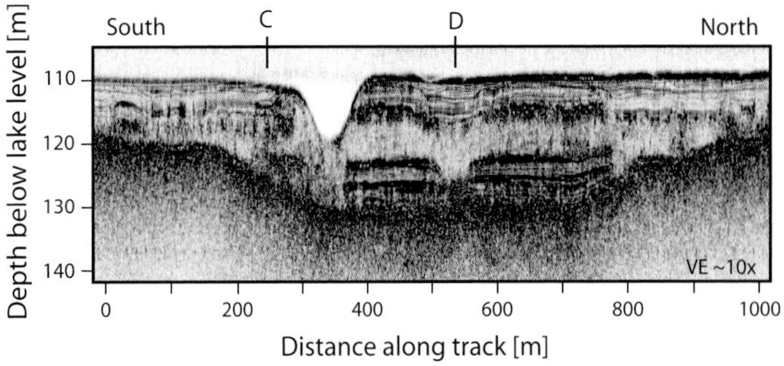

Figure 11.11. Subbottom profile across the sinkhole in figure 11.10. Shown with less vertical exaggeration than in figure 11.9.

dimensions of the northern sinkhole along line A-B are about 50 m in diameter by 5 m deep. The dimensions of the third sinkhole are about 100 m in diameter by 5 m deep (figures 11.10 and 11.11). In other sinkholes that were imaged, but are not shown here, the dimensions ranged from close to 400 m in diameter and up to 20 m deep to less than 25 m in diameter and only a few meters deep. Some would be better described as pockmarks instead of sinkholes. In addition, some of the sinkholes were nearly perfectly round in shape and others were very elongated, or elliptical in shape. Therefore, there was high variability in sinkhole shapes and sizes across the region. Lastly, most were found in water depths between about 80 and 120 m. However, if surveying had continued into deeper water along the same west–northwest to east–southeast trend as the sinkholes, it is quite possible that more would have been discovered.

The interpretation of the subbottom images of the sinkholes and surrounding subsurface geology is given in figures 11.12 and 11.13. The positions of prominent reflective sedimentary units are interpreted based on details given in Moore et al. (1994) and Rea et al. (1994). These authors identified a number of seismic reflectors in Lake Huron subbottom profiles and correlated them to sediment cores that contained datable horizons. The acoustic character of these reflective units was used to interpret the profiles collected over the sinkholes (figures 11.9 and 11.11). The top of each unit was assigned a different color, which also represents a different stratigraphic age. This same color scheme was adopted in the sinkhole interpretations (figures 11.12 and 11.13). A description of the acoustic characteristics between horizons and the radiocarbon date of fossils found just

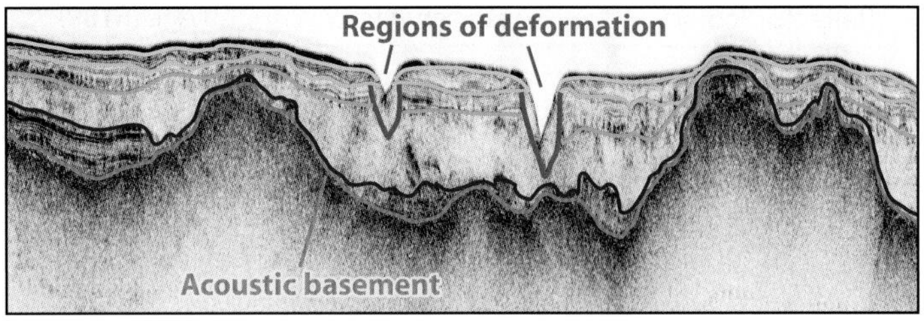

Figure 11.12. Interpretation of seismic stratigraphic sequences in cross section A-B as in figures 11.8 and 11.9 (based on Huron seismic stratigraphy given by Moore et al. [1994] and Rea et al. [1994]). It appears the sinkholes formed prior to deposition of light green and light blue reflectors. Refer to text for detail.

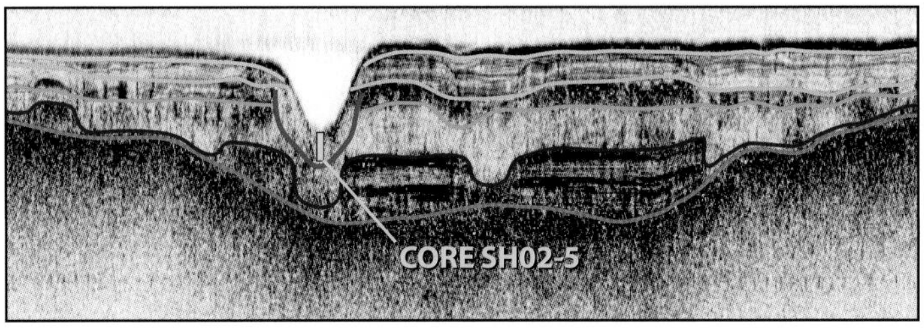

Figure 11.13. Interpretation of seismic stratigraphic sequences in cross section C-D as in figures 11.10 and 11.11 (based on Huron seismic stratigraphy given by Moore et al. [1994] and Rea et al. [1994]). Location of Core SH02-5 is shown on the seismic section.

above each horizon is given in figure 11.14 (Rea et al. 1994). The unique features are the sinkholes. The region of deformation is interpreted on each cross section. These regions of deformation can be represented by a V-shaped "fault" in cross section (a ring in plan view). These "faults" are actually the boundary within which the sinkholes have formed. Outside this boundary, the sediments have not been deformed and retain their primary depositional character. By carefully examining the uppermost strata adjacent to the sinkholes, it appears these layers are conformably overlying the regions of deformation. Just above the dark green horizon, the "faults" truncate the strata (including the dark green unit). This indicates that the sinkholes formed sometime following deposition of the dark green stratigraphic unit. Since the orange stratigraphic unit does not appear to be deposited inside the sinkhole, the sinkhole probably formed after that unit was deposited, but prior to deposition of later stratigraphic units. According to the interpretation by Rea et al. (1994), this would be between the yellow and orange depositional regime, around 10,000 BP. Sediments that have been deposited since their formation have sunk into the sinkhole, but this Holocene sequence is too thin to be resolved by the subbottom profiler.

The 110-m-deep sinkhole depicted in cross section C-D (figure 11.13) was explored by the ROV and video imagery was collected. The sediment surrounding the sinkhole was primarily very fine cohesive mud, with scattered cobbles mixed in with the deposit. As the lip of the sinkhole was approached, a large mass of clay could be observed slowly slumping into the sinkhole. This mass of clay sloped down toward the middle of the depression and in some places was undercut by clay that had already calved off into the sinkhole. The clays deposited

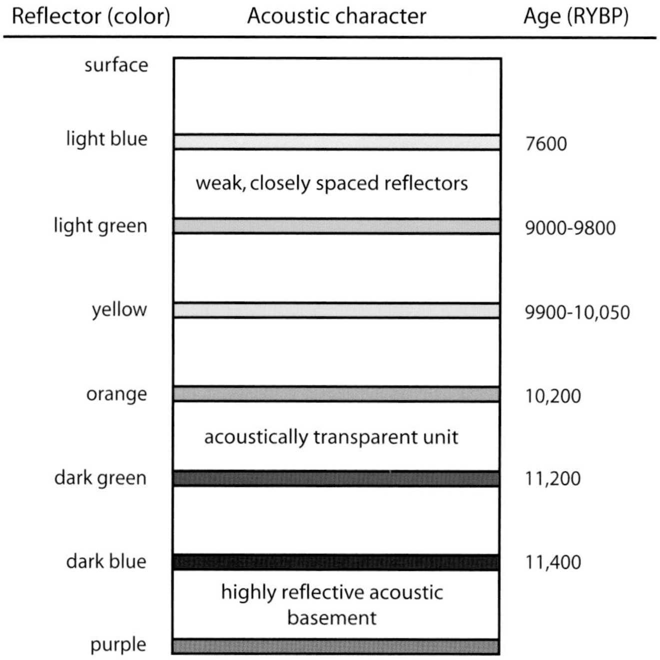

Figure 11.14. Radiocarbon dating results for each stratigraphic sequence (after Rea et al. 1994). Age dates are in radiocarbon years before present (RYBP). No stratigraphic units younger than the orange sequence are observed near the sinkholes.

within the sinkhole appeared identical to the deposits around its perimeter. Most of the rim of the sinkhole and the deeper middle of the sinkhole was explored to look for any interesting associated features, but nothing noteworthy was found.

Piston core SH02-5 (figure 11.15) was collected in the third sinkhole (figure 11.13). This core was 182 cm long and consisted primarily of massive, light gray-colored, nondescript clay. Very fine laminations could be seen throughout the lower portions of the core. The uppermost 17 cm is characterized by a slight change in lithology to a massive, dark gray-colored clay. The sediments represented by this core were probably deposited soon after the sinkhole formation and their lithologic character suggests they may be correlative with varved clay deposits observed in cores elsewhere from around 10,000 BP, representing a time interval between the yellow and orange reflectors (Mike Lewis, 2003, personal communication). An interesting component of this core was discovered when the split core was being subsampled for paleomagnetic analysis. At a depth of 8 cm from the top of the core, a very dense, 1-cm-diameter, spherical, highly magnetic stone was recovered (John King, 2003, personal communication). Its appearance is anomalous compared with the surrounding fine clay sediment, and does not reflect what would be expected for ice-rafted coarse material. The possibility that the stone is a small meteorite or possibly a human artifact has not been ruled out, but this would require further investigation.

Another sinkhole, referred to here as the isolated sinkhole (figure 11.16), was also explored by the ROV. This sinkhole was several kilometers inshore from the sinkholes that were imaged by the subbottom profiler, in about 93 m water depth. As the ROV approached this sinkhole a hazy fluid was first noticed that created a layer overlying the entire depression. Near the rim of the sinkhole,

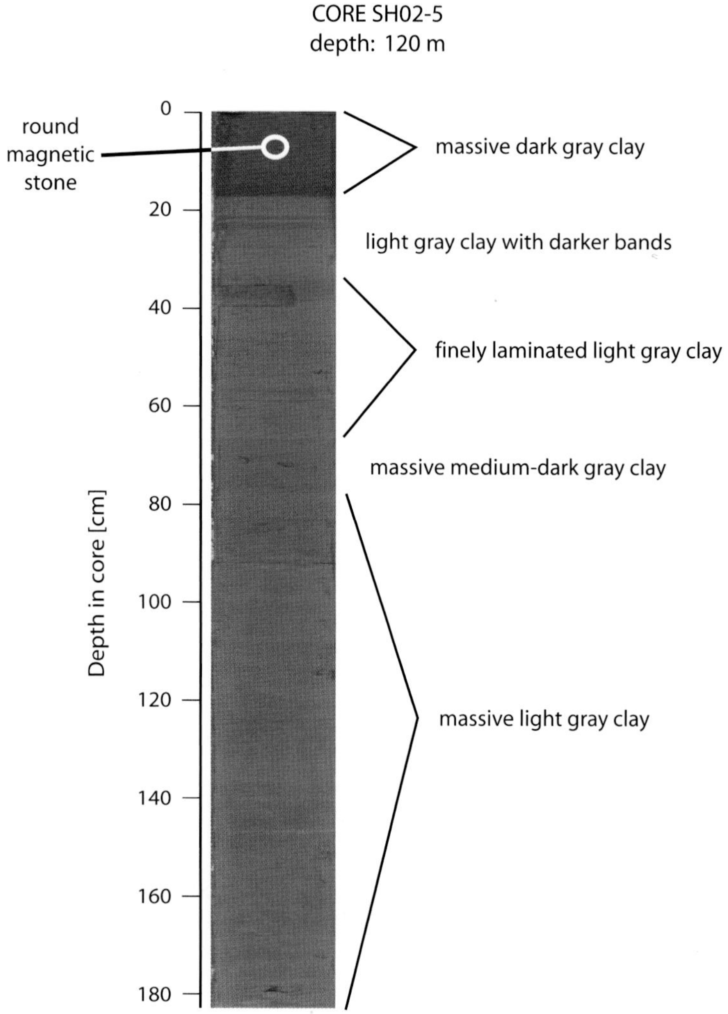

Figure 11.15. Whole-core photograph of core SH02-5 with description. Core was collected from a sinkhole that was acoustically imaged (figures 11.10 and 11.11). Note the sharp transition at about 17 cm from massive dark clay to laminated lighter-colored clay. One anomalous stone was recovered from a depth of about 8 cm in the core. The stone was very dense and highly magnetic.

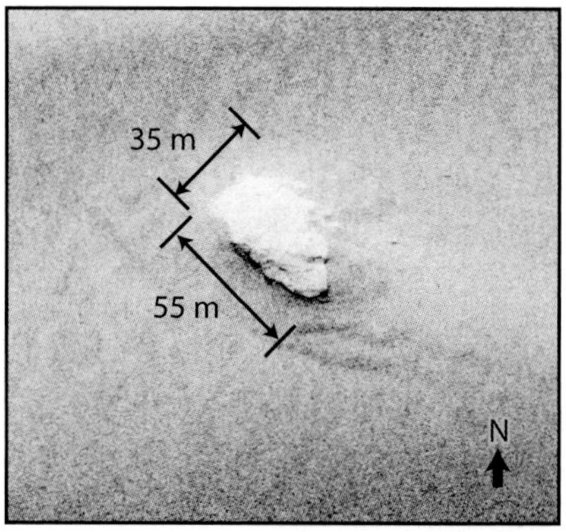

Figure 11.16. Geocorrected side-scan sonar image of an isolated sinkhole from 93 m water depth. Refer to figure 11.5 for location. This was the only significant sinkhole within several hundred meters radius and stood out prominently on the side-scan sonar record. Sinkhole measures about 55 m long by 35 m wide and was about 4 m deep. This image was created by the sonar being towed from west to east north of the sinkhole, resulting in poor reflectance on the northern downsloping side, and strong reflectance on the southern upsloping side.

Figure 11.17. Video still image taken by the ROV during a dive in the isolated sinkhole. Scale is approximately 20 cm across the bottom of the image. This particular item appears to be a rounded stone with a hole. This is not confirmed, however, and could represent benthic biology or an object that had drifted in to the sinkhole. (Copyright Institute for Exploration)

several interesting deposits were immediately imaged, including what appears to be a rounded stone containing a hole (figure 11.17). This is only one interpretation, however. Other large boulders were observed surrounding the sinkhole. As the ROV proceeded in toward the center of the depression, the pilot noticed some buoyancy differences created by the hazy fluid layer. Within the center of the sinkhole, a number of other unique-looking and perplexing deposits were imaged (figure 11.18). In one place, what appears to be a preserved tree trunk was imaged (upper image). At first the deposit had the appearance of bone material, but upon closer examination, that was ruled out. If it is a fossil tree, it needs to be determined whether it is in situ or was transported and deposited in the sinkhole. Another unique deposit was imaged near the first site (middle image). Here, small boulders were found deposited with wood and sediment. At first, the deposit looked like it had been arranged in a particular fashion. The lower images in figure 11.18 reveal wood debris from within the aforementioned deposits that have the appearance of being cut by a tool, not broken naturally. Of course, all this discourse is speculation at this time, and this sinkhole warrants further investigation; sampling and analysis of the organic material and stone will confirm or refute this interpretation. One observation that was confirmed visually was that in several locations throughout this sinkhole, evidence for spring-fed groundwater venting was obvious. Also, some organic deposits, possibly bacteria, were found around these vents. This venting explains the existence of the hazy fluid layer. As groundwater travels through the subsurface karst formations, it picks up dissolved salts, then gets emitted through the vents as a dense briny fluid that is warmer than the surrounding lake bottom water. This dense, warm fluid layer lies in suspension above the sinkhole, slowly diffusing upward and outward into the surrounding water. The isolated

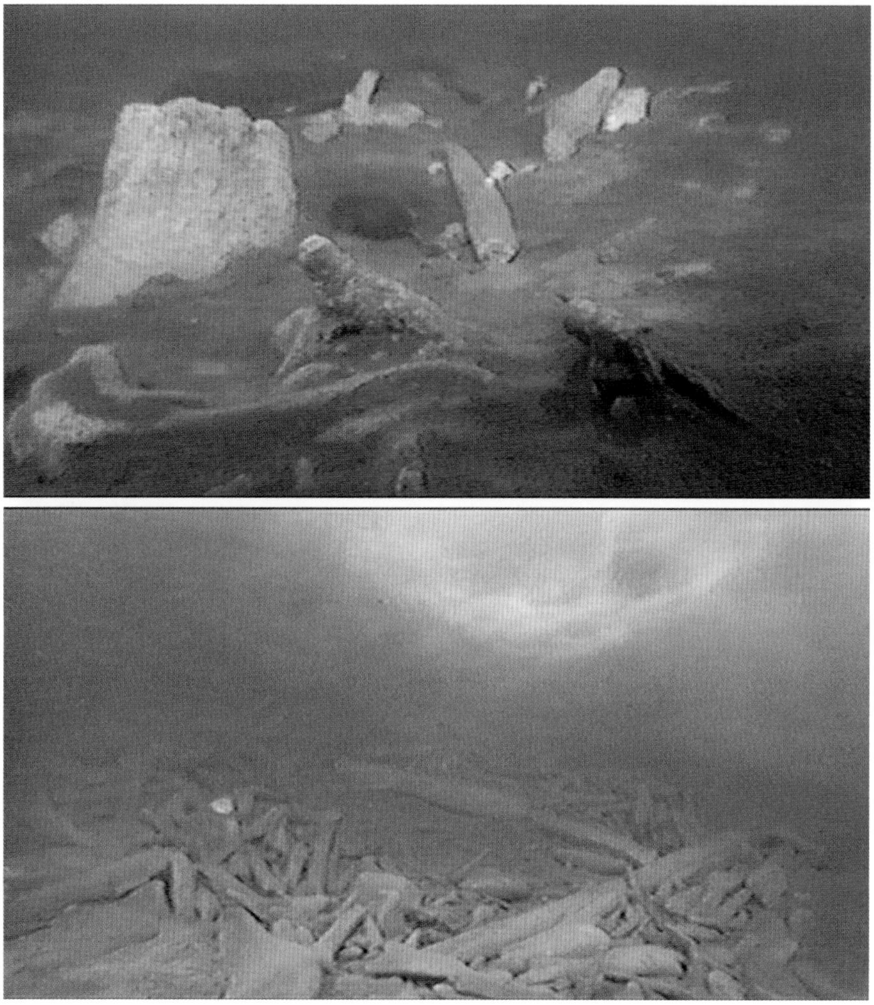

Figure 11.18. Video still images collected from the isolated sinkhole. Scale is approximately 2 m across the bottom of the images. The top figure reveals possible remains of a tree trunk with associated roots and branches, which could exist in situ and be preserved by the chemical conditions inside the sinkhole. The middle figure reveals a unique deposit of stone and wood material mixed together. Also note the hazy fluid layer that sits above the sinkhole that is caused by the emission of warmer briny water into the bottom of the lake through the sinkhole. (Copyright Institute for Exploration)

sinkhole was revisited during an expedition in 2003. Additional video imagery of the venting was collected, as were water samples and CTD data (Ruberg et al. 2005; Biddanda et al. 2006).

The last sinkhole that was studied during the 2002 expedition was off Middle Island where a CTD cast was made to the bottom of the sinkhole (figure 11.19). The temperature profile plot shows depth in meters on the left vertical axis and temperature on the right. The horizontal axis represents a distance equating to time as the vessel drifted over the sinkhole. The temperature profile shows a sharp thermocline from about 5 to 17 m depth where the temperature drops from about 18 to 8°C (figure 11.19, upper plot). The temperature then gradually lowers to about 6°C about 1 m above the bottom. A thermal anomaly

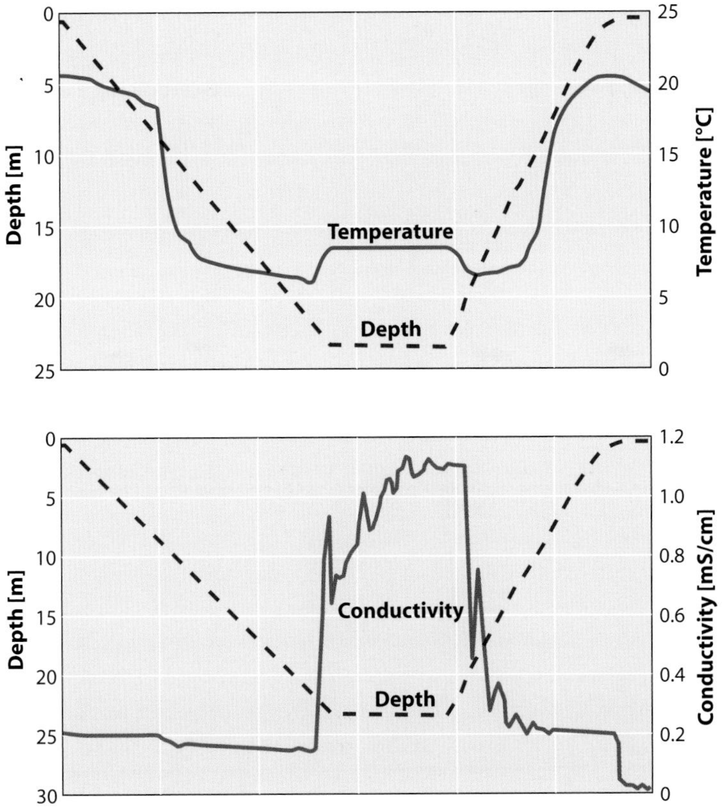

Figure 11.19. Temperature (*above*) and conductivity (*below*) profiles collected from the sinkhole north of Middle Island (see figure 11.7 for location). The horizontal scale represents a distance of about 100 m. The CTD was launched upcurrent from the sinkhole site for the cast, then lowered down into the sinkhole as the vessel drifted with the current, then brought back up to the surface. Significant thermal and salinity anomalies exist at the bottom of the sinkhole. (Data collected and compiled by Steve Ruberg)

of about 2–3°C exists immediately above the deepest part of the sinkhole. In a similar fashion, the depth/conductivity plot indicates a conductivity increase just above the bottom of the sinkhole that is 4–5 times higher than ambient lake conductivity (figure 11.19, lower plot). This simultaneous increase in temperature and conductivity is probably associated with an influx of groundwater into the lake through the bottom of the sinkhole.

Discussion and Conclusions

A plot of the elevation uplift of the isolated sinkhole is given in figure 11.2, which follows the uplift trend of the Alpena Sill given by Lewis and Anderson (1989), but has been corrected to calendar years according to Anderson and Lewis (2002). This plot indicates that the isolated sinkhole was always below the level of Lake Huron following the last glaciation. If the deposits in the sinkhole prove to be related to an in situ tree, or if evidence of human activity is found at the site, then this would be more data to be added that will modify the lake level curve in figure 11.2. Otherwise, the deposits in the sinkhole could have been transported there. Sampling and analysis of the material in the sinkhole is necessary to confirm or disprove these theories.

The presence of submerged karst formations in deep portions of Lake Huron had not been reported previously until this investigation and the subsequed work reported elsewhere. Moore et al. (1994) reported similar depression features, although not nearly as deep relative to the surrounding lake floor, in results from a side-scan sonar survey in Lake Huron. They described the features as shallow, circular or elliptical depressions on the lake floor, which were common in the side-scan records. Although they did not offer a detailed explanation, these authors did observe that a relationship between the occurrence of the depressions and the near-surface stratigraphy indicated that the depressions may not have formed in modern times. Based on the interpretation of subbottom profiler data across several deep sinkholes, a conclusion has been drawn that indicates the sinkholes formed sometime prior to 9000 BP and have been evolving since that time. Modern sediment is being deposited both in and surrounding the sinkholes. Near the rim of the sinkholes, the uppermost surface sediment is slumping in toward the center of the depressions.

If the sinkholes contain evidence of ancient human activity, this could potentially be significant for a number of different reasons. Firstly, the bottom sediments in the sinkholes are probably anoxic, and will therefore preserve any organic material associated with the sites. Secondly, late Paleo-Indian and early Archaic sites are rare throughout the Great Lakes region, and actually throughout eastern North America, so the discovery and investigation of any site could significantly add to the present knowledge of the nature of past human activity and could shed light on some controversial theories about the peopling of the New World. Lastly, no significant prehistoric underwater sites have been documented from midcontinent region lakes in North America, so a new discovery could be very important and add to our archaeological knowledge for this region.

In shallow water, sinkholes near Middle Island and in Misery Bay and even up to the depths of the isolated sinkhole (93 m), show a relationship with the groundwater hydrologic system, as there is evidence for fluid venting in the bottom of these features. During a return expedition to the isolated sinkhole in 2003, using an ROV from the University of Michigan, a team of researchers was able to measure the salinity (from conductivity) and temperature of the venting groundwater (Ruberg et al., 2005). In addition, the team was able to use the ROV to collect pumping samples from inside the sinkhole to analyze the geochemistry and microbiology of the venting fluid (Biddanda et al., 2006). Unfortunately, the team was unable to collect any geological, biological, or potential archaeological samples from within the sinkhole. Clearly more work is required at these sites to be able to draw conclusions about the submerged prehistoric archaeology, but the potential is there.

References

Adovasio, J. M., J. Donahue, J. E. Guilday, R. Stuckenrath, J. D. Gunn, and W. C. Johnson (1983). Meadocroft rock shelter and the peopling of the new world. In *Quaternary Coastlines and Marine Archaeology*, eds. P. M. Masters and N. C. Flemming, 275–302. London: Academic Press.

Anderson, T. W., and C.F.M. Lewis (2002). Upper Great Lakes climate and water-level changes 11 to 7 ka: effect on the Sheguiandah archaeological site. In *The Sheguiandah Site: Archaeological, Geological and Paleobotanical Studies at a Paleoindian Site on Manitoulin Island, Ontario*, ed. P. Julig, 195–234. Canadian Museum of Civilization, Archaeological Survey of Canada, Mercury Series Paper 161.

Biddanda, B. A., D. F. Coleman, T. H. Johengen, S. A. Ruberg, G. A. Meadows, H. W. Van Sumeran, R. R. Reoiske, and S. T. Kendall (2006). Exploration of a submerged sinkhole ecosystem in Lake Huron. *Ecosystems* 9:1–16.

Black, T. J. (1983). Selected views of tectonics, structure, and karst in Northern Lower Michigan. In *Tectonics, Structure, and Karst in Northern Lower Michigan*, ed. R. E. Kimmel, 11–35. Michigan Basin Geological Society Field Trip Guidebook.

Black, T. J. (1984). Tectonics and geology in karst development of Northern Lower Michigan. In *Sinkholes: Their Geology, Engineering and Environmental Impact*, ed. B. F. Beck, 87–91. Proceedings of the First Multidisciplinary Conference on Sinkholes. Boston: A. A. Balkema.

Black, T. J. (1997). Evaporite karst of Northern Lower Michigan. *Carbonates and Evaporites* 12:81–83.

Broecker, W. S. (1989). Routing of meltwater from the Laurentide Ice Sheet during the Younger Dryas cold episode. *Nature* 341:318–21.

Butzer, K. W. (1971). *Environment and Archaeology: An Ecological Approach to Prehistory*. Chicago: Aldine Atherton.

Chappell, J., and N. J. Shackleton (1986). Oxygen isotopes and sea level. *Nature* 324: 137–40.

Clark, P. U., and A. C. Mix (2002). Ice sheets and sea level of the Last Glacial Maximum. *Quaternary Science Reviews* 21:1–7.

Clausen, C. J. (1975). The early man site at Warm Mineral Springs. *Journal of Field Archaeology* 2:191–213.

Clausen, C. J., and J. B. Arnold (1976). The magnetometer and underwater archaeology: magnetic delineation of individual shipwreck sites, a new control technique. *International Journal of Nautical Archaeology* 5:159–69.

Clausen, C. J., A. D. Cohen, C. Emiliani, J. A. Holman, and J. J. Stipp (1979). Little Salt Spring, Florida: a unique underwater site. *Science* 203:609–14.

Cleland, C. E. (1992). *Rites of Conquest: The History and Culture of Michigan's Native Americans*. Ann Arbor: University of Michigan Press.

Coleman, D. F. (2002). Underwater archaeology in Thunder Bay National Marine Sanctuary, Lake Huron—preliminary results from a shipwreck mapping survey. *Marine Technology Society Journal* 36:33–44.

Coleman, D. F., J. B. Newman, and R. D. Ballard (2000). Design and implementation of advanced underwater imaging systems for deep sea marine archaeological surveys. In *Oceans 2000 Conference Proceeding*. Columbia, MD: Marine Technology Society.

Culliton, T. J., M. A. Warren, T. R. Goodspeed, D. G. Remer, C. M. Blackwell, and J. J. McDonough (1990). *50 Years of Population Change Along the Nation's Coasts*. Rockville, MD: National Oceanic and Atmospheric Administration.

Denton, G. H., and T. J. Hughes (1981). *The Last Great Ice Sheets*. New York: Wiley.

Dikov, N. N. (1983). The stages and routes of humanoccupation of the Beringian land bridge based on archaeological data. In *Quaternary Coastlines and Marine Archaeology*, eds. P. M. Masters and N. C. Flemming, 347–64. London: Academic Press.

Dunbar, J. S., and S. D. Webb (1996). Bone and ivory tools from submerged Paleoindian sites in Florida. In *The Paleoindian and Early Archaic Southeast*, ed. D. G. Anderson and K. E. Sassaman. Tuscaloosa, AL: University of Alabama Press.

Dunnel, R. C., and W. S. Dancey (1983). The siteless survey: a regional scale data collection strategy. In *Advances in Archaeological Method and Theory*, vol. 6, ed. M. B. Schiffer, 267–87. New York: Academic Press.

Dyke, A. S., J. T. Andrews, P. U. Clark, J. H. England, G. H. Miller, J. Shaw, and J. J. Veillette (2002). The Laurentide and Innuitian ice sheets during the Last Glacial Maximum. *Quaternary Science Reviews* 21:9–31.

Erickson, J. (1990). *Ice Ages: Past and Future*. Blue Ridge Summit, PA: TAB Books.

Eschman, D. F., and P. F. Karrow (1985). Huron basin glacial lakes: a review. In *Quaternary Evolution of the Great Lakes*, ed. P. F. Karrow and P. E. Calkin, 79–93. Geological Association of Canada Special Paper 30.

Fairbanks, R. G. (1989). A 17,000-year glacio-eustatic sea level record: influence of glacial melting rates on the Younger Dryas event and deep-ocean circulation. *Nature* 342:637–42.

Flemming, N. C. (1985). Ice ages and human occupation of the continental shelf. *Oceanus* 28:18–25.

Halsey, J. R. (1991). *Beneath the Inland Seas: Michigan's Underwater Archaeological Heritage*. Lansing, MI: Bureau of History, Michigan Department of State.

Hough, J. L. (1962). Lake Stanley, a low stage of Lake Huron indicated by bottom sediments. *Geological Society of America Bulletin* 73:613–20.

Lewis, C.F.M., and T. W. Anderson (1989). Oscillations of levels and cool phases of the Laurentian Great Lakes caused by inflows from glacial Lakes Agassiz and Barlow-Ojibway. *Journal of Paleolimnology* 2:99–146.

Lewis, C.F.M., T. C. Moore, D. K. Rea, D. L. Dettman, a. M. Smith, and L. A. Mayer (1994). Lakes of the Huron basin: their record of runoff from the Laurentide ice sheet. *Quaternary Science Reviews* 13:891–922.

Mann, M. E., R. S. Bradley, and M. K. Hughes (1999). Northern Hemisphere temperatures during the past millennium: Inferences, uncertainties, and limitations. *Geophysical Research Letters* 26:759–62.

Mason, R. J. (1981). *Great Lakes Archaeology*. New World Archaeological Record. New York: Academic Press.

Mayer, L. A., and R. Leblanc. The Chirp sonar: a new quantitative high-resolution profiling system. In *Acoustics and the Seabed*, ed. N. G. Pace. Bath, UK: Bath University Press.

Moore, T. C., D. K. Rea, L. A. Mayer, C.F.M. Lewis, and D. M. Dobson (1994). Seismic stratigraphy of Lake Huron—Georgian Bay and postglacial lake level history. *Canadian Journal of Earth Science* 31:1606–17.

Moore, T. C., J.C.G. Walker, D. K. Rea, C.F.M. Lewis, L.C.K. Shane, and A. J. Smith (2000). Younger Dryas interval and outflow from the Laurentide ice sheet. *Paleoceanography* 15:4–18.

Mörner, N.-A. (1980). Eustasy and geoid changes as a function of core/mantle changes. In *Earth Rheology, Isostasy, and Eustasy*, ed. N.-A. Mörner. New York: Wiley.

Morse, D. F., D. G. Anderson, and A. C. Goodyear (1996). The Pleistocene–Holocene transition in the eastern United States. In *Humans at the End of the Ice Age: The Archaeology of the Pleistocene–Holocene Transition*, ed. L. G. Straus, B. V. Eriksen, J. M. Erlandson, and D. R. Yesner, 319–38. New York: Plenum Press.

Peltier, W. R. (2002). On eustatic sea level history: Last Glacial Maximum to Holocene. *Quaternary Science Reviews* 21:377–96.

Penland, S., R. Boyd, and J. R. Suter (1988). Transgressive depositional systems of the Mississippi delta plain; a model for barrier shoreline and shelf sand development. *Journal of Sedimentary Petrology* 58:932–49.

Rea, D. K., T. C. Moore, C.F.M. Lewis, L. A. Mayer, D. L. Dettman, A. J. Smith, A. J., and D. M. Dobson (1994). Stratigraphy and paleolimnological record of lower Holocene sediments in northern Lake Huron and Georgian Bay. *Canadian Journal of Earth Science* 31:1586–1605.

Reeves, B.O.K. (1983). Bergs, barriers, and Beringia: reflections on the peopling of the New World. In *Quaternary Coastlines and Marine Archaeology*, ed. P. M. Masters and N. C. Flemming, 413–39. London: Academic Press.

Ruberg, S., D. Coleman, T. Johengen, G. Meadows, H. Van Sumeran, G. Land, and B. Biodanda (2005). Groundwater plume mapping in a submerged sinkhole in Lake Huron. *Marine Technology Society Journal* 39:65–9.

Smith, W.H.F., and D. T. Sandwell (1997). Global seafloor topography from satellite altimetry and ship depth soundings. *Science* 277:1957–62.

U. S. Department of Commerce. NOAA National Marine Sanctuaries Division (1999). Thunder Bay National Marine Sanctuary final environmental impact statement and management plan. (Report).

Warthin, A. S., and G. A. Cooper (1943). Traverse rocks of Thunder Bay Region, Michigan. *Bulletin of the American Association of Petroleum Geologists* 27:571–95.

PART FIVE

Telepresence and Submerged Cultural Sites

12

Long-term Preservation and Telepresence Visitation of Cultural Sites beneath the Sea

Robert D. Ballard and Mike J. Durbin

Recent advances in undersea robotics and telepresence technology now make it possible to consider the long-term preservation of cultural sites beneath the sea as well as the opportunity to visit such sites without getting wet, advances that will have a profound impact on the future handling and management of submerged cultural sites. One of the more popular arguments made by salvagers is that they are recovering objects from the bottom of the ocean to protect them from destruction and/or to permit them to be seen by others who otherwise would not see them, since most shipwrecks are lost in remote and inaccessible regions of the world's oceans. Were it not for the fact that their primary reason for recovering valuable objects is to make money, one could enter into a true debate as to the long-term strategy for managing a submerged cultural site in an effort to maximize its benefit to society. But, in fact, this argument is more of an excuse by salvagers seeking moral cover than a genuine concern as to what the best long-term strategy is for any particular submerged cultural site.

Telepresence technology, however, is beginning to weaken their argument, as it becomes increasingly easier to provide in situ public access to submerged cultural sites in the oceans of the world. The development of remotely controlled undersea vehicles makes it possible to visit cultural sites in the deep sea without coming into contact with the site while delivering high-quality video images to the public thousands of miles away.

Use of Satellite Technology for Limited Visitation to Submerged Cultural Sites

During the last several years, our team has conducted a number of expeditions to submerged cultural sites around the world and used shipboard high-bandwidth satellite technology to provide public access to these sites. The first such effort took place in 1989 in the central Mediterranean Sea when our team of archaeologists and oceanographers investigated an ancient deep-water trade route connecting ancient Carthage with Rome. The program focused on a 5th-century CE. Roman trading ship that had been lost in approximately 1000 m of water. During the course of that program, the Jason Foundation for Education produced and distributed a series of live educational programs to a network of students and teachers around the world via a high-bandwidth, gyro-stabilized satellite antenna that connected the ship to a shore-based receiving center. A similar program took place in Lake Ontario in 1990 when a team of archaeologists and oceanographers were investigating two ships from the War of 1812 that were lost in 100 m of water. Using a barge anchored above the site, a series of live educational programs was distributed once again to the Jason network of students and teachers around the world.

Recently, in the summer of 2004, our team returned to the RMS *Titianic* with our remotely operated vehicle Hercules to investigate its condition close to 20 years after our discovery (figure 12.1), during which live educational programs were transmitted from the ship. The team carried out its investigation from within a command/control center aboard the NOAA *Ronald H. Brown*. Digital images and computer displays within the shipboard command/control center (figures 12.2 and 12.3) were also transmitted via a high-bandwidth satellite system (figure 12.4) back to shore. At the University of Rhode Island's Graduate School of Oceanography, a team of historians was able to watch, listen, and direct the team at sea from a similar command/control console within the Graduate School's Inner Space Center (Ballard, 2004) (figure 12.5). At the same time scientists were participating in the expedition ashore in real time via

Figure 12.1. Hercules inspecting the starboard rail of *Titanic*'s bow section. Image taken from the Argus support vehicle to which Hercules is attached. (Copyright Institute for Exploration)

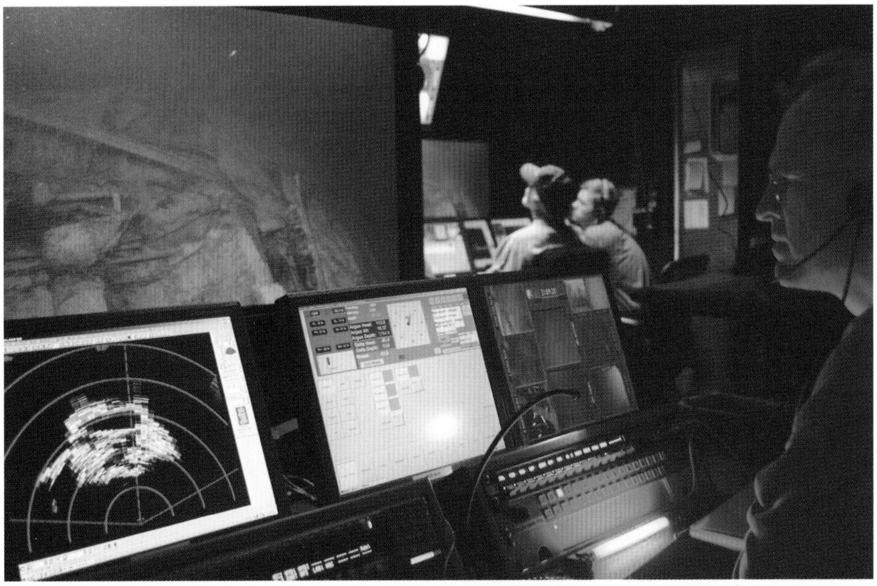

Figure 12.2. Scientist (*foreground*) directs a pilot and navigator (*background*) as they maneuver Hercules around stern section of *Titanic*. (Copyright Institute for Exploration)

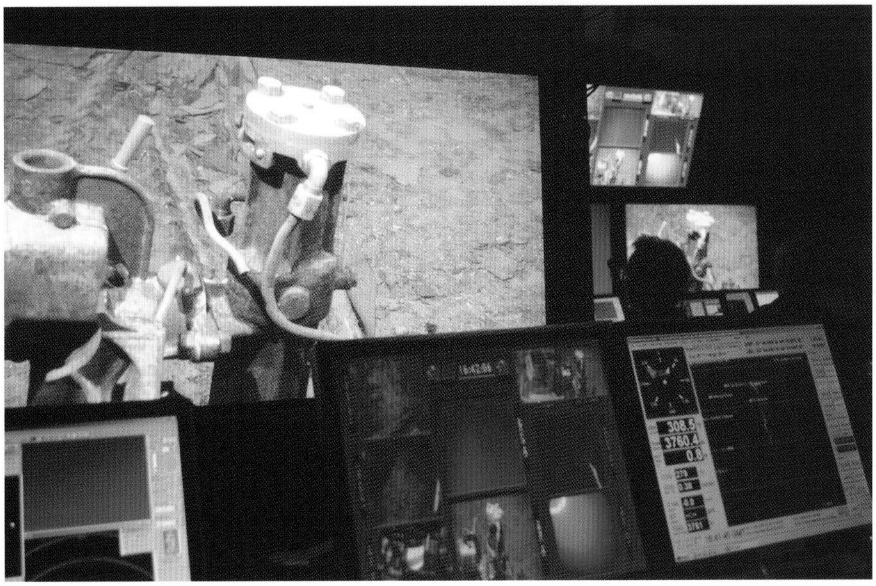

Figure 12.3. Closeup image of *Titanic*'s telemotor as Hercules hovers within centimeters of the ship's superstructure. (Copyright Institute for Exploration)

telepresence from the URI/GSO Inner Space Center, a television production team, associated with the Immersion Presents program, was able to take the same video feeds to produce a series of educational outreach programs to selected sites associated with the Boys and Girls Clubs of America and a network of museums, science centers, and aquariums across the country in real time (figure 12.6).

Newly Emerging Developments in Satellite Technology

For remote locations around the world and especially ships at sea, high-speed data communications on a real time basis, have always been a dream. Thanks,

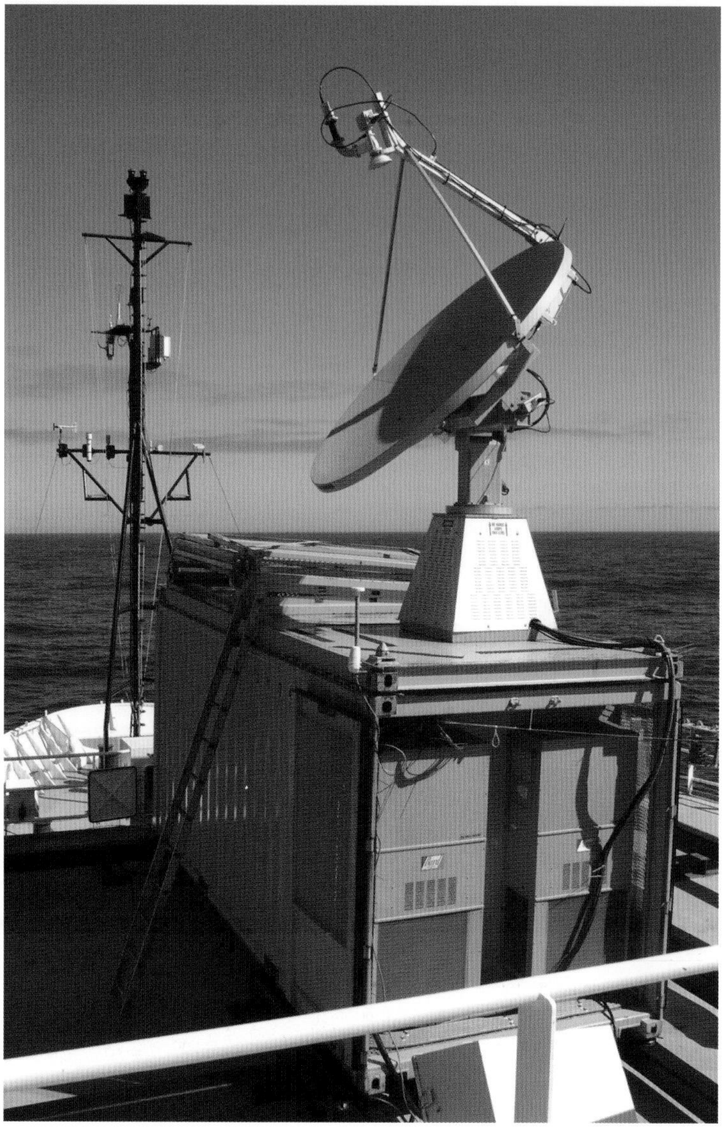

Figure 12.4. Satellite antenna mounted on the deck of the NOAA ship *Ronald H. Brown* relays high-resolution images and digital data stream back to shore. (Copyright Institute for Exploration)

however, to new IP technology, MPEG video, data compression technologies, and high-speed Internet, access to these normally land-based technologies can be offered to archaeologists and historians at sea. As previously stated, this new technology can supply high-quality video and data, and even enable real-time conferencing with experts who are based on shore. This will allow a single expert to be part of many expeditions without ever leaving his or her own environment and also provide public access to very remote submerged cultural sites. Commercial satellite communications are mostly carried in the super-high-frequency range from just above 1 GHz to around 60 GHz. These frequencies are divided into bands, those most used for satellite communications being the L, S, C, X, Ku, and Ka bands. The bands at the lower end of the spectrum generally have better propagation properties, while those at the higher end suffer from atmospheric attenuation problems such as rain fade. The higher

Figure 12.5. Team of historians participates in the 2004 expedition to *Titanic* inside the University of Rhode Island's Graduate School of Oceanography's Inner Space Center. (Copyright Institute for Exploration)

Figure 12.6. Immersion Presents production team creating live programming within the URI/GSO Inner Space Center (*upper left*). Satellite truck at URI/GSO transmits programming to a domestic satellite (*upper right*), which is received by an Immersion site (*lower left*) and presented to children watching the broadcasts (*lower right*). (Copyright Institute for Exploration)

the bandwidth, the greater the need for antenna pointing accuracy. This is an important consideration in marine applications where yaw, pitch, and roll are significant concerns. "L-band," according to Inmarsat's Philip Van Bergen, "is more forgiving of pointing errors than, say, C-band." This is one of the reasons why Inmarsat uses L-band as its operating frequency for the satellite-to-mobile terminal link. C-band, on the other hand, allows for more flexible systems and more choices of satellite providers. The new designs in satellite tracking equipment and advances in satellite coverage area have allowed the development of high-speed communications to ships at sea over a much larger area of the earth than ever before. This technology has brought speeds of 15 Mbps or more to ships at sea, allowing full-motion, high-quality video and data traffic to be accomplished in a real-time environment. With the use of Internet high-speed interconnectivity, scientists around the world can go on science expeditions from their own PC. Now scientists can call upon experts in all fields as needed, instantly from remote locations. The months required (not to mention expense) of planning a visit to remote locations with these experts to help can be eliminated.

An increasing number of satellite communications companies are offering LEO (low Earth orbiting) and MEO (medium Earth orbiting) constellations of satellites. Essentially, the closer a satellite constellation is to Earth, the smaller the antenna needs to be, making it possible to buy a satellite phone that's approximately the size of a regular cell phone. However, the closer the constellation to Earth, the more satellites are required for suitable coverage. The MEO systems need at least a dozen; the LEO systems need at least four dozen. There are also other factors that add to the complexity and cost of LEO systems. *Iridium*, one of the first of the LEOs, has built a $5 billion global communications system, with 66 satellites linked via ground stations to existing wireless infrastructure, allowing transmission of phone calls and paging worldwide. Iridium consortium members include Lockheed Martin, Raytheon, SK Telecom, and Sprint. Motorola is the chief manufacturer of the satellites and Iridium's handheld phones.

The system shown in figure 12.4 was designed to be moved from ship to ship as necessary and all communications equipment is in a nearby van. When installed permanently on a ship the system will not require the van, as the communications equipment will be installed in the ship's communications area.

C-band/K-band

C-band and the newer K-band systems to date have all been designed to use a dedicated land-based station for support of individual ships at sea. The future will bring a network of at least 3 land-based stations to give global coverage to a fleet of scientific and research ships or moored buoys at sea or on land anywhere in the world. Scientists from one ship in the Sea of Cortez could communicate and send high-quality digital video to a ship exploring similar sites in the Atlantic Ocean. These same research programs could be provided to the world in a real-time television broadcast. A good example of this technology in action was our 2004 expedition to *Titanic*.

MPEG, which stands for Moving Picture Experts Group, is the name of family of standards used for coding audiovisual information (e.g., movies, video, music) in a digital compressed format. The major advantage of MPEG compared to other video and audio coding formats is that MPEG files are much smaller for the same quality. This is because MPEG uses very sophisticated compression techniques. This is the technology that allows high-quality digital video to be sent with telephone, and internet traffic on a common carrier. Multiplexing video, data, and voice on a common carrier using MPEG compression lowers the total bandwidth required to send and receive large volumes of information.

True Bandwidth on Demand

The transmission speed required for any MPEG-2 broadcast varies according to the nature of the video material. The MPEG-2 encoder located at the satellite uplink station has a finite time to make encoding decisions. Prerecorded movies and other taped material do not push the time constraints of the encoder to the limit; the encoder can select at its leisure the most efficient method for encoding at the lowest possible data rate. Live sports and other live action materials require a higher data rate because the encoder is forced to make immediate coding decisions and must also transmit complex, rapid-motion changes without introducing high levels of distortion.

REPRESENTATIVE MPEG-2 TRANSMISSION RATES

Type of Video Service Data Rate

- Movies (VHS quality) 1.152 Mb/s
- News/entertainment 3.456 Mb/s
- Live sports event 4.608 Mb/s
- 16:9 wide-screen TV 5.760 Mb/s
- Studio-quality broadcast TV 8.064 Mb/s
- High-definition television 14.00 Mb/s
- Monaural sound 0.128 Mb/s
- Stereo sound (L + R) 0.512 Mb/s
- Digital data 9.6 kb/s

MPEG Transmissions

TV programmers use a transmission format called multiple channel per carrier (MCPC) to multiplex two or more program services into a single unified digital bit stream. With MCPC, a package of program services can use the same conditional access and forward error correction systems, thereby economizing on the overall bandwidth and transmission speed requirements. What's more, programmers can dynamically assign capacity within the digital bit stream of any multiplexed transmission, so that more bits are available to a live sporting event broadcast and fewer bits to a news report or interview program consisting of "talking heads." At the conclusion of a live basketball game, for example, the digital capacity previously used to relay a single sporting event could be used to transmit multiple separate movies.

It is clear from this brief discussion that new advances in telecommunications are now being used to provide archaeologists, historians, and the general public short-time access to submerged cultural sites in more and more remote locations and in increasing water depths.

Initial Experimentation in Providing Long-term Public Access to Submerged Culture Sites

Since 2003, the Institute for Exploration has been experimenting in telepresence access to undersea habitats through a project known as the Immersion Institute. Its goal has been to work with the National Oceanic and Atmospheric Administration's (NOAA) National Marine Sanctuary Program to provide the public access to marine sanctuaries using telepresence technology. The initial focus of this program has been the National Marine Sanctuary in Monterey Bay, California, and its kelp forest habitat. The project began when two metal stanchions were installed in approximately 15 m of water, 25 m apart. A nonconducting steel cable connected the two stanchions to support a small remotely operated vehicle (ROV) named Orpheus (figure 12.7). A fiber-optic tether, which carried power and telemetry to the ROV extended and contracted in an accordion-like fashion as the vehicle traveled back and forth along its 25-m cable.

Figure 12.7. The Orpheus ROV, which operated in the Monterey Bay National Marine Sanctuary in California for more than 3 years, was controlled from the Immersion Theater in Mystic, Connecticut. (Copyright Institute for Exploration)

In addition to thrusters and lights, the ROV carried a 3-chip pan, tilt, and zoom color video camera, which transmitted its signal via its tether back to the shoreward stanchion and via an undersea cable back to a shore-side hub at a facility on historic Cannery Row. A second camera was mounted on the shoreward stanchion, making it possible to see the ROV as it traversed a portion of its cable. Two additional cameras were mounted above water: one at a nearby California sea lion "haul-out" area on a rock jetty, while the second camera was mounted on a pier at the Cannery Row facility. The latter was connected via a subaerial cable to the station, while the second camera's signal was transmitted via a two-way microwave communication link. Both cameras could be controlled remotely and had pan, tilt, and zoom functions.

During the first two years of this program, the shore station was connected to the University of California at Santa Cruz campus, located about 25 nautical miles to the north, via a microwave link that connected it to the Internet2 backbone. In 2004, the City of Monterey, California provided direct access to Internet2, eliminating the need for the microwave link across the Bay, which commonly was plagued by atmospheric interference. Internet2 is a consortium of more than 200 universities working in partnership with government and industry to create a new high-bandwidth nationwide network. The high bandwidth capabilities of Internet2 make it possible to transmit multiple channels of video, including the four camera systems that were in the Monterey Bay National Marine Sanctuary.

Any member of the Internet2 network could access the system, including the Institute for Exploration at Mystic Aquarium in Mystic, Connecticut. At Mystic, an interactive theater was constructed that let a team of interpreters conduct regular public programming for visitors who wished to visit this Sanctuary.

Figure 12.8. Artist rendering of the Immersion Theater in Mystic, Connecticut, where interpreters conducted regular live tours of the Monterey Bay National Marine Sanctuary, by controlling the ROV Orpheus. (Copyright Institute for Exploration)

Using a control station, the interpreters were able to operate all four cameras systems in Monterey Bay as well as drive the ROV along its 25-m path through the kelp forest. Audio links also made it possible for visitors in Connecticut to hear natural sounds of the sea as well as communicate with undersea divers who regularly supported the live tours.

Based on the success of this pilot program, NOAA's National Marine Sanctuary Program authorized similar telepresence networks to be established in other National Marine Sanctuaries, including their Thunder Bay National Marine Sanctuary in Lake Huron. Also known as "Shipwreck Alley," this region of Lake Huron has seen numerous ships lost in the Bay due to collisions in the fog, gales, and rocky shoals. In 2001, the Institute for Exploration carried out a systematic survey of the Sanctuary using its Echo side-scan sonar system to precisely locate both known and unknown shipwrecks inside and outside the Sanctuary. In 2002, IFE returned with its ROV Little Hercules and imaging sled Argus to document several of the shipwrecks in detail (see chapter 11 on Lake Huron). Based in part on that survey, certain shipwrecks are being considered to be equipped with similar telepresence technologies as those installed in Monterey Bay.

The overall goal of the National Marine Sanctuary telepresence network is to build a network of museums, science centers, and aquariums that will help to support the establishment of other remotely accessed natural and cultural sites beneath the sea. Once a cultural site is established for telepresence access, it will permit constant monitoring of the site to insure its long-term preservation. As this concept grows, and technology continues to improve, more remote sites will become accessible. The long-term goal of this program is to establish telepresence access to more underwater cultural sites in deeper water and in more remote locations such as the *Titanic* and the submerged battlefields of Iron Bottom Sound. Given this long-term strategy, it is critical that efforts be made to protect such sites from those who lack this vision and think the only way to appreciate cultural sites beneath the sea is to strip them of their artifacts, frequently damaging the ship in the process.

Long-term Preservation of Deep-water Sites

The science of deep-sea conservation and preservation is in its infancy but given the natural preserving characteristics of that environment, long-term preservation of shipwrecks in the deep sea appears promising. Major advances, for example, have been made in the development of underwater vehicles that can clean the hull of large ships while they remain in water. And equally important advances have been made in the development of antifouling epoxy paints and polymer waxes that can be applied underwater to provide long-term protection. We were struck by the appearance of *Titanic* during our initial dives in 1986, including the pinkish tinge on the surface of hull below the waterline where antifouling paint had been applied years before. It seems perfectly possible that our remotely operated vehicle Hercules could clean and then paint a section of *Titanic's* hull to prolong its life, and epoxy or polymer paint could be laid down on its deck.

While such experimentation is taking place, legal steps need to be taken to insure that important historical and archaeological wreck sites are placed under national or international protection. Rules should be established through the enactment of international treaties that govern the regular visitation of such sites to insure that damage is not caused, for example, by a manned submersible, to a shipwreck that lies beyond the protection of sovereign states. When the *Titanic* was discovered in 1985, attempts were made to protect the ship from subsequent salvage operations but the resulting law, the Titanic Memorial Act of 1986, was unsuccessful, since other countries refused to join the United States in the creation of bilateral agreements as the Act requested. Ironically, it was largely American citizens who then went to other countries, namely France and Russia to carry out a number of activities at the *Titanic* wreck site that have seriously and irreversibly altered the site.

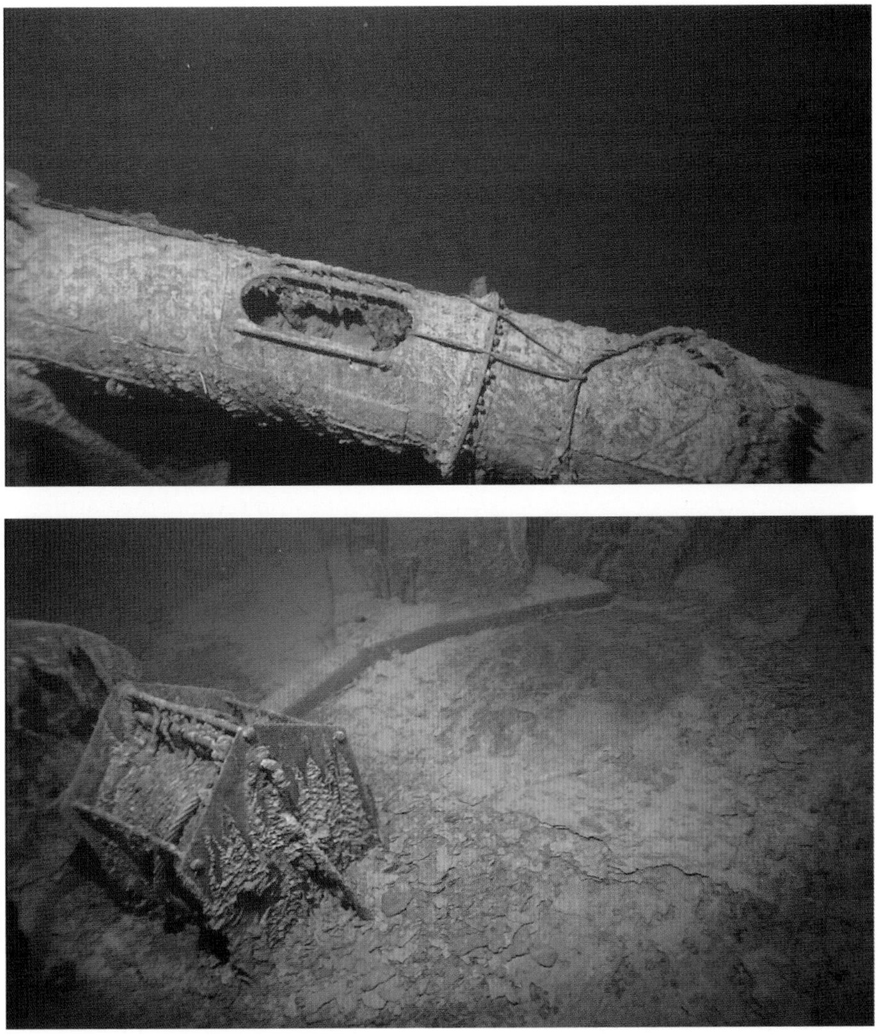

Figure 12.9. Foremast (*top*), which lost its crow's nest during a time when a submersible expedition was underway. One of two large oval impressions (*bottom*) in *Titanic*'s deck at the base of the forward mast caused by "something" crushing the deck. (Copyright Institute for Exploration)

Recently, Great Britain signed a treaty to protect the *Titanic* but other nations have yet to follow, including the United States. In the summer of 2004, our expedition to the RMS *Titanic* attempted to bring national and global attention to the site in hopes of the establishment of further international laws to protect it from human-induced damage and to document the change that has occurred since human intervention began in 1997 when large-scale salvage operations began. During our 2004 visit to *Titanic* it was clear that submersibles visiting *Titanic* to retrieve artifacts, conduct filming operations, or simply bring visitors to the site had inflicted damage to the ship. Since such damage

Figure 12.10. Collapsed bulkhead (*top*) of the First Officer's cabin, which has occurred since the 1986 expedition. Holes in the deck (*bottom*) frequently used as submersible landing site above the Marconi Room and forward off the Grand Staircase. (Copyright Institute for Exploration)

is seldom documented or acknowledged by the submersible operations, it is impossible to differentiate between the natural processes of chemical, biological, and physical deterioration and that done by human intervention (figures 12.9–12.12).

Given these advances in technology and the natural preserving conditions of the deep sea, cultural sites that have historic, archaeological, and anthropological importance should become undersea museums accessed via telepresence as discussed earlier. In summary, the technology exists to develop methods for in situ preservation and remote visitation to these underwater museums.

Figure 12.11. Shells (*top*) of dead wood-boring mollusks that consumed wooden deck planks. Hole (*bottom*) in starboard boat deck also used frequently as a submersible landing site. (Copyright Institute for Exploration)

Figure 12.12. A pair of shoes (*top*) and a "slicker" worn by a crewmember (*bottom*) have survived long periods of submergence in the deep sea. (Copyright Institute for Exploration)

GLOSSARY

The definitions below are drawn from the following sources: *American Heritage Dictionary*; *American Heritage Science Dictionary*; Robert L. Bates and Julia A. Jackson, *Glossary of Geology*, 2d ed. (Falls Church, VA: American Geological Institute, 1980).

Abiotic process A chemical or physical change in the natural world that occurs independently of biological agents.

Acoustic imaging The propagation of sound waves and analysis of their reflections to create an image.

Acoustic pulse A short burst of sound energy.

Acoustic shadow An area of a sonar record showing no return energy due to the relief of the target impeding the path of the acoustic signal.

Active margin The boundary between the oceanic and continental crust that coincides with the juncture between two tectonic plates. This tectonically active region is often characterized by a subduction zone or transform fault.

Aerobic An environmental condition where air, either in its native form or as dissolved gasses, is present.

Amidships The middle part of a ship, between the bow and stern.

Amphora A type of ceramic vessel with two handles and a neck narrower than the body which was used for storage and shipping of bulk materials from the 15th Century BCE until the 13th Century CE.

Anaerobic An environmental condition where air, either in its native form or as dissolved gasses, is absent.

Anoxic An environmental condition where oxygen, either in its native form or as a dissolved gas, is absent.

Anthropogenic Caused, produced, or influenced by human activity.

Anthropology The scientific study of humans, especially of their origin, their behavior, and their physical, social, and cultural development.

Archaeology The study of human culture and history through the examination of material culture and its context

Archaeo-magnetism The measurement of the position of the virtual geomagnetic pole recorded in archaeological samples. Magnetic minerals in archaeological samples align with the orientation of the earth's magnetic field, which changes over time. Therefore, the signature recorded in the sample gives an indication of the time when it formed

Archaic Period The period of human occupation of the Americas that began around 8000 BCE and ended with the adoption of sedentary farming around 1000 BCE. The Archaic period is characterized by larger communities of mobile hunter gatherers who participated in trade and a specific form of lithic production.

Artifact In the field of archaeology, any physical object made, modified, or used by humans.

Assemblage A group of artifacts found in association with one another, that is, in the same context, often representing a limited time period or a discrete human activity.

Asthenosphere The ductile portion of the earth's upper mantle that lies around 100 to 200 km below the surface beneath the rigid lithosphere. The temperature and pressure conditions in the asthenosphere allow the material to move plastically beneath the tectonic plates.

Attenuation A reduction of amplitude of an electrical or acoustic signal.

Backfill The process of filling in an excavated archaeological site with sediment in order to protect it from further degradation.

Bandwidth In acoustics, the range of frequencies of a propagating acoustic signal. In network communications, the rate at which data can be transferred over a computer network.

Barrier beach A narrow, elongate sand ridge rising slightly above the high-tide level and extending generally parallel with the shore, but separated from the shore by a lagoon or marsh.

Barrier lagoon A shallow pond or lake that is roughly parallel to the coast and is separated from the open ocean by a strip of land, barrier beach, or barrier reef.

Bathymetry A measurement of the depth within bodies of water, or a series of measurements to make, for example, a nautical chart or a map of seafloor topography.

BCE Before the Common Era. Refers to the number of years prior to the start of the Common Era.

Benthic Relating to, or living in or on the bottom of a body of water.

Biocide A chemical agent that is used to destroy micro-organisms or to retard their growth.

Biodeterioration Degeneration and disintegration due to biological processes.

Biological accretions Any solid material that has been chemically precipitated or otherwise developed on a surface as a by-product of biological activity or growth.

Box core Rectangular-shaped sample taken by insertion of a hollow box into a surface; in oceanography, a core that typically consists of the uppermost seafloor sediment.

BP Before Present. Refers to the number of years before present day, either in calendar years or radiocarbon years.

Bronze Age Time period in the development of civilization when the technology to smelt copper and tin to produce bronze tools and weapons becomes significant. This technology is adopted at different times in various areas but spanned 3000 BCE to 1100 BCE in ancient Greece.

Byzantine Relating to the period of the Byzantine Empire. The Byzantine Empire gradually evolved from the eastern Roman Empire around 300 CE and is considered to have ended with the conquest by the Ottoman Turks in 1453.

Byzantium The ancient Greek city that became the capital of the Byzantine Empire around 200 CE and was renamed Constantinople after the death of Emperor Constantine. The city flourished as a dominant trading port between the Mediterranean and Black Seas and became the modern city of Istanbul in 1930.

Calcium carbonate A soluble chemical compound, $CaCO_3$, which occurs naturally in solid form in the minerals calcite and aragonite, as well as in bone and shell material. It is commonly found dissolved in natural bodies of water.

CE The Common Era. Refers to the number of years following the start of the Common Era.

Cenozoic Era The geologic time period that began around 65 million years ago with a mass extinction that included the dissapearance of the dinosaurs and is characterized by the diversification of mammals, the formation of modern continents, and the onset of glaciations.

Chalcolithic The period in history that begins with the first appearance of copper metallurgy at a particular locality, which varies significantly from region to region.

Chirp A sonar signal in which the acoustic pulse changes frequency throughout its duration.

Clovis culture An ancient American culture whose distinct tool kit was first classified at a site near Clovis, New Mexico. The Clovis people are traditionally considered the earliest inhabitants of the Americas, arriving around 11,000 years ago.

Coccolith A minute plate of calcium carbonate secreted by single-celled algae called coccolithophores that is deposited on the seafloor after the organism dies.

Concretion In underwater archaeology, a type of corroded metal artifact encased in a surrounding shell of dense corrosion products. During the corrosion process, some or all the original metal is dissolved leaving an internal cavity that takes the form of the original artifact.

Contact Period The period in North American history after Europeans arrived and began influencing local culture. The beginning of this period varies according to region and the spread of European colonization.

Continental margin The region that separates the continental landmass from the deep ocean floor and consists of the continental shelf, slope and rise. (see also *Active margin*, *Passive margin*)

Continental shelf The part of the ocean floor that is characterized by a very gentle slope and lies between the shoreline and the beginning of the continental slope, which represents the steeper portion of the continental margin.

Cordilleran ice sheet A large permanent body of ice that covered the western coast of North America during glacial episodes of the Quaternary Period.

Core The innermost layer of a planet. The earth's core is extremely dense and consists of the elements iron and nickel. It is divided into the liquid outer core and solid inner core.

Corrosion The breakdown or alteration of a material, especially a metallic substance, through chemical or biochemical reactions with its environment.

Cretaceous Period The geologic time period that spanned around 145 to 65 million years ago. It is the last period of the Mesozoic Era and its conclusion correlates to the mass extinction best known for the disappearance of the dinosaurs.

Crust The outermost layer of a planet. The earth's crust is less dense than the core and mantle. Oceanic crust is basaltic in composition and is about 10 km thick on average. Continental crust is granitic in composition and is about 30 km thick on average.

Cultural resource The physical products associated with art and manufacturing and the intangible aspects of culture such as folklore and dance. Cultural resource management is the practice of managing, maintaining, and studying cultural sites, artifacts, and traditions.

Deformation Alteration of the shape, size, or position of a rock as a result of stresses caused by various forces within the earth.

Dendrochronology The study of growth rings in trees for the purpose of analyzing past climate conditions and used in archaeology for the dating of wooden artifacts and the structures they comprise.

Deposition The constructive process of accumulation of any kind of loose rock material by any natural agent, such as the mechanical settling of sediment from suspension in water, the chemical precipitation of mineral matter by evaporation from solution, or the accumulation of organic material following the death of plants and animals.

Depressor weight A weight used between two lengths of underwater cable and vehicle tether to decouple the motion of the ship from the vehicle at the far end of the cable system.

Desalination The removal of soluble salts from a porous artifact through a process of soaking in changes of fresh or distilled water.

Dynamic positioning system A shipboard propulsion system that employs computer driven thrusters to stabilize and maneuver a ship with very high precision.

Echosounder An acoustic instrument that measures the depth of water below a ship.

Electrical resistivity A measure of how strongly a material opposes the flow of electric current. The measurement of this property can be used in archaeological surveys to identify features buried beneath the surface because electricity flows differently through archaeological materials and the surrounding sediment matrix.

Erratic A boulder or rock carried by glacial ice and deposited a large distance from its source.

Eustasy The variation in the globally averaged absolute elevation of the sea.

Evaporite Any water-soluble mineral sediment produced by chemical precipitation following the evaporation of water.

Excavation In archaeology, the systematic documentation and removal of sediment or soil to expose artifacts and features from an archaeological site in order to analyze past human behavior.

Fennoscandian ice sheet A large permanent body of ice that spread from Scandinavia and the British Islands into northern Europe during glacial episodes of the Quaternary Period.

Flocculent A substance that causes fine particles in aqueous suspension, like colloids, to coalesce into larger particles which then settle or fall out of solution.

Fluted point A manufactured, sharp lithic material used with a spear, dart or arrow that has one or more long narrow grooves extending from the basal edge along the face.

Fungicides A chemical substance that destroys or inhibits the growth of fungus.

Gantry A mechanical frame for supporting the weight of another object.

Garum A strongly flavored condiment or sauce, made by aging and salting fish, that was popular in the ancient Greece and Rome.

Geoarchaeology The study of the techniques and subject matter of geography and the earth sciences applied to archaeological investigations.

Geochemistry The study of the chemical composition of the earth and other planets and chemical processes and reactions that govern the distribution of chemical elements in minerals, rocks, soils, waters and atmosphere.

Geographic Information System (GIS) A computer system for capturing, storing, integrating, managing and displaying data that are spatially referenced to the earth. This tool allows individuals to analyze the spatial information, edit data and maps, and create georeferenced data products.

Geomorphology The study of the characteristics, origin and evolution of landforms.

Geophysics The study of the earth by quantitative physical methods such as seismic, electromagnetic, and radioactive techniques.

Geotextile A large woven or non-woven fabric engineered to be applied in or on soils or sediments to achieve specific goals such as sediment isolation, consolidation, or the prevention of erosion.

Glacial drift Rock material transported and deposited by glaciers, icebergs, or meltwater emanating from a glacier.

Glacial erratic See *Erratic.*

Glacial period The period of time during an ice age when temperatures drop, glaciers advance, sea level falls and ice sheets expand.

Glacial rebound Isostatic uplift in response to the removal of a load of continental ice due to melting.

Glacial till See *Till.*

Glacier A large mass of ice that forms from the compaction of snow in areas where snow accumulation exceeds melting and sublimation. Glaciers slowly deform and flow in response to gravity and can move large masses of sediment and rock.

Global positioning system (GPS) A system of multiple satellites used to determine absolute positions on the earth's surface (latitude and longitude). A GPS receiver uses radio time signals sent from the satellite to calculate and record the positions. A differential GPS compares a fixed reference to the calculated position to correct and improve the reported position.

Gravity corer A heavy, cabled, tubular device that is deployed from a surface ship and uses gravity to descend to the bottom to penetrate the surface sediment layers and extract core samples.

Ground-penetrating radar (GPR) An electronic device that emits electromagnetic pulses toward a surface and records the pattern of reflections from subsurface features. GPR has many applications but is most commonly used to identify either manmade or geologic features buried beneath the surface of the earth.

Heave The vertical motion of a vessel on water, generally due to wave action.

Holocene The geologic time period spanning from about 11,000 years ago to the present time. It is characterized by the rise in human civilization and is the last epoch in the Neogene period of the Cenozoic Era.

Homo erectus An extinct species of humans who migrated out of Africa around 2 million years ago and colonized most of Eurasia.

Homo sapiens The scientific name of modern humans. It is widely believed that Homo sapiens first appeared in Africa around 200,000 BP, migrated into Eurasia and Oceania by 40,000 BP and the Americas around 10,000 BP.

Hunter-gatherer Members of a community who subsist primarily on plants and animals directly obtained from the wild by foraging and hunting.

Hydrocarbon An organic compound that contains hydrogen and carbon and frequently occurs in natural gas, petroleum, coal and bitumen.

Hydrothermal vent An outlet on the sea floor through which water that has been heated inside the earth flows. The water usually contains dissolved minerals that precipitate out upon contact with the colder seawater and provides a foundation for biological ecosystems.

Hypsometry The measure of the amount of land at particular elevations relative to sea level.

Ice Age Any time period of long-term reduction in global temperatures during which ice sheets cover continental crust. Commonly used to refer to periods of extensive glaciation when ice sheets covered the North American and Eurasian continents. When noted in capital letters, refers the most recent ice age, which ended around 11,000 years ago, at the beginning of the Holocene.

Inclusion Any foreign material in a larger matrix. In archaeological ceramic materials, inclusions are often in the form of minerals or organic matter in a clay matrix.

In situ Situated in the original place of deposition or discovery; in context.

Interglacial Any warmer period of time during ice ages when temperatures and sea levels rise and ice sheets and glaciers retreat.

Intrusive techniques Archaeological methods that disturb the surface features of a site. Any act which involves disrupting the stratigraphy, moving artifacts, or altering the original context of the site.

Inundation The process of begoming covered with water; submergence.

Iron Age Time period in the development of a culture during which iron is adopted as a prominent material for the production of implements. The Iron Age is traditionally considered the last phase for classifying prehistoric cultures, preceded by the Bronze Age and the Stone Age. In classical studies, the Iron Age is considered to have begun in the Mediterranean in the 12th century BCE and ended with the beginning of the Hellenistic tradition in the 4th century BCE.

Insonified When an object or area of the seafloor is scanned by a sonar system.

Isostatic uplift The positive vertical adjustment of the earth's lithosphere in response to the removal of a load (e.g., water, ice, sediment, or rock).

Isostasy The pressure equilibrium between the earth's rigid lithosphere and the plastic asthenosphere. Changes in the thickness and density of the lithosphere over time produce uneven stresses on the asthenosphere, which flows to restore equilibrium.

Isotope One of two or more forms of an element having the same atomic number but different atomic mass. Isotopes of the same element have the same number of protons, giving them the same atomic number, but have differing numbers of neutrons, giving them each a unique atomic mass.

Karst terrain A region characterized by the dissolution of bedrock, usually a carbonate such as limestone or dolomite, to produce erosional features such as sinkholes, underground streams and caverns.

Kettle pond A water-filled basin that forms when a detached body of ice is left behind by a retreating glacier.

Kinematically Having to do with pure motion, not the masses or forces involved in the motion.

Lacustrine Relating to a system of inland wetlands and deep-water habitats associated with freshwater lakes and reservoirs

Lanceolate Narrow and tapering to a pointed apex; shaped like the head of a lance.

Land bridge Any body of land that connects two landmasses and allows plants and animals to migrate and colonize.

Last Glacial Maximum (LGM) A period of time around 20,000 years ago when the continental ice sheets were at their maximum extent, and covering the greatest area across the northern continents.

Late Roman Amphora 1 A specific shipping jar shape of Late Antiquity (4th–7th century CE) found throughout the Mediterranean Sea and in the Black Sea.

Laurentide ice sheet A large permanent body of ice that covered much of North America, including Canada and a large portion of the United States during glacial episodes of the Quaternary Period.

Layback Horizontal distance from the ship to a towed vehicle, usually calculated from the known length of cable deployed and depth of the towed vehicle.

Lithic Stone, or made from stone.

Lithology The macroscopic quality or description of the physical characteristics of a rock or rock layer.

Lithosphere The solid outermost portion of the earth's crust and upper mantle that lies from the surface to around 100 to 200 km depth. The brittle lithosphere makes up the tectonic plates, which lie on top of the denser, ductile asthenosphere.

Lithostratigraphic unit A body of rock that is unified by consisting dominantly of a particular lithology. A rock layer characterized by similar physical properties and features, and contains a substantial degree of overall homogeneity.

Little Ice Age A brief, cold episode from about the 16th to 19th centuries CE when the global average temperature was about 1 degree colder than today, spawning a short re-advance of glaciers.

Littoral drift Also known as longshore drift, the process by which sediments move along a beach or shoreline due to coastal wave action.

Longshore current Also known as littoral current, the continuous, directed movement of water caused by the approach of waves toward a coast at an angle that causes water to flow parallel to and near the shoreline.

Magnetometer An instrument that measures the magnitude and direction of a magnetic field, usually the earth's magnetic field. In archaeological survey, a magnetometer can detect variations in the field caused by man-made objects.

Mantle The thick layer within the earth that lies between the crust and the core, containing approximately 70% of the planet's volume. The mantle is stratified by changes in its elastic properties. Variations in temperature at the boundary layers drive convection within the mantle that moves material and redistributes heat.

Marine gravimetry The measurement of the gravitational field of the earth over the oceans. Data collected by marine and satellite gravimeters can be used to determine the depth of the ocean floor.

Marine transgression A rise in sea level that inundates the land and causes the shoreline to move to higher elevations on the terrestrial landscape.

Mass wasting The process of movement of material down slope due to gravity.

Matrix In archaeology, the physical environment surrounding an artifact, i.e. the sediment in which it is embedded. The matrix and its components are studied to better understand the age and alteration of original artifacts and archaeological features.

Megafauna A qualitative term used to describe large animals, often referring to those that have become extinct in geologically recent times.

Mesolithic The prehistoric period characterized by the development of stone technology to include small composite tools called microliths and the beginning of settled communities. The Mesolithic Period follows the Paleolithic Period and precedes the Neolithic Period; its timing varies according to region.

Midden A mound or deposit containing domestic refuse such as shells and animal bones that usually indicates the site of a prehistoric human settlement.

Mid-ocean ridge An underwater mountain range formed by the uplift of oceanic crust at a divergent plate boundary where convection in the mantle causes magma to form and rise to create new oceanic crust.

Mineralization In archaeology, a process of alteration or corrosion, in which a metal or organic object is converted to a mineral substance by reaction with its environment.

Miocene The geologic time period spanning from about 25 to 5 million years ago. The fourth epoch of the Tertiary period in the Cenozoic Era.

Moraine See *Terminal moraine.*

Mortar A vessel with a depression in which substances are ground and mixed. Used in conjunction with a pestle.

Multibeam sonar A sonar system that emits multiple simultaneous sound pulses and measures the timing of their reflections to determine distance to a number of points. Normally it is configured in a fan-shaped pattern perpendicular to the direction of travel of the sensor and produces a detailed bathymetric profile of a swath of seafloor.

Multiplex The process of combining input signals from multiple sources onto a single communications path.

National Marine Sanctuary A protected marine ecosystem managed by the National Oceanographic and Atmospheric Administration (NOAA), which serves to conserve, safeguard, and enhance the biodiversity, ecological integrity, and cultural legacy of these areas.

Neanderthal An extinct species of humans who inhabited Europe and western Asia during the late Pleistocene epoch until about 24,000 years ago.

Neolithic The last period of the Stone Age, characterized by the rise of domestication and agriculture. It occurred in the Middle East around 10,000 BCE, though its timing varies according to region.

New World The new lands discovered by the European explorers in the fifteenth century; the American continents.

North Atlantic Deep Water (NADW) Highly saline, dense water from the North Atlantic that sinks to depth and travels south along the coast of North and South America and plays a significant role in the thermohaline circulation of the world ocean.

Oceanography The exploration and scientific study of oceans, the life that inhabits them, and their physical characteristics.

Old World The world as it was known by early Europeans before the discovery of the Americas; the Eurasian and African continents.

Outwash sediment Stratified detritus removed from a glacier by meltwater streams and deposited in front or beyond the end moraine or the margin of an active glacier.

Oxygen isotope curve A plot of the variations in the ratio of oxygen-18 to oxygen-16, as measured in oceanic sediments, through specified time intervals. The ratio of oxygen isotopes in the ocean can be used as a proxy for sea level and temperature. Oxygen-16 is preferentially evaporated from the ocean and during periods of increased glaciation, and it becomes locked in continental ice after being precipitated as snow, leaving the oceans heavier in oxygen-18.

Overburden Sediment that covers an archaeological site or feature of interest, usually void of archaeological material.

Paleo-Indian The earliest known people to have populated the Americas, the Paleo-Indians banded in small communities of hunter-gatherers who are believed to have migrated across the Bering land bridge around 11,500 years ago. Recent evidence suggests that they could have migrated as early as 13,000 years ago, however this is still debated in the scientific community.

Paleolithic The earliest period in human history showing evidence of technology, which began with the introduction of the first stone tools. Also known as the "old stone age", dates for the Paleolithic vary from region to region. The earliest stone tools were discovered in Africa and date to about 2 million years ago.

Paleontology The scientific study of life in the geologic past, especially through the study of animal and plant fossils.

Paleoshoreline A relic surface associated with the shore of an ancient body of water.

Pallet A portable platform used for storing or moving cargo or freight.

Palynology The scientific study of spores and pollen, both living and fossilized. Palynology helps improve knowledge of ecosystems in both the recent and distant past, since pollen and spores are extremely durable, unlike many other plant parts.

Parrel A mechanism for fastening the yard of a sailing ship to the mast, ranging from a simple rope to a system of beads and ribs that facilitate raising and lowering the sail.

Passive margin The boundary between the oceanic and continental crust that does not coincide with a plate boundary and is not tectonically active.

Peat An unconsolidated deposit of semi-carbonized plant remains in a water-saturated environment.

Pelagic Relating to, or living in or on the open ocean.

Pestle An instrument for pounding and grinding used in conjunction with a mortar.

Petrology The field of geology that focuses on the study of rocks and the conditions in which they form by examination of their composition and structure.

Phoenician A maritime culture that was located around modern Lebanon and Syria and flourished from about 1200 BCE to 900 BCE. The Phoenicians were known for their advanced seamanship and skills in manufacture and trade.

Photogrammetry A measurement technology in which the three-dimensional coordinates of points on an object are determined by measurements made in two or more photographic images taken from different positions.

Photomosaic A set of photographs that overlap and are stripped together to produce a composite image of a large region or feature that cannot be imaged by a single photograph.

Ping See *Acoustic pulse.*

Plate tectonics The theory of geology that describes the movements of the earth's crust and upper mantle, which is broken into large lithospheric plates, over the ductile lower mantle, or asthenosphere.

Pleistocene The geologic time period spanning between about 1.806 years ago and the beginning of the Holocene Epoch about 11,000 years ago. The third epoch of the Neogene period, it is characterized by repeated glaciations and its end correlates to the end of the Paleolithic archaeological time period.

Pneumatic Describing the use of pressurized gas, usually air, to transfer energy.

Polycarbonate A high-impact plastic used in tubular form in underwater archaeology to extract tube cores of archaeological sediment from the seabed.

Polyethylene A low-density, inexpensive, and chemically stable plastic.

Post glacial rebound See *Glacial rebound.*

Posthole A depression into which a vertical timber is placed for the construction of a fence or house. In archaeology, once the post has decayed, postholes can be identified by circular stains left in the sediment.

Prehistory The period of time before written history; the earliest phase in human development.

Preservation In archaeology, the act of maintaining for future generations the durability, integrity and accessibility of cultural property through scientific study, education and legislation.

Provenance In general, the original source location of an artifact or feature of archaeological significance. Specifically, the three-dimensional location of an artifact or feature within an archaeological site, measured by two horizontal dimensions, and a vertical elevation.

Pseudomorph In archaeology, the preserved form of a perishable organic substance resulting from the slow mineralization of the original.

Pyroclastic eruptions Volcanic eruptions characterized by explosive energy and pyroclastic flows that are destructive, fast-moving, fluidized bodies of hot gas, ash and rock.

Quaternary The geologic time period that represents the most recent period of repeated glaciations to the present. It includes the Pleistocene and Holocene Epochs.

Radiocarbon dating A method that uses the ratio of the naturally occurring radioactive isotope carbon-14 to the stable isotope carbon-12 and the rate of decay of carbon-14 to determine the age of organic material.

Radioisotopic methods Any technique that measures the ratios of atoms with unstable nuclei that decay radioactively, usually to determine a unique chemical signature or age of formation.

Reconnaissance survey An initial survey of a previously unexplored region intended to identify likely targets for more detailed investigation.

Remotely operated vehicle (ROV) A tethered, unoccupied, highly maneuverable underwater robot operated by a person aboard a surface vessel.

Satellite A device that has been placed in orbit around the earth and is used for receiving and transmitting communications signals.

Satellite remote sensing The use of a satellite to gather data from a distance via radar or infrared photography.

Scraper A unifacial stone tool used to scrape hides, bone or other materials in the preparation of food, clothing and shelter.

Sector-scanning sonar A sonar system in which the acoustic path is sequentially moved around the arc of a circle, or part of a circle.

Sedimentation The deposition or accumulation of solid fragmented material, such as silt, sand, gravel, chemical precipitates, and fossil fragments, that is transported and deposited by water, ice, or wind, or that accumulates through chemical precipitation or secretion by organisms, and that forms layers on the earth's surface.

Sedimentology The study of modern depositional processes of sediments and the understanding of these processes applied to the formation of sedimentary rocks.

Seismic stratigraphic profiling The use of sound waves to determine the internal structure of sediment deposits.

Shoreface The narrow, rather steeply sloping zone seaward or lakeward from the low-water shoreline. A region permanently covered by water, over which beach sands and gravels actively oscillate with changing wave conditions.

Side-scan mosaic A georeferenced composite image that merges multiple side-scan images.

Side-scan sonar A sonar system that emits sound pulses, aimed to hit the sea floor at a shallow angle, and records the timing and amplitude of their reflection patterns to create an image of the sea floor. Side-scan sonar emits fan-shaped pulses out to the side, perpendicular to the direction of travel of the sensor, to produce a swath of coverage with a fixed width depending on the signal frequency.

Sill In geology, an intrusion (such as a volcanic lava flow) into the earth's crust that forms a flat sheet-like deposit between weak layers of sedimentary rock.

Sinkhole A natural depression in a land surface produced by the dissolution and collapse of soluble bedrock.

Site formation processes The cultural and natural events that form and transform archaeological material and its matrix over time to produce an archaeological site.

Slant range The slanting distance between two points (e.g., a sonar and a target) at differing depths.

Sonar An acoustic system for making measurements or detecting objects under water. (see also *Multibeam sonar*, *Sector-scanning sonar*, *Side-scan sonar*)

Specific gravity The density of a material or object relative to that of standard pure water.

Stone Age The period in human history before the development of metal implements when stone was the primary material for tool production.

Stratigraphy In geology, the classification, analysis and correlation of layers, or strata, in sedimentary rocks. In archaeology, the sequence of layered sediments and cultural deposits in an archaeological site that is used to determine relative ages of artifacts and to identify context within the site.

Structural geology The study of the three dimensional distribution of rock bodies, surfaces, and their internal fabrics to make inferences about tectonic history, past geological environments, and deformation events.

Subbottom profiler A sensor that emits a sound pulse toward the seafloor and records the reflections to identify and characterize the layers of sediment or rock under the seafloor. The sound pulse emitted by the subbottom profiler penetrates the seafloor and is reflected off different layers in the substrate to provide information on what is below the surface.

Submersible An underwater vehicle that must rely on human pilots who either occupy the vehicle or control the vehicle remotely from a support ship.

Subsidence A geological term which describes the relative lowering of the earth's surface elevation. Subsidence can occur in many ways: when a load, such as sediment, is deposited on the crust; when the crust thins due to extension; or when material is removed from the crust, such as in the extraction of water or natural gas.

Supersaturated The state of substance having a higher concentration in a chemical solution than is normally possible under given conditions.

Tectonic Relating to the movement of the earth's crust and the structures that result from these forces.

Telemetry Transmission of information, data, and measurements over long distances, such as through fiber optic tethers that connect ROVs to the surface.

Telepresence The use of video and other information to give a remote observer a sense of being present as an active participant at a local site.

Terminal moraine A large deposit of glacial sediment pushed in place by an advancing ice sheet and built upon by outwash sediment.

Thermocline A layer within a body of water (or air) in which the temperature gradient changes rapidly with depth (or altitude).

Till Unsorted sediment that is directly deposited by a glacier.

Topography The detailed surface and relief features of a region, determined by elevation measurements.

Towfish A piece of equipment, that is usually hydrodynamically streamlined, and is towed under water behind a vessel.

Transducer A device that converts electrical energy into sound waves.

Transgression See *Marine transgression.*

Trawling A fishing technique that utilizes a strong net that is dragged along the bottom of the sea to capture deep-water organisms.

Triage In archaeological conservation, the essential initial steps taken to insure the survival of newly excavated artifacts.

Trireme A type of warship that was propelled by both oars and sails and was used to ram enemy ships. These ships were used for nearly a thousand years beginning in the Classical period.

Tube cores Extracted stratified sediment samples, collected by forcing a hollow plastic tube downward into the seabed.

Ultra-short baseline positioning An acoustic positioning system in which a signal is received by multiple hydrophones located very close together (less than one acoustic wavelength apart) that uses a phase measurement to calculate the angle of the received signal path.

Unconformity An erosional surface separating two rock units of different age indicating a gap in the depositional sequence.

Uranium-series analysis The measurement of the radioactive decay of Uranium, which is commonly used to determine the age of carbonate materials.

Varve A layer or series of layers of sediment deposited in a body of still water in one year.

Vibracorer A long hollow cylindrical coring device that inserts a tube into sediment to retrieve a sample. The cylinder is vibrated to loosen the sediments and ease its insertion into the solid surface.

Volcanology The study of volcanoes, lava, magma and related geological phenomena.

Wisconsin glaciation The latest Pleistocene glaciation in North America, which ended about 11,000 to 10,000 years ago.

Woodland period The cultural period in North America that spanned about 1000 BCE to 1000 CE and marked the beginning of pottery production, permanent settlements, elaboration of burial practices, and, in later times, the cultivation of maize, beans and squash.

Würm glaciation The latest Pleistocene glaciation in Eurasia, which ended about 11,000 to 10,000 years ago.

X-radiography Production of an image on a radiosensitive surface by x-radiation.

Younger Dryas A brief period of very cold, dry climate that occurred about 11,000 years ago and caused a return to glacial conditions in the higher latitudes of the Northern Hemisphere.

LIST OF CONTRIBUTORS

Robert D. Ballard
President, Institute for Exploration, Mystic, CT; Director, Institute for Archaeological Oceanography, University of Rhode Island Graduate School of Oceanography, Narragansett, RI

Ali Can
General Electric Medical Systems Research Center, Schenectady, NY

Dwight F. Coleman
Marine Research Scientist, University of Rhode Island Graduate School of Oceanography, Narragansett, RI; Director of Research, Institute for Exploration, Mystic, CT

Michael J. Durbin
Advisory Engineer, Verizon Business

Ryan Eustace
Assistant Professor, Department of Naval Architecture & Marine Engineering, University of Michigan, Ann Arbor, MI; Guest Investigator, Department of Applied Ocean Physics and Engineering, Woods Hole Oceanographic Institution, Woods Hole, MA

Brendan Foley
Department of Applied Ocean Physics and Engineering, Woods Hole Oceanographic Institution, Woods Hole, MA

Cathy Giangrande
Affiliate, Centre for Maritime Archaeology, University of Southampton, Southampton, UK

Todd S. Gregory
Marine Development Engineer, University of Rhode Island Graduate School of Oceanography, Narragansett, RI; Adjunct Research Scientist, Institute for Exploration, Mystic, CT

Rachel L. Horlings
Department of Anthropology, Florida State University, Tallahassee, FL

Jonathan Howland
Department of Applied Ocean Physics and Engineering, Woods Hole Oceanographic Institution, Woods Hole, MA

Kevin McBride
Department of Anthropology, University of Connecticut, Storrs, CT

James B. Newman
Chief Engineer, Institute for Exploration, Mystic, CT; President, Woods Hole Marine Systems, Woods Hole, MA

Dennis Piechota
Conservator, Fiske Center for Archaeological Research, University of Massachusetts at Boston, Boston, MA

Oscar Pizarro
Australian Center for Field Robotics, University of Sydney, Sydney, Australia

Christopher Roman
Assistant Professor, University of Rhode Island Graduate School of Oceanography, Narragansett, RI

Hanumant Singh
Department of Applied Ocean Physics and Engineering, Woods Hole Oceanographic Institution, Woods Hole, MA

Cheryl Ward
Associate Professor, Department of Anthropology, Florida State University, Tallahassee, FL

Sarah Webster
Engineer II, Department of Applied Ocean Physics & Engineering, Woods Hole Oceanographic Institution, Woods Hole, MA; presently with Department of Mechanical Engineering, Johns Hopkins University, Baltimore, MD

INDEX

Page numbers in *italics* refer to images